Basic Anatomy and Physiology

H.G.Q. Rowett
M.A., F.R.S.A., M.I.BIOL.

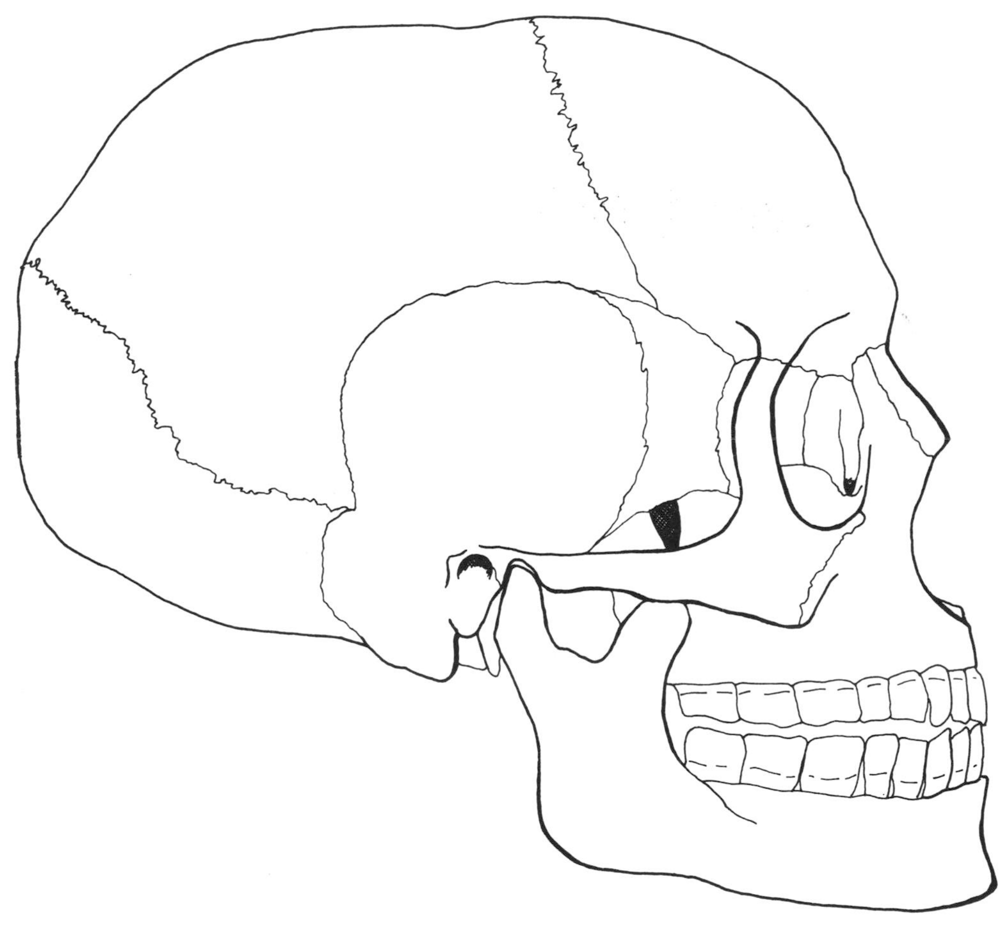

third edition

John Murray

Also by H. G. Q. Rowett:

Dissection Guides
I The Frog
II The Dogfish
III The Rat
IV The Rabbit
V Invertebrates

Guide to Dissection
(Dissection Guides I-V in one volume)

First published 1959 by
John Murray (Publishers) Ltd
50 Albemarle Street, London W1X 4BD

Reprinted 1962 (revised), 1966 (revised), 1968

Second edition 1973
Reprinted 1975 (revised), 1977, 1979, 1983 (revised), 1986, 1987

Third edition 1988
Reprinted 1988, 1990, 1991, 1992, 1993, 1994, 1996

Printed in Great Britain by
Butler & Tanner Ltd, Frome and London

British Library Cataloguing in Publication Data

Rowett, H. G. Q.
Basic anatomy and physiology. – 3rd ed.
1. Anatomy, Human 2. human physiology
I. Title
611 QM23.2

ISBN 0-7195-4395-9

Introduction

The human body is made up of an enormous number of tiny units of living material called **cells**, plus intercellular non-living substance called **matrix**.

Groups of cells and the matrix associated with them form **tissues**. The tissues are used to construct **organs** concerned with special functions. For convenience the organs may be considered to be grouped into **systems**, but the functioning of each system is closely related to that of others with overall concern for stability or **homeostasis**.

The study of the form and arrangement of organs is called **anatomy**, and of their methods of functioning, **physiology**. This book deals with the anatomy and physiology of the various parts of the body, system by system, with cross references to indicate the integration of the whole. The study of cells and tissues is called **histology**. Special features of the cells are closely bound up with the functions of the organs in which they are found, but there are certain common characteristics of all living cells, to understand which some basic knowledge of **chemistry** is essential.

Diagrams are used as far as possible with the mininum of written description.

General Plan

Essential Chemistry

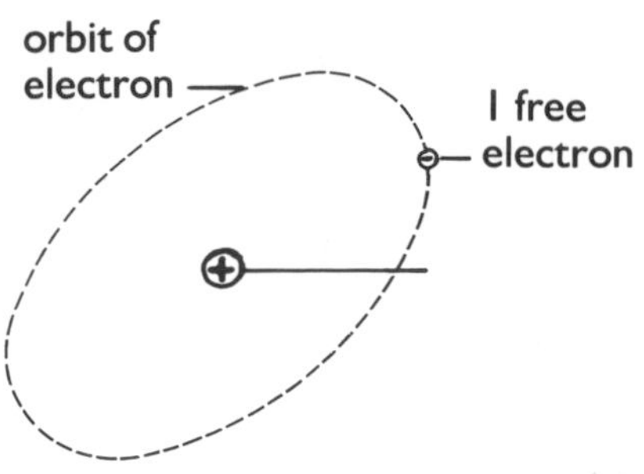

Hydrogen atom— *diagram of structure*

Every living body contains an enormous number of different substances, many with complex chemical structures. The study of these is called **biochemistry**. To understand basic physiology we need only consider the commonest of these substances.

ELEMENTS

About 24 different elements are found in the body, the most abundant being carbon (C), hydrogen (H), oxygen (O) and nitrogen (N). For convenience and brevity symbols will be used for these and also for many of the commonly occurring compounds.

The **atom** (the smallest unit) of any element is electrically neutral because it has an equal number of positively charged particles (**protons**) in its core (**nucleus**) and of negatively charged electrons **orbiting** outside the nucleus. The nuclei of atoms (except H) also contain uncharged particles (**neutrons**) which contribute to the atomic mass.

Some elements and their symbols

Metallic		**Non-metallic**	
Calcium	Ca	Carbon	C
Iron	Fe	Chlorine	Cl
Potassium	K	Fluorine	Fl
Sodium	Na	Iodine	I
Magnesium	Mg	Nitrogen	N
Non-metallic, replaced by metals in salts		Oxygen	O
Hydrogen	H	Phosphorus	P
		Sulphur	S

COMPOUNDS

Compounds are formed when atoms are **bonded** (joined) together. The smallest unit of a compound is a **molecule**. Molecules have very different characteristics from the atoms of which they are composed. Bonding to form compounds occurs when electrons in the outermost orbit of an atom are transferred to another atom—**ionic** bonding, or are shared—**covalent** bonding. The number of electrons available or needed to form a stable bonding is the **valency** of the element concerned.

Ionic bonding

Electrons are transferred. The atom (or group of atoms) which loses one or more electrons becomes a positively charged **ion**, e.g. H^+, Na^+, K^+, Ca^{2+}. The atom (or group of atoms) which gains the electrons becomes a negatively charged ion, e.g. Cl^-, OH^- (hydroxyl), CO_3^{2-} (carbonate), SO_4^{2-} (sulphate). The number before the + or − sign indicates the valency, which must be satisfied to hold the oppositely charged ion in the molecule, e.g. sodium chloride will be NaCl but sodium sulphate will be Na_2SO_4. The numbers below the line indicate the number of atoms of each element involved in the molecule.

Covalent bonding

Electrons are shared and neither of the combining atoms loses or gains. The bonding may be between identical atoms or between atoms of different elements. The gases hydrogen, oxygen and nitrogen usually exist as bonded pairs. Organic compounds contain carbon atoms covalently bonded to other carbon atoms to form chains, rings and branching molecules. Each carbon atom is capable of making four covalent bonds. Those which are not occupied by other carbon atoms are often occupied by hydrogen, oxygen, the hydroxyl group, nitrogen, or, through oxygen, such groups as phosphate. Nitrogen has three valency bonds. It never makes chains or rings of its own, but can become involved in carbon rings. Of special importance is the amino group $-NH_2$ which, with the organic acid group $-COOH$, occurs in all amino acids.

Elements covalently bonded

Hydrogen	Single bond	H−H	H_2
Oxygen	Double bond	O=O	O_2
Nitrogen	Triple bond	N≡N	N_2

−C− carbon atom with 4 available valencies

−C−C−C−C−C−C− chain as in saturated fatty acid

−C=C−C−C=C−C− chain with some double bonds as in unsaturated fatty acids

ring as in the hexose sugar, glucose

branching as in some amino acids

characteristic amino acid group $R-C\begin{smallmatrix}COOH\\NH_2\end{smallmatrix}$ or $R-C\begin{smallmatrix}C(=O)-OH\\N-H_2\end{smallmatrix}$

Shapes of some organic molecules

Compounds are classified into two groups:

1. **Inorganic compounds**. These, with the exception of carbon monoxide (CO), carbon dioxide (CO_2), carbonic acid, carbonates and bicarbonates, do not contain carbon. Water and the electrolytes found throughout the body are inorganic compounds.
2. **Organic compounds** invariably contain carbon. Most also contain hydrogen and oxygen, but many other elements may be included, e.g. nitrogen, sulphur, phosphorus and iron. Organic compounds are characteristic of living material, but some have been made (**synthesised**) artificially.

Water

A large proportion of the body (about 60%) is water, in which other substances are dissolved or suspended. Pure water is neither acid nor alkaline. Its molecule, H_2O, is formed of H^+ (hydrogen ion) and OH^- (hydroxyl ion), but the bonding is very firm and there is very little dissociation of water into its ions. However, it provides an environment in which many dissolved substances can readily ionise and upset the balance of H^+ to OH^- present, thus producing acidic or alkaline conditions.

Note. Fatty (lipid) substances are insoluble in and immiscible with water.

Electrolytes

These are compounds whose molecules can dissociate into their component ions when in solution in water, yielding positively charged **cations** and negatively charged **anions**. In **acids** the cations are H^+. In **salts** the hydrogen ions of the acid are replaced by metallic ions. A **base** or **alkali** is a compound in which metallic cations are paired with hydroxyl ions.

$$H^+Cl^- + Na^+OH^- \rightleftharpoons Na^+Cl^- + H_2O$$

hydrochloric acid ionised in solution; sodium hydroxide ionised in solution; sodium chloride ionised in solution; water unionised

Some acids and their salts

Acids		**Salts**	
Hydrochloric	HCl	Chloride	e.g. NaCl
Sulphuric	H_2SO_4	Sulphate	e.g. Na_2SO_4
Phosphoric	H_3PO_4	Phosphate	e.g. $Ca_3(PO_4)_2$
Carbonic	H_2CO_3	Carbonate	e.g. Na_2CO_3
		Bicarbonate	e.g. $NaHCO_3$

Not all electrolytes dissociate equally readily. A weak acid like carbonic acid which does not ionise easily can reduce the ionisation of strong alkalis, while similarly weakly alkaline basic salts like sodium bicarbonate can reduce the free hydrogen ions of strong acids. This reaction is called **buffering** and, in the body, prevents extreme fluctuations of pH.

pH is a measure of the relative concentrations of hydrogen ions and hydroxyl ions. It is calculated as the logarithm (to base 10) of the reciprocal of the concentration of hydrogen ions, therefore the more acidic the solution the lower the pH value. pH 7 is neutral.

Physiological functions of electrolytes

1. Buffering—see above.
2. Establishment of correct osmotic gradients—see pages 99, 119 and 120.
3. Assistance in enzyme activity—see pages 38, 99 and 102.
4. Production of the excitable state in cell membranes—see pages 61, 69 and 119.

CHEMICAL REACTIONS

Any change in bonding is a **chemical reaction**.

In the case of electrolytes, most of the changes take place in solution and with very little energy involved. Sometimes in the exchange of ion partners insoluble substances are formed and deposited (precipitated) or gases are released.

Changes to organic compounds may take place in a number of ways, the most important of which are:

1. **Rearrangement** of atoms within the molecule, e.g. glucose⇌fructose.

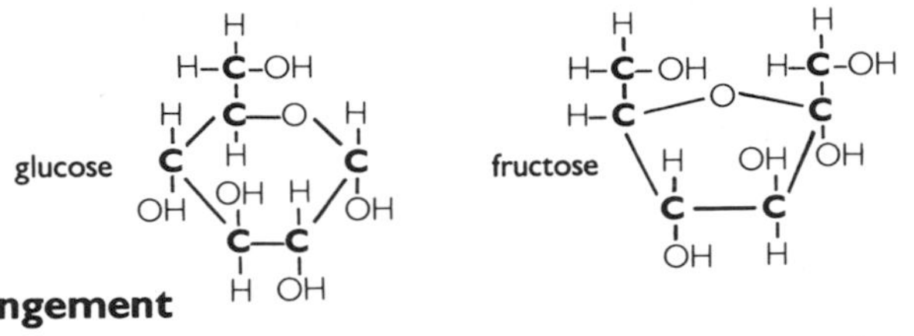

glucose   fructose

Rearrangement

2. **Hydrolysis/dehydration synthesis,** i.e. the addition or removal of water during destruction or formation of a bond. This is the way that fatty acids are joined to glycerol to form fats; hexose sugars (monosaccharides) are joined together to form disaccharides and further joined to form polysaccharides such as starch in plants and glycogen in animals including man; and amino acids are joined by peptide bonds to form first dipeptides, then tripeptides, polypeptides and eventually proteins. **Phosphorylation** is a dehydration synthesis in which one of the units is phosphate.

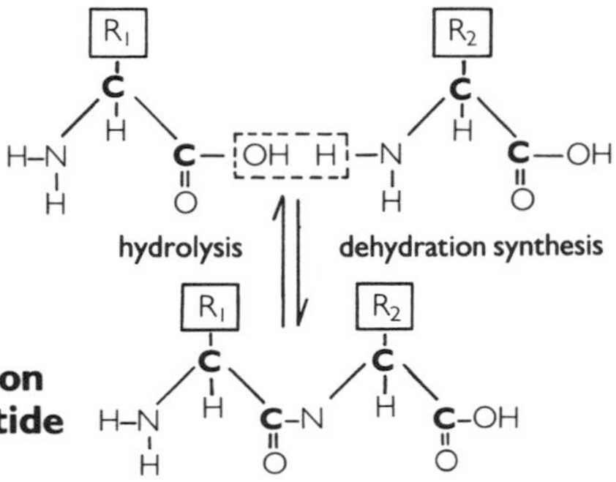

hydrolysis   dehydration synthesis

Formation of a peptide link

3. **Oxidation/reduction**. Oxidation literally means addition of oxygen, but more commonly in organic chemistry the reaction involves removal of hydrogen ions and electrons, i.e. **dehydrogenation**. Reduction is the opposite of oxidation. Frequently one substance is reduced as another is oxidised. The series of events by which glucose ($C_6H_{12}O_6$) is oxidised to $6CO_2+6H_2O$ involves a large number of dehydrogenation processes.

4. **Hydrogen bonding**. Hydrogen sometimes bonds with an oxygen or nitrogen atom on a fluctuating basis which produces a weak, temporary link or 'bridge' between large molecules or parts of the same molecule.

Note. **Binding sites** are positions in a complex organic molecule which are available for bonding with other molecules or ions. A **ligand** is a molecule or ion whose shape is complementary to the binding site of a particular molecule.

ENERGY AND METABOLISM

Metabolism comprises all chemical reactions which take place within the living body and the energy changes which accompany them.

Energy may take a number of forms, which are interchangeable in appropriate circumstances. The energy changes involved in the vital processes are measured in **joules** (J), or the more old-fashioned heat units, **calories**—1 cal = 4.19 J. In any chemical reaction energy is needed to break one or more bonds and is released during formation of others. The important quantity is the net result, i.e. the change in energy, whether **intrinsic** (retained internally in the bonds of the molecules) or released as **heat** or as **mechanical** energy, often with a change in volume.

Anabolic reactions, whereby smaller molecules are built up into larger ones, require the addition of energy.

Catabolic reactions, whereby larger molecules are broken down, release energy.

In both anabolic and catabolic activity there are chains of reactions called **metabolic pathways**, which may include internal rearrangements of atoms, some requiring and some yielding energy.

Activation of reactions

Energy is needed to start reactions. The function of **enzymes** is to lower the energy threshold, so that reactions within the body can proceed at lower temperatures. The enzymes are **catalysts**.

Enzymes are proteins and are therefore damaged by heat (heat coagulation). The optimum temperature for human enzyme action is about 37 °C i.e. normal body temperature. During the catalysed reaction, the molecules concerned (substrate) become temporarily bonded to the catalyst enzyme. After the reaction is complete the enzyme molecules are released for re-use. Sites of attachment are highly specific and, in some cases, **co-enzymes** are also needed as **carriers**. Named enzymes frequently carry the suffix **-ase**. The reaction is usually written:

$$\text{substrate} \xrightarrow{\text{enzyme}} \text{products}$$

Reactions are also influenced by environmental conditions and concentrations of the reacting substances and products. Some are readily reversible, others always proceed one way.

Energy transfer

Energy is needed by the living body for a large number of purposes, e.g. muscular activity, nerve action, pumping of substances in and out of cells and anabolic processes of many kinds, including growth and repair. The most abundant substance from which free energy is rapidly available is **adenosinetriphosphate (ATP)**, composed of adenine–ribose–3phosphate units.

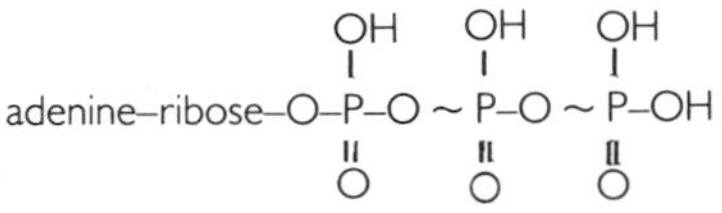

adenine–ribose–O–P(OH)(=O)–O ~ P(OH)(=O)–O ~ P(OH)(=O)–OH

The symbol ~ signifies that when this bond is **hydrolysed** relatively large amounts of free energy are liberated. Some of this is in the form of heat, but the rest can do useful work. The reaction is catalysed by **ATP-ase** and **adenosinediphosphate (ADP)** is produced. ADP can be **rephosphorylated** to ATP when energy is available from catabolic processes inside the cells. The most important of these involves oxidation of glucose, but fats and proteins can also be used.

$$ATP + H_2O \rightleftharpoons ADP + \text{phosphate} + \text{energy}$$

Thus ATP is an intermediary reserve substance, between catabolic sources of energy and energy-using activities.

SOURCES OF ENERGY

Carbohydrates, **proteins** and **fats** taken in the food—see page 94—can all be used as sources of energy. For nutritional calculations the average amount of energy available from 1 g of carbohydrate or protein is 18 J (4.3 cal) and from 1 g of fat 39 J (9.3 cal). The energy is released by **oxidation (dehydrogenation)**. Glucose is the basic carbohydrate raw material.

The oxidation of glucose

The oxidation of glucose ($C_6H_{12}O_6$) takes place within every living cell and results in the liberation of large amounts of energy, 34% of which is used for the conversion of **ADP** to **ATP** for storage till required, while the rest is in the form of heat.

The complete oxidation process takes place in three stages—**glycolysis**, the **Krebs cycle** and **electron transport**—and results in the production of six molecules each of **carbon dioxide** and **water**.

1. **Glycolysis**. This is a chain of 10 reactions, each catalysed by a specific enzyme, at the end of which each 6-carbon molecule of glucose produces two 3-carbon molecules of **pyruvic acid** plus 4H which are transferred to a co-enzyme known as **NAD (nicotinamide adenine dinucleotide)** to form **2NADH$_2$**. Though some energy is used in the early stages of the chain, there is a net gain of 2ATP per glucose molecule plus a small amount of heat energy which helps to maintain optimum conditions for enzyme action. Summarised:

$$1 \text{ molecule glucose} \rightarrow 2 \text{ molecules pyruvic acid} + 2ATP + 2NADH_2$$

If conditions in the cell are **anaerobic** (lacking in oxygen) the pyruvic acid is reduced to **lactic acid** using the $NADH_2$. Lactic acid can be converted back to pyruvic acid either in the cells when conditions become **aerobic** or in the liver. Pyruvic acid then passes on to the Krebs cycle.

2. The **Krebs cycle** or **citric acid cycle**. Before the actual cycle starts the pyruvic acid is **decarboxylated** with the loss of CO_2 and 2H transferred to form $NADH_2$. The remaining 2-carbon unit or **acetyl group** becomes attached to a carrier substance known as **co-enzyme A (CoA)**. This combines with **oxaloacetic acid** from the end of a previous cycle, **citric acid** is produced and CoA released for re-use. A series of molecular changes then occurs, some involving incorporation of water as a result of which 2 more molecules of CO_2 are released and 8H are transferred to co-enzymes, 6 to NAD to form $3NADH_2$ and 2 to **FAD (flavine adenine dinucleotide)** which becomes **FADH$_2$**. A high-energy compound **GTP (guanine triphosphate)** is also produced from which energy can be transferred to produce ATP. Summarised:

$$2 \text{ pyruvic acid } (C_3H_4O_3) \rightarrow 6CO_2 + 8NADH_2 + 2FADH_2 + 2ATP$$

Note. When proteins and fats are used as sources of energy the preliminary processing produces acetic acid which then enters the Krebs cycle and thereafter follows the same oxidation path as carbohydrates.

3. **Electron transport**. This is the final series of oxidation–reduction reactions. Hydrogen from $NADH_2$ (2 from glycolysis and 8 from the Krebs cycle) is transferred to FAD forming $FADH_2$, liberating NAD for re-use and forming 10ATP. $FADH_2$ reduces **co-enzyme Q** (CoQ). Reduced CoQ then liberates **2H⁺** and **2 electrons**. The latter are transferred via a series of iron-containing co-enzymes called **cytochromes**. Eventually the electrons are transferred to **oxygen** molecules, forming **oxygen ions**, O^{2-}, which combine with the $2H^+$ released from CoQ. Water is formed.

During the cytochrome transfer 2ATP is produced. For the $12FADH_2$ from the original glucose molecule this means 24ATP from this part of the system; $6O_2$ is used and $12H_2O$ produced. ($6H_2O$ is equatable to the water used during the Krebs cycle).

Summarised: $10NADH_2 + 2FADH_2 \rightarrow 12FADH_2 + 10ATP$

$12FADH_2 + 6O_2 \rightarrow 12H_2O + 24ATP$

Overall summary: $C_6H_{12}O_6 + 6O_2 \rightarrow 6CO_2 + 6H_2O + 38ATP$

Note. All the enzymes and co-enzymes are re-usable and recycled freely so long as oxygen is ultimately available. Anaerobic periods hold all reactions back except those of glycolysis. Once glucose stores are used up, a shortage of ATP soon follows and the functions of the cell are inhibited. If shortage is prolonged the cell will die.

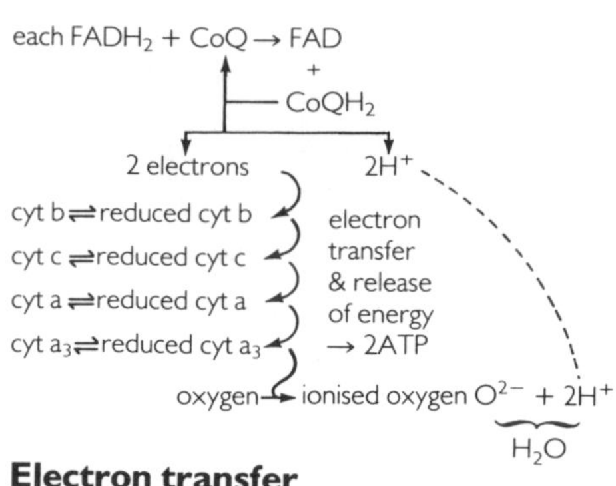

Electron transfer

PROTEINS

Proteins are of two types:
1. **Fibrous proteins** in which the molecules are filamentous and relatively insoluble. They are mainly structural, e.g. **collagen** of the connective tissues, **keratin** of skin and **muscle proteins**.
2. **Globular proteins** in which the molecules are folded into complex shapes and much more soluble, e.g. **haemoglobin** of blood, **albumens** and most **enzymes**.

Note. **Conjugated proteins** are those which are linked to other substances, e.g. lipoproteins (with lipids) or phosphoproteins (with phosphate).

Amino acids

Required in the diet	Can be synthesised
Arginine*	Alanine
Histidine*	Aspartic acid
Isoleucine	Cysteine
Leucine	Cystine
Lysine	Diiodotyrosine
Methionine	Glutamic acid
Phenylalanine	Glycine
Threonine	Hydroxylysine
Tryptophane	Hydroxproline
Valine	Proline
	Thyroxine
*can be made in	Tyrosine
small quantities	Serine

Structure of proteins

Proteins are formed of long chains of **amino acids**. 23 different amino acids are commonly involved in living material. Eight of these cannot be synthesised in the human body and two cannot be made in adequate quantities. These must be present in the diet. The remaining 13 can be manufactured, using amino groups from other amino acids and various complex synthetic pathways.

The peptide link which joins the amino acids together is formed by dehydration synthesis—see page 3—only in conjunction with minute cell granules called **ribosomes**—see page 6. The order in which the amino acids occur in any protein is critical to the functioning, particularly of enzymes, and is templated by **nucleic acids**, so called because they were first identified in the nuclei of cells—see page 7.

NUCLEIC ACIDS

Nucleic acids are made up of units called **nucleotides**, each of which contains a **pentose** (5-carbon) sugar, **phosphate** and a **nitrogenous base**. There are two types of nucleic acids, known as **DNA (deoxyribonucleic acid)** and **RNA (ribonucleic acid)**.

1. **DNA** occurs as linked double strands like a coiled ladder. The uprights of the ladder are formed of alternating sugar (**deoxyribose**) and **phosphate** groups while the rungs are formed of paired bases: **adenine–thymine**, **guanine–cytosine**. The bases occur in varied order, sections of about 1000 base pairs forming the average length of a protein template (**gene**) responsible for a particular developmental feature.

During cell division—see page 124—the **chromosomes**—see page 7—on which the genes lie become identifiable. There are about 20 000 genes on each of the 23 pairs of human chromosomes responsible for hereditary continuity. This continuity depends on accurate **replication** of the DNA strands. Replication starts from one end of the double coil (double helix). Each base attracts the appropriate partner, already combined with sugar and phosphate, and a link to the preceding nucleotide is made under the influence of the enzyme **DNA-polymerase**.

2. **RNA** is like a single strand of DNA but with **ribose** as the sugar and a row of protruding nucleotides in which **uracil** replaces thymine. RNA is synthesised by **transcription** from sections of one strand of DNA which temporarily separates from its partner. The RNA is a mirror image of the section of active strand except for the uracil opposite each adenine.

RNA is synthesised in the nucleus—see page 7—and carries the genetic instructions to other parts of the cell to regulate protein synthesis.

Protein synthesis

Translation is the mechanism whereby the instructions transcribed from the DNA to the RNA are interpreted to produce the correct sequence of amino acids in the proteins. It involves three types of RNA: **mRNA** (messenger RNA), **tRNA** (transfer RNA) and **rRNA** (ribosomal RNA found in ribosomes—see page 6).

mRNA consists of long strands, each of which templates a complete protein. Units of three nucleotides along a strand form **codons** (coding units) which match **anticodons** borne on small strands of tRNA. The tRNA molecule is twisted into a cloverleaf shape so that the anticodon section is correctly exposed. With some energy input the appropriate amino acid is picked up, but the anticodon can only link with the mirror image codon if a ribosome with its rRNA is also present. The ribosome becomes attached to the mRNA, reads the codon, attracts the anticodon and promotes the formation of a peptide bond before it moves on to read the next codon. The protein grows lengthwise till a **termination codon** is reached. The assembled protein is then complete and is released. The ribosome breaks into its two sub-units until next required.

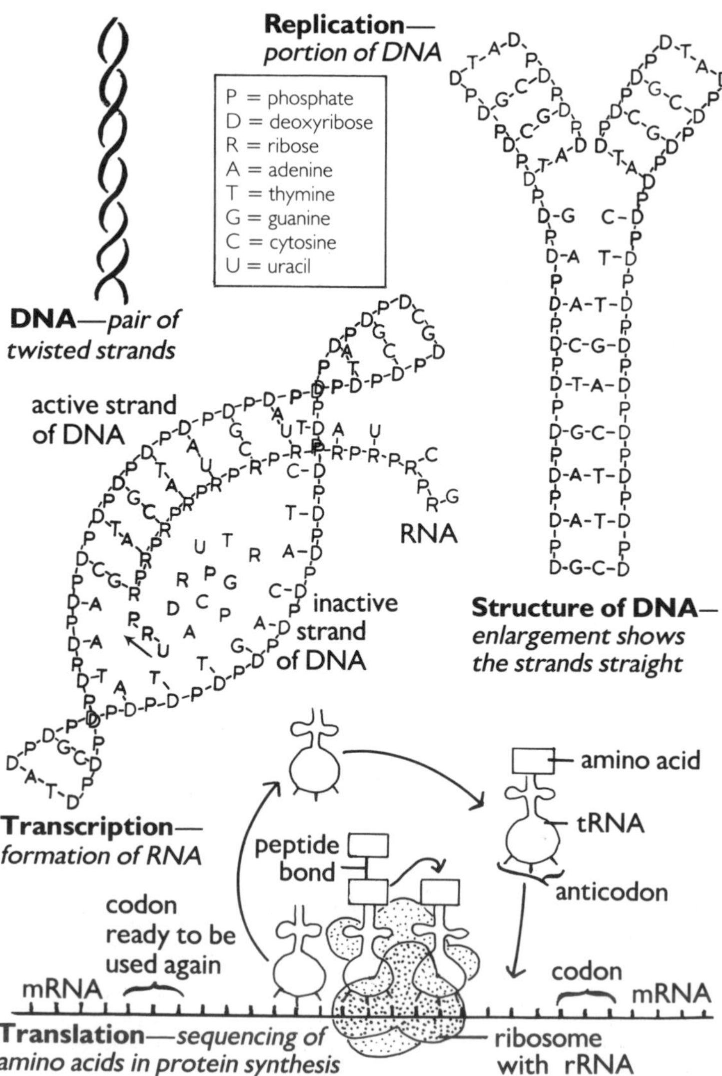

Cells

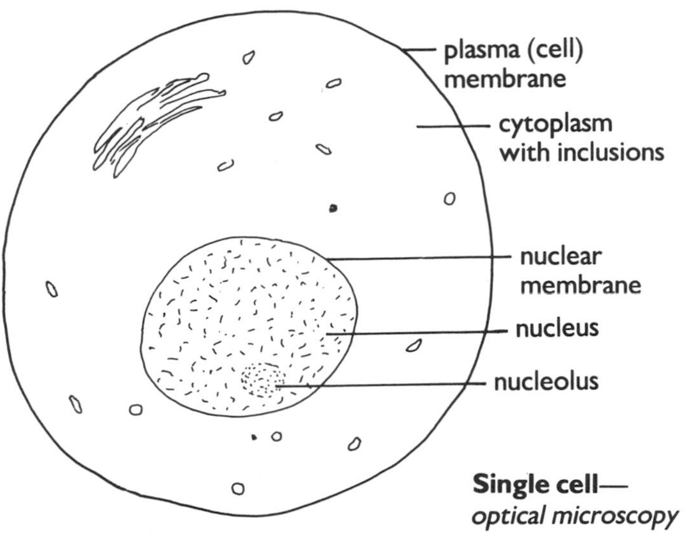

Single cell— *optical microscopy*

The body is composed of many millions of living cells and non-living intercellular substance (**matrix**) between them. Each cell consists of an outer **cell membrane (plasma membrane)**, **cytoplasm** and **organelles** including at least one **nucleus**. It may also contain non-living **inclusions**.

Cells vary in size from about 7.5 to 300 μm in diameter (1 μm = 1/1 000 000 m). Some nerve cells may have processes extending to over 1 m in length and muscle fibres vary in length according to the state of contraction. Shape may also be affected by juxtaposition of other cells. With a good optical microscope, capable of magnification up to ×1000, and suitable stains, it is possible to identify cell outline, nucleus or nuclei and some granulations and fibres, and also the gross appearance and quantity of intercellular matrix. With an electron microscope, capable of magnification up to ×200 000, details measurable in Ångström units can be identified (1Å = 1/10 000 μm). This and the use of radioactive tracers has clarified the distribution and functions of many of the cell components.

PLASMA MEMBRANE

This is a pliable surface membrane, 65–100 Å thick, which separates the cell contents from the surroundings. It is constructed of a double layer of **phospholipid** molecules with **integral protein** and **peripheral protein** molecules. Some of the integral proteins can form minute temporary channels, while others combine with carbohydrates to form **glycoproteins**. Some of the peripheral proteins serve as enzymes while others are believed to be involved as scaffolding or to assist changes of membrane shape.

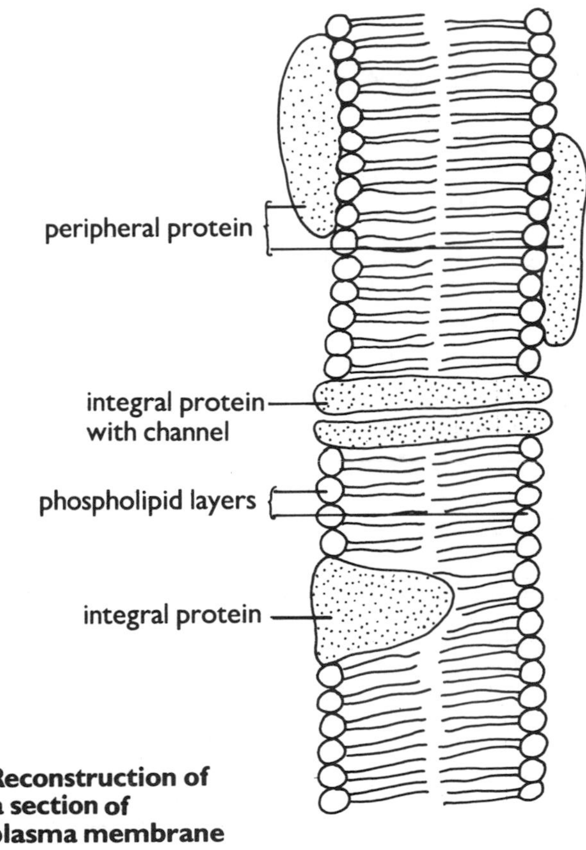

Reconstruction of a section of plasma membrane

Functions of the plasma membrane

1. **Selective permeability**. This depends on:
(a) **size of molecules**—water and amino acids pass readily while proteins are too large for easy passage;
(b) **solubility**—substances which dissolve in lipids, e.g. other lipids, oxygen and CO_2, pass more readily than those which are only water soluble;
(c) **ionisation**—the proteins of the membrane can be ionised and attract ions of opposite charge which then pass through;
(d) **carrier molecules**—e.g. the transport of glucose in a combined form which makes it lipid soluble.

2. **Passive transport**. This is movement across the membrane without use of energy from the cell. It depends on the potential energy of the concentration gradients. Three processes are involved:
(a) **diffusion**, which is the tendency to equalise the concentration of any dissolved substance (or gas);
(b) **osmosis**, which occurs when the membrane is impermeable to a dissolved substance but permeable to water;
(c) **dialysis**, which results in separation of larger molecules to which the membrane is impermeable from smaller ones which can pass through provided there is a suitable diffusion gradient.

3. **Active transport**. This uses energy derived from the breakdown of ATP. It is responsible for absorption of such substances as glucose from the alimentary canal—see page 88, or from the kidney tubules back into the blood—see pages 38, 61 and 119, and also for the establishment of ionic gradients—see page 119. Active intake is helped by temporary pores in the integral protein units.

4. **Recognition**. Glycoproteins in the cell membrane are capable of recognising similar proteins, hormones and nutrients and also of identifying foreign cells bearing different glycoproteins and thus initiating protective responses—see page 101.

The plasma membrane and shape

As all exchanges between the contents of the cell and the environment take place at the surface, the ratio of the area of surface to the cell volume is important. A basic globular shape produced by surface tension alone gives a minimum surface/volume ratio for the size of the cell. All but the very smallest cells tend to be flattened or elongated or to have folds and processes to increase the extent of the plasma membrane and reduce the distance from the outside of the cell to its core. Special examples include:
(a) **cups** of goblet cells—see page 87;
(b) **microvilli** of intestinal cells—see page 87;
(c) **pseudopodia** which project to engulf solid particles into vacuoles, a process called **phagocytosis**;
(d) **pockets** which grow inwards to surround liquid droplets into vacuoles, a process called **pinocytosis**.

Note. The vacuoles formed by phagocytosis and pinocytosis are lined with plasma membrane. After material in the vacuoles has been processed, any unwanted residues are voided to the exterior and the membrane components are recycled.

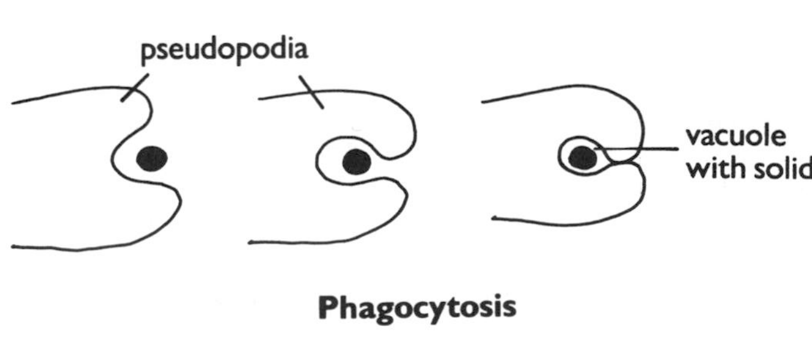

Phagocytosis

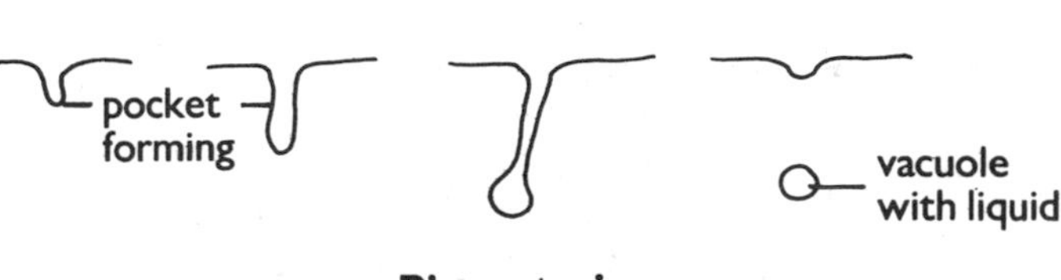

Pinocytosis

CYTOPLASM

Cytoplasm is a semitransparent fluid bounded by the plasma membrane. It is up to 90% water with inorganic substances and sugars in true solution, but most other organic compounds in a colloidal state, i.e. dispersed, but not truly dissolved. This gives the cytoplasm **viscosity** (stickiness).

The shape of the cytoplasmic unit is influenced by juxtaposition of neighbouring cells and by the presence of the **cytoskeleton** of minute filaments and tubules, which also help to support and disperse other cytoplasmic **organelles**. Of these organelles the **centrosome** and **ribosomes** are free in the cytoplasm, but the **endoplasmic reticulum**, **Golgi apparatus**, **lysosomes**, **mitochondria** and **nucleus** have outer membranes of the same structure and selective permeability as the plasma membrane.

Cytoplasm is the site of numerous chemical reactions on which life depends. The differentiated parts of the cell have specific roles and keep critical chemical activities from interfering with one another.

A cytoplasmic unit with more than one nucleus is said to be **syncytial**.

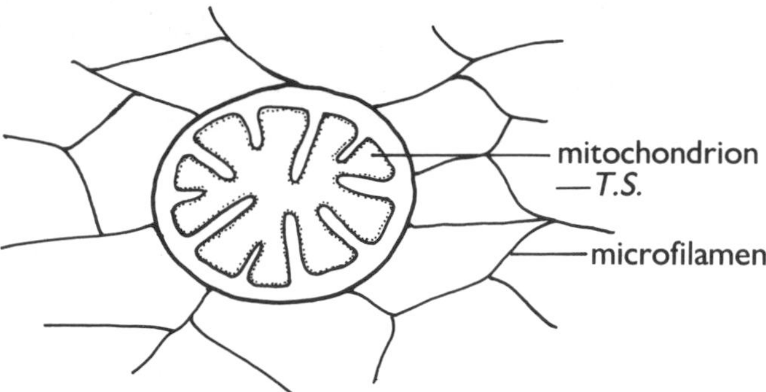

Microfilament lattice holding a mitochondrion

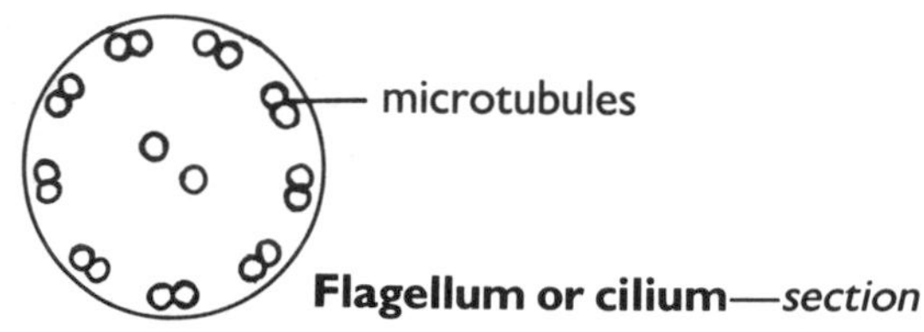

Flagellum or cilium—*section*

Cytoskeleton

The cytoskeleton includes microfilaments and microtubules.

1. **Microfilaments** form a lattice which supports the other organelles. They are 30–120 Å thick, vary in length, and are scattered or regularly arranged according to the type of cell. They are formed of proteins and help movement within the cell. They are involved in phagocytosis, pinocytosis and the contractility of muscle—see page 38.
2. **Microtubules** average about 240 Å in diameter. They are largely made of the protein **tubulin**, and while they may act as conducting channels, their main function is supportive, assisting movement of pseudopodia, flagella and cilia. **Flagella** form the tails of spermatozoa—see page 126. **Cilia** are shorter and more numerous vibratile projections lining the respiratory tract—see page 95. In both flagella and cilia the whip-like motion is influenced by the asymmetry of the nine pairs of microtubules around two core tubules in each flagellum or cilium.

Centrosome

This is a dense region of cytoplasm close to the nucleus. It contains DNA—see page 4, and is self-replicatory. It also contains two **centrioles**, each of which is made up of nine bundles of three microtubules. These are important in spindle formation during cell division—see page 124. The absence of centrosomes from mature nerve cells makes replacement after damage impossible.

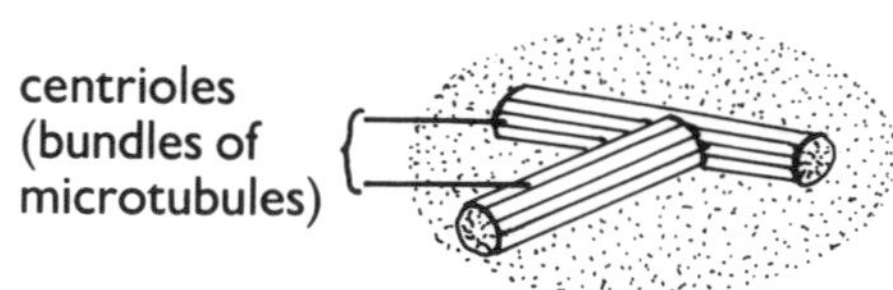

Centrosome

Ribosomes

Ribosomes are so called because they contain **ribonucleic acid** (RNA)—see page 4, which is bonded to protein to form granules of maximum size 250 Å. Each ribosome has two parts, one about twice the size of the other. These unite temporarily during the cycle of active manufacture of protein—see page 4. The small sub-unit binds to the messenger RNA (mRNA) while the large sub-unit contains the enzymes needed to make the peptide bond. Several ribosomes can follow one another along the mRNA strand when rapid protein synthesis is required.

Some ribosomes remain free in the cytoplasm, either in clusters or in chains, where they are loci for the manufacture of enzymes and other proteins needed within the cell, but many are associated with the endoplasmic reticulum and assist synthesis of enzymes and other proteins for export from the cell.

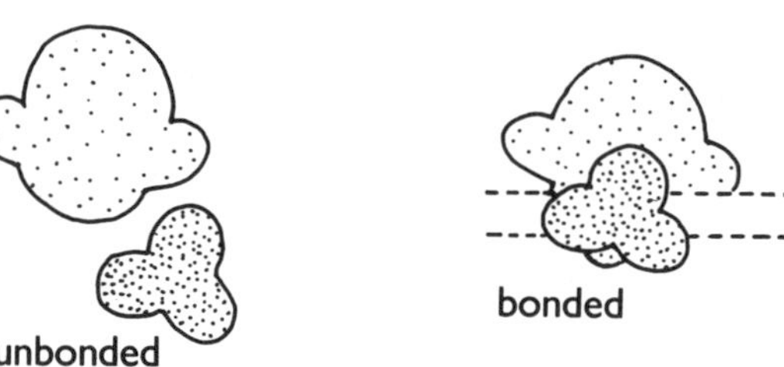

Ribosomes

Endoplasmic reticulum

This consists of pairs of parallel membranes enclosing narrow cavities and thus forming a network of channels called **cisternae**. The membranes are continuous with the plasma membrane and the nuclear membrane and also connect with the Golgi apparatus.

The endoplasmic reticulum contributes to the mechanical support of the cytoplasm and conducts intracellular impulses, e.g. in muscle cells where it is called the sarcoplasmic reticulum—see page 38. Its large surface area allows it to be actively involved in chemical reactions, while the cisternae provide spaces in which internal circulation can take place and in which substances harmful to the cytoplasm can be stored.

The endoplasmic reticulum is of two types:

(a) **rough**—with numerous ribosomes attached to it and synthesising proteins;

(b) **smooth**—without ribosomes and believed to be concerned with synthesis of certain lipids, including those of the plasma membranes, and also of some carbohydrates.

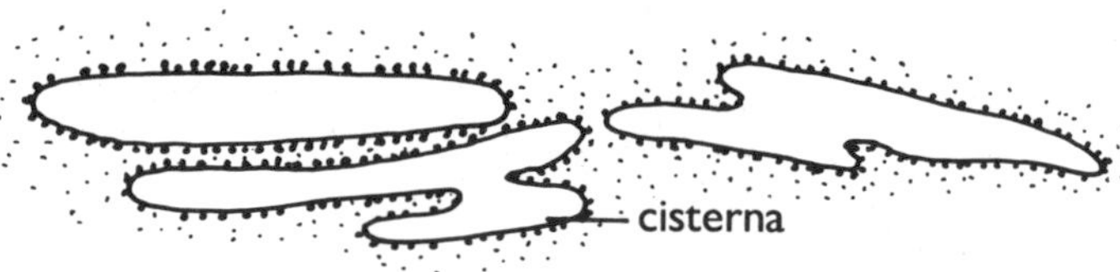

Rough endoplasmic reticulum

cisternae

Smooth endoplasmic reticulum

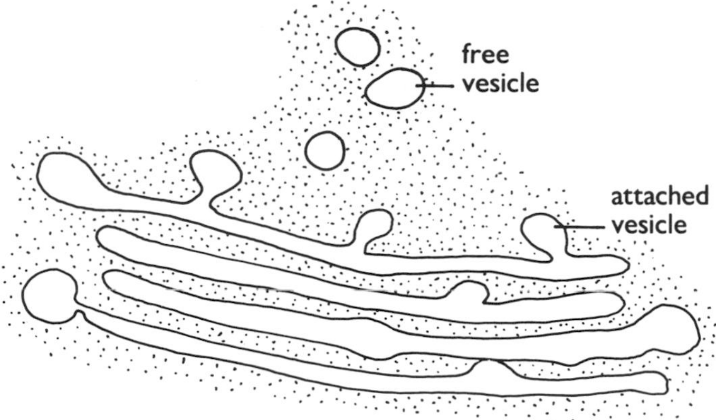

Golgi apparatus—*section*

Mitochondria

These are minute, usually sausage-shaped structures and 1.5 μm long. Each mitochondrion has a double wall of similar structure to plasma membrane. The outer wall forms a simple sac, but the inner wall is much folded. The folds, called **christae**, protrude into the fluid-filled cavity and bear numerous round knobs attached by stalks. These knobs contain the enzymes involved in the energy-releasing reactions which convert ADP to ATP—see page 3. Active cells such as those of the liver may have more than 1000 mitochondria, while inactive, short-life cells like spermatozoa have only about 25.

Mitochondria contain DNA which differs from that in the rest of the cell. They are self-replicatory in response to need for ATP, i.e. for energy-giving oxidative processes. During cell division they are shared rather than regenerated under genetic instructions, thus the supply of energy is uninterrupted.

Nucleus

A nucleus is present in all cells except mature red blood cells where its loss limits activity and life span. It is usually spherical or oval, but may become lobulated, as in some white blood cells—see page 100. It is limited by a double-layered nuclear membrane of plasma membrane type. The outer layer is continuous with the endoplasmic reticulum, while the inner layer is produced by the nucleus itself. The membranes are fused at intervals to form pores up to several hundred Å in diameter, which communicate with the endoplasmic cisternae and permit passage of large RNA molecules.

The **nucleoplasm** or **karyolymph** is gel-like. It contains one or more **nucleoli** and threads of **chromatin**. The nucleoli consist mainly of RNA, stored until required. They are especially large in active secretory cells such as those of the pancreas. The chromatin (so called because of its affinity for stains) appears granular because its threads are made up of thicker units called **nucleosomes** joined by continuous DNA strands. Each nucleosome has about 200 base pairs of DNA associated with proteins called **histones**. The DNA is responsible for templating the RNA which in turn templates the arrangement of amino acids in proteins—see page 4. In order to maintain the accuracy of these instructions, accurate replication of DNA strands is necessary before any of the cell divisions which take place within the body, whether for growth or repair or for hereditary continuity.

Prior to any cell division the chromatin strands in the nucleus become twisted into dense helical coils and shortened to form rods known as **chromosomes**. Replication of DNA strands precedes the shortening so that each chromosome is formed of two double strands called **chromatids**, twisted round one another and adherent to one another at a point called the **centromere**. Human cells, with the exception of eggs and sperm, have 23 pairs of chromosomes. The full details of cell division are described later—see page 124.

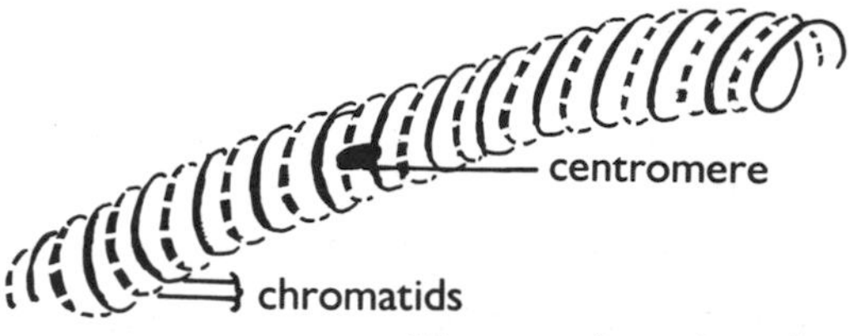

Chromatin twisted and shortened to form a chromosome

Golgi apparatus

This is like the smooth endoplasmic reticulum (ER) but with four to eight wider **cisternae** and **vesicles** at the edges. The latter swell, become pinched off and migrate into the cytoplasm. They may contain:
(a) **proteins/enzymes** formed in the rough ER and concentrated in the Golgi apparatus;
(b) **lipids** formed by smooth ER;
(c) **glycoproteins**, **collagen** or **mucus** formed in the Golgi apparatus itself.

The vesicles may discharge into the cytoplasm or be carried to the surface of the cell for secretion. The Golgi apparatus is best developed in cells with special secretory functions, e.g. the pancreas or salivary gland, but is present and active in many others. It keeps secretions, especially powerful enzymes, away from the cytoplasm.

Lysosomes

These are produced as vesicles from the Golgi apparatus and contain powerful enzymes capable of breaking down any bacteria which may enter the cell, e.g. during phagocytosis by white blood cells—see page 100. When a cell is damaged they bring about destruction or **autolysis**, thus removing cell parts or unwanted extracellular material. Bone reshaping by osteoclasts—see page 8—is a form of autolysis.

Peridoxosomes are small vesicles abundant in liver cells and containing, among other enzymes, catalase which destroys hydrogen peroxide.

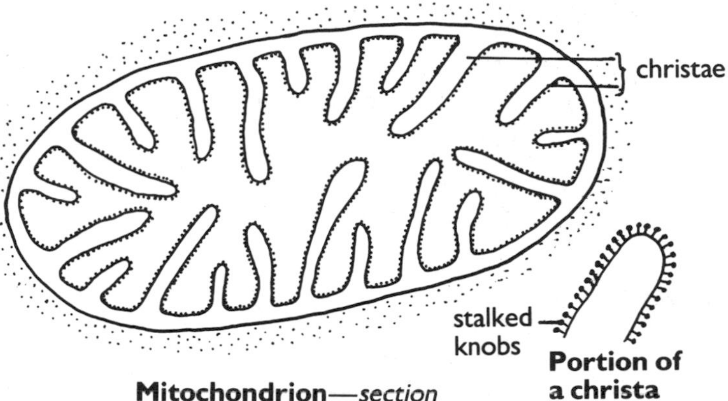

Mitochondrion—*section*

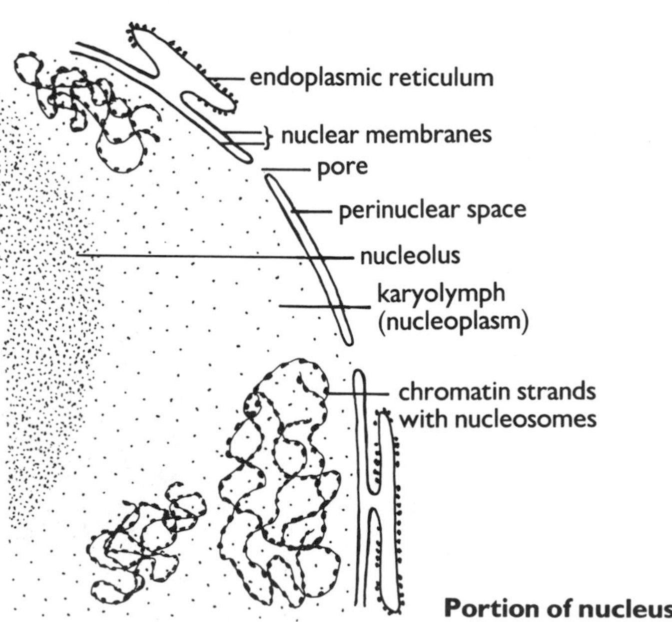

Portion of nucleus

CELL INCLUSIONS

These are a diverse group of substances found in cells, but taking no direct part in cell metabolism. They include:
(a) **stored carbohydrate**—chiefly glycogen, e.g. in liver and muscle cells;
(b) **stored fat**—e.g. in adipose cells—see page 81;
(c) **mucus**—in cells lining organs where lubrication is required, e.g. respiratory passages and alimentary canal;
(d) **melanin**—the pigment of skin, hair and eyes where it acts as a screen, particularly against ultraviolet light.

The Skeletal System

The skeleton is composed of two types of tissue, **cartilage** and **bone**, both of which are stiffened forms of **connective tissue**.

1. **Cartilage** is softer than bone, less rigid and slightly elastic. It has a clear (**hyaline**) matrix of **chondrin**, a gelatinous protein, strengthened by varying amounts of tough **collagen fibres** and **yellow elastic (elastin) fibres**, also proteins. Spaces in the matrix, called **lacunae**, are occupied by **chondrocytes**, which lie singly or in groups and give rise to new cartilage in the young. Later growth is from the surface membrane called the **perichondrium**.

Cartilage forms the temporary skeleton of the developing foetus, but is gradually replaced almost entirely by bone. It is retained throughout life on the articular (joint) surfaces of most bones and as the costal, nasal, laryngeal, tracheal, and bronchial cartilages—see pages 25, 74 and 96.

2. **Bone** is a rigid, non-elastic tissue. Its matrix is 67% **calcium salts** (chiefly phosphate with some carbonate) and 33% organic matter, mainly **collagen**. The cells, **osteocytes**, possess many fine processes which lie in minute **canaliculi** in the bone matrix. **Hard bone** is dense, with comparatively narrow **Haversian canals** around which the osteocytes are regularly arranged. It forms the surface layer of all bones and the whole of the tubular shaft of long bones. Spongy or **cancellate bone** is found inside hard bone and has irregular spaces filled with **red bone marrow**. The shafts of long bones have large marrow cavities filled with **yellow bone marrow**. Haversian canals and marrow cavities are penetrated by blood vessels from the **periosteum** around the bone.

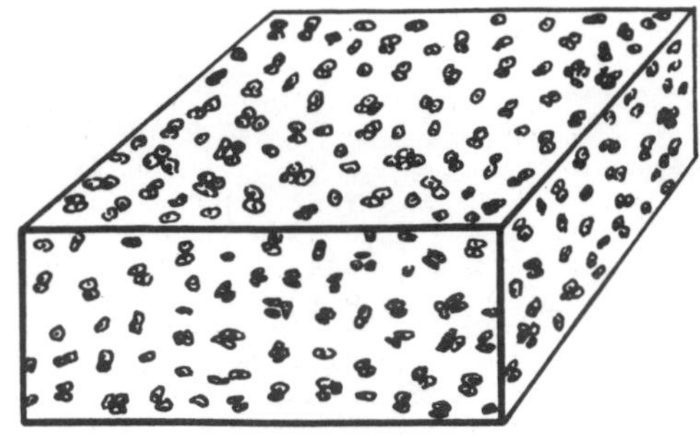

Block of cartilage

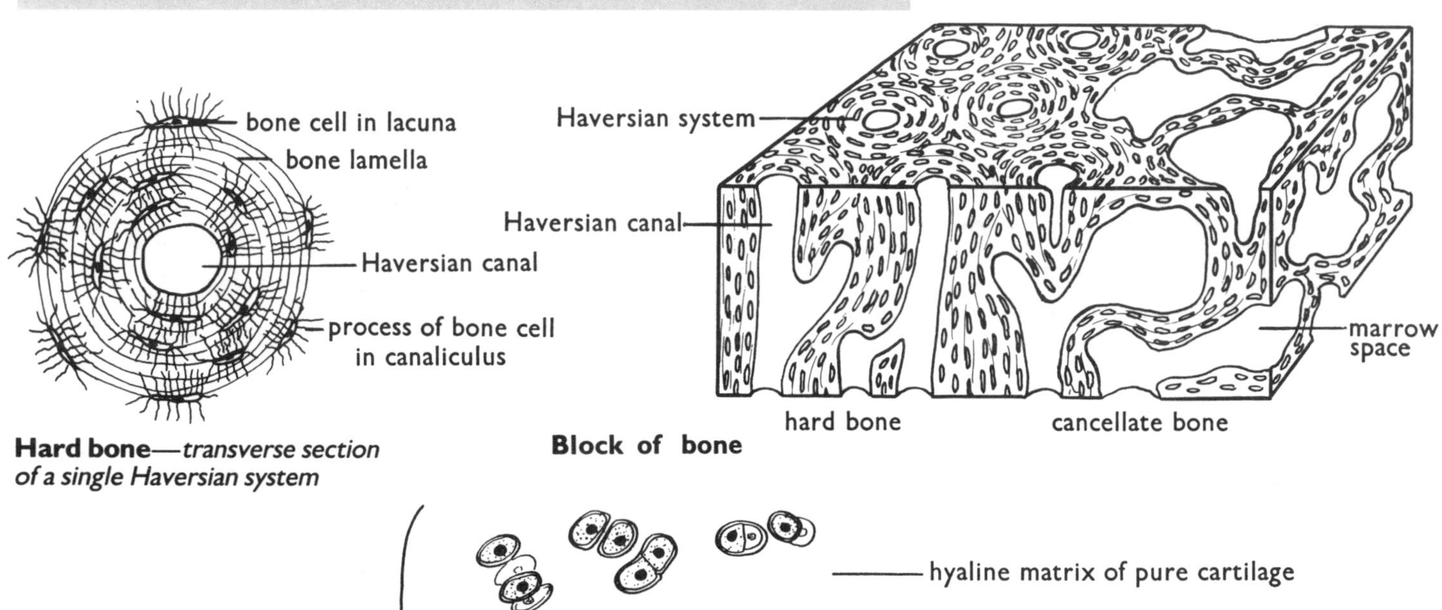

Hard bone—*transverse section of a single Haversian system*

Block of bone

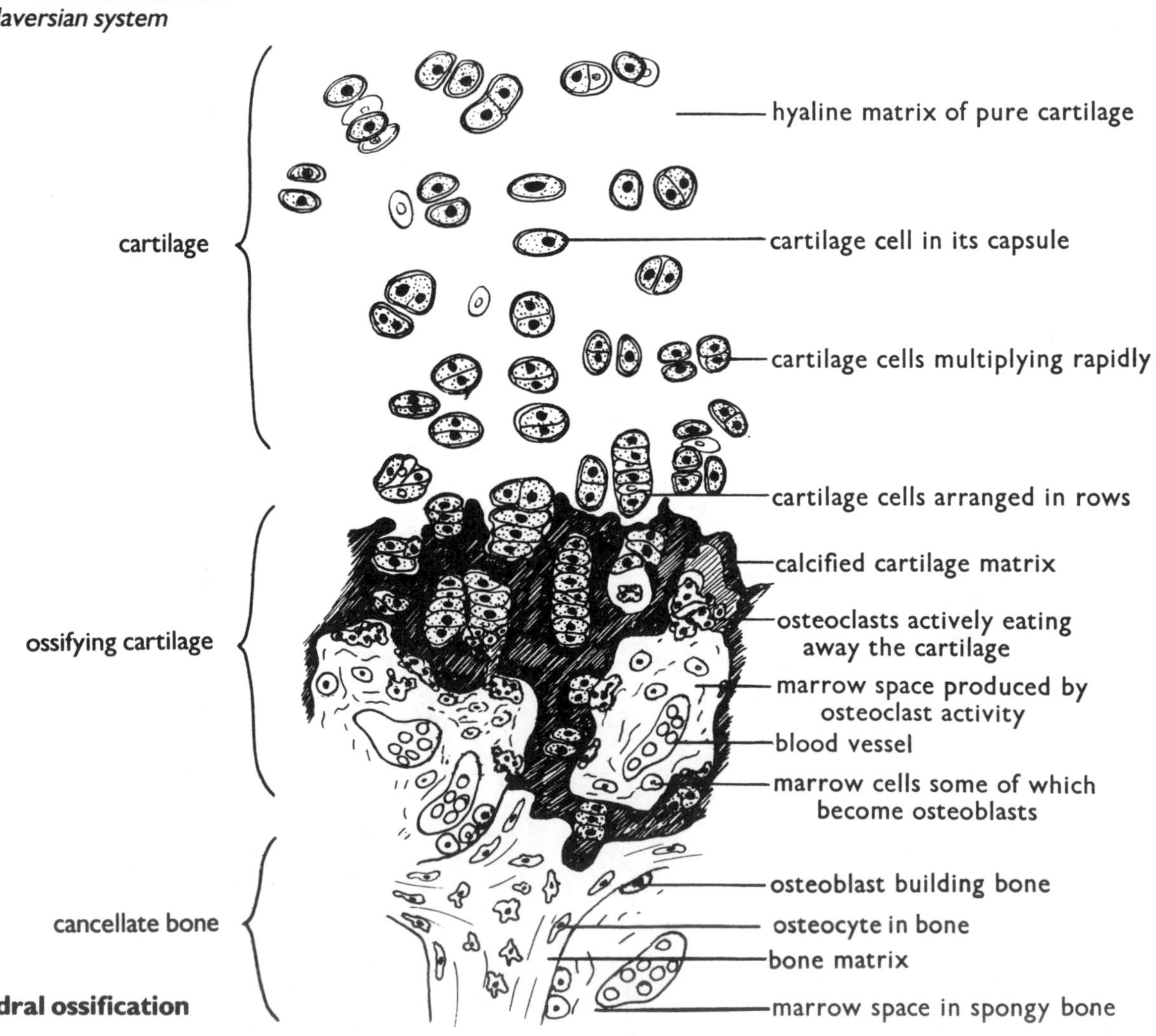

Endochondral ossification

OSSIFICATION

Ossification is the process by which bone is formed.

Some bones, e.g. the flat bones of the skull, parts of the mandible (lower jaw) and the clavicles (collar bones), form directly in membranes. The **osteoblasts** (bone-building cells) secrete a framework of **collagen** fibres in which **calcium salts** are deposited. As a lattice is built up the trapped **osteoblasts** become **osteocytes** and the surrounding connective tissue becomes the **periosteum**. Eventually the surface layers are reconstructed to form hard bone, but variable amounts of cancellate bone remain inside.

Most bones are formed originally by replacement of cartilage, though bone is added from the surrounding membrane during growth. The stages of the **endochondral ossification** are shown in the diagrams. **Osteoclasts** destroy unwanted cartilage by **autolysis**—see page 7, while osteoblasts build bone around themselves and become osteocytes. The **perichondrium** becomes the **periosteum**. There is continual remodelling, with additional bone from the periosteum, until ossification is complete.

By about the age of 25 the final size and shape is assumed. Thereafter there is still phased destruction and replacement, which enables worn or damaged bones to be repaired and also enables bone to act as a reservoir for the calcium needed for proper functioning of the other tissues, e.g. nerves—see page 69, muscles—see page 38, and blood—see pages 99 and 102.

Development of a long bone

The following series of diagrams illustrates the stages in transformation of a cartilage rudiment of a developing baby into a long bone and the subsequent growth pattern.

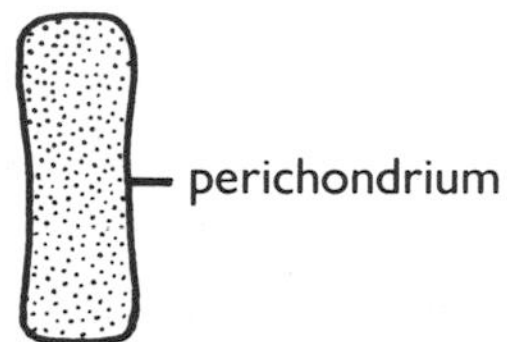

1. Cartilage.

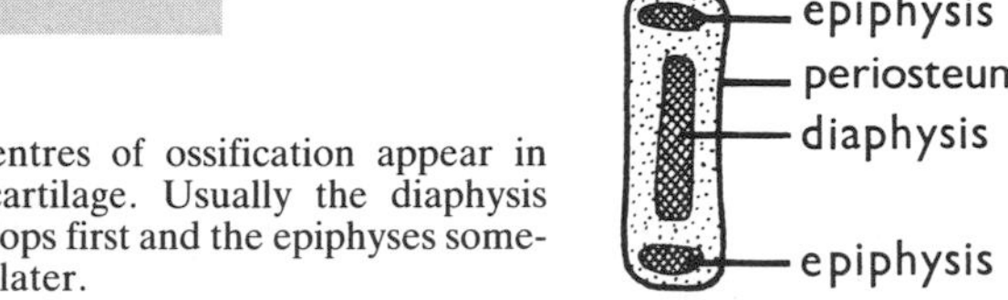

2. Centres of ossification appear in the cartilage. Usually the diaphysis develops first and the epiphyses somewhat later.

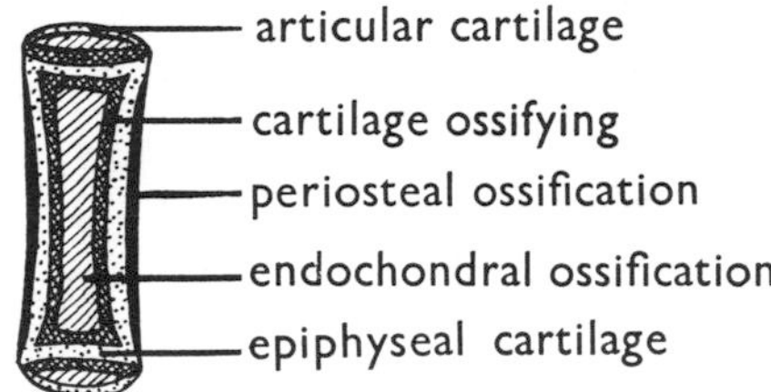

3. Endochondral ossification is the formation of bone from cartilage in the epiphyses and diaphysis. Periosteal ossification is the formation of bone from the periosteum surrounding the cartilage and the endochondral ossification. Epiphyseal and articular cartilages are retained.

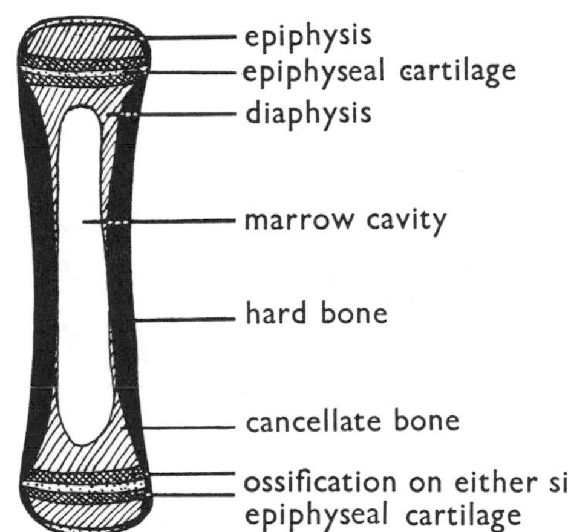

4. Continuous growth of the epiphyseal cartilages with ossification on both sides of each of them produces increase in length. Periosteal ossification produces increase in thickness. A marrow cavity is excavated in the shaft.

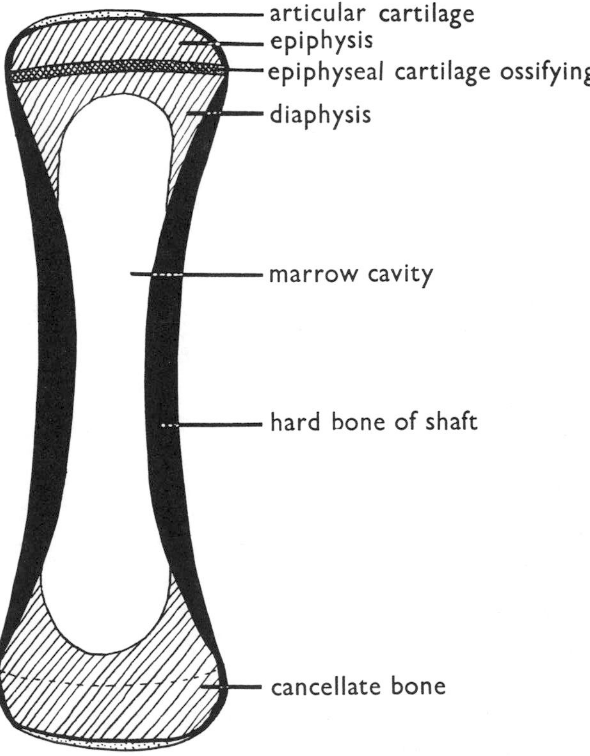

5. When growth in length is complete the epiphyseal cartilage ossifies and the epiphyses fuse with the diaphysis. The shaft is hard bone but cancellate bone with red bone marrow remains in the ends. The marrow cavity is filled with pulpy yellow bone marrow.

Terms used when describing features of bones

Shape—**long**, e.g. arm, leg, fingers and toes;
short, e.g. wrist and ankle;
flat, e.g. cranium, ribs, sternum (breast bone) and scapulae (shoulder blades);
irregular, e.g. vertebrae (spine bones) and facial bones;
Wormian—small irregular bones sometimes found in cranial sutures (joins);
sesamoid—small bones formed in tendons, e.g. patella.

Openings—**fissure** = narrow cleft;
foramen = hole;
meatus = tube.

Depressions—**sulcus** = groove;
sinus = air-filled space.

Processes (projections) forming a joint surface may have a
—**facet** = smooth flat surface;
head = rounded surface;
condyle = curved but not rounded surface.

Processes to which tendons and ligaments are attached may be a
—**line** = slight ridge;
crest = prominent ridge;
spine or spinous process = sharp knob;
tubercle = small blunt knob;
tuberosity = large blunt knob;
trochanter = large tuberosity of the femur (thigh bone);
epicondyle = process projecting above a condyle.

FUNCTIONS OF THE SKELETON

The functions of the skeleton are:
1. To **support** the soft parts of the body. All the tissues except cartilage and bone are soft. Without the skeleton the body would be flabby and could not stand up. The arrangement of the bones gives shape to the body as a whole.
2. To give **attachment** to the muscles. The skeleton is jointed to allow movement. The movements are brought about by muscles which are attached to the bones and so pull them into different positions. The bones act as levers.
3. To **protect** the more delicate parts of the body, e.g. the cranium protects the brain, the neural arches of the vertebral column surround the spinal cord, and the heart and lungs lie within the thoracic cage.

THE PARTS OF THE SKELETON

The skeleton may be considered in two sections: the **axial** skeleton which supports the parts on the main axis of the body, viz. the head, neck and trunk; and the **appendicular** skeleton which supports the appendages or limbs and gives them attachment to the rest of the body.

Note. In describing the parts of the skeleton, superior = above or upper; inferior = below or lower; anterior = in front; posterior = behind; lateral = side.

1. The **axial skeleton** is composed of the following parts:
skull—cranium 8 bones; face 14 bones; auditory ossicles 6 bones;
hyoid bone—1 bone;
vertebral column—33 vertebrae but some of them fused so that there are usually 26 bones;
sternum—3 bones;
ribs—12 pairs (24 bones).

2. The **appendicular skeleton** is composed of the following parts:
shoulder girdle—2 scapulae and 2 clavicles;
arm bones—1 humerus, 1 radius and 1 ulna in each arm;
wrist bones—8 carpal bones in each wrist;
hand bones—5 metacarpal bones in each hand;
finger bones—14 phalanges on each hand, 2 in each thumb and 3 in each of the other fingers;
hip girdle—2 pelvic bones each formed of an ilium, an ischium and a pubis;
leg bones—1 femur, 1 tibia and 1 fibula in each leg;
ankle and foot bones—7 tarsals and 5 metatarsals in each foot;
toe bones—14 phalanges on each foot, 2 in each great toe and 3 in each of the other toes.

Note. In addition there is a sesamoid bone, the **patella**, capping each knee and a variable number of other small sesamoid bones, especially in the tendons of the wrists.

The skull

The skull consists of two parts, the **cranium** or brain-box and the **face**, and has three small bones enclosed in the cavity of each ear.

1. The **cranium** is a complete box formed by the following bones:
1 **occipital** bone at the back of the skull;
2 **parietal** bones forming the hind part of the roof of the skull;
1 **frontal** bone forming the front part of the roof of the skull, the forehead, and the upper parts of the eye sockets or orbits;
2 **temporal** bones at the sides, above and around the ears;
1 **sphenoid** bone at the base of the cranium, with wings on either side forming the temples;
1 **ethmoid** bone between the frontal and the sphenoid, forming the roof of the nasal cavities.
Note. There may be small Wormian bones in some of the cranial sutures.

2. The **face** is supported by 14 bones:
2 **zygomatic** or **malar** bones in the cheeks;
2 **maxillae** forming the upper jaw, most of the side walls of the nose and front part of the hard palate;
2 **palatine** bones forming the rest of the hard palate and part of the side walls of the nose;
2 **lacrimal** bones, one in each orbit;
2 **nasal** bones forming the bridge of the nose;
2 **turbinate** bones inside the nose;
1 **vomer** forming part of the nasal septum;
1 **mandible** forming the lower jaw.

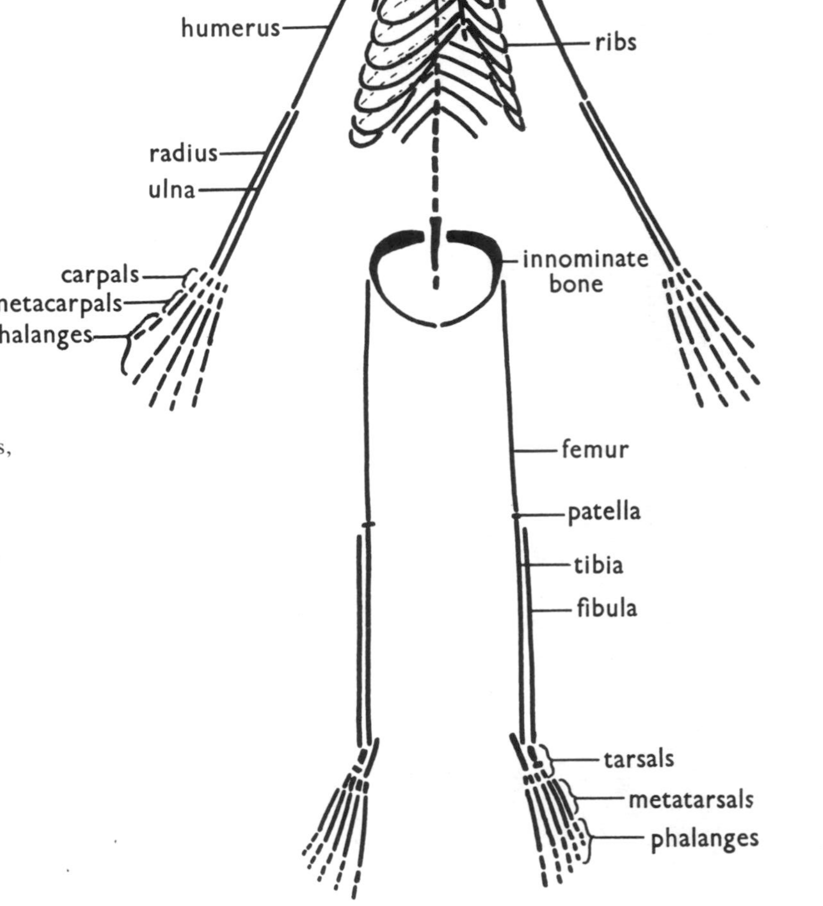

The arrangement of the parts of the skeleton

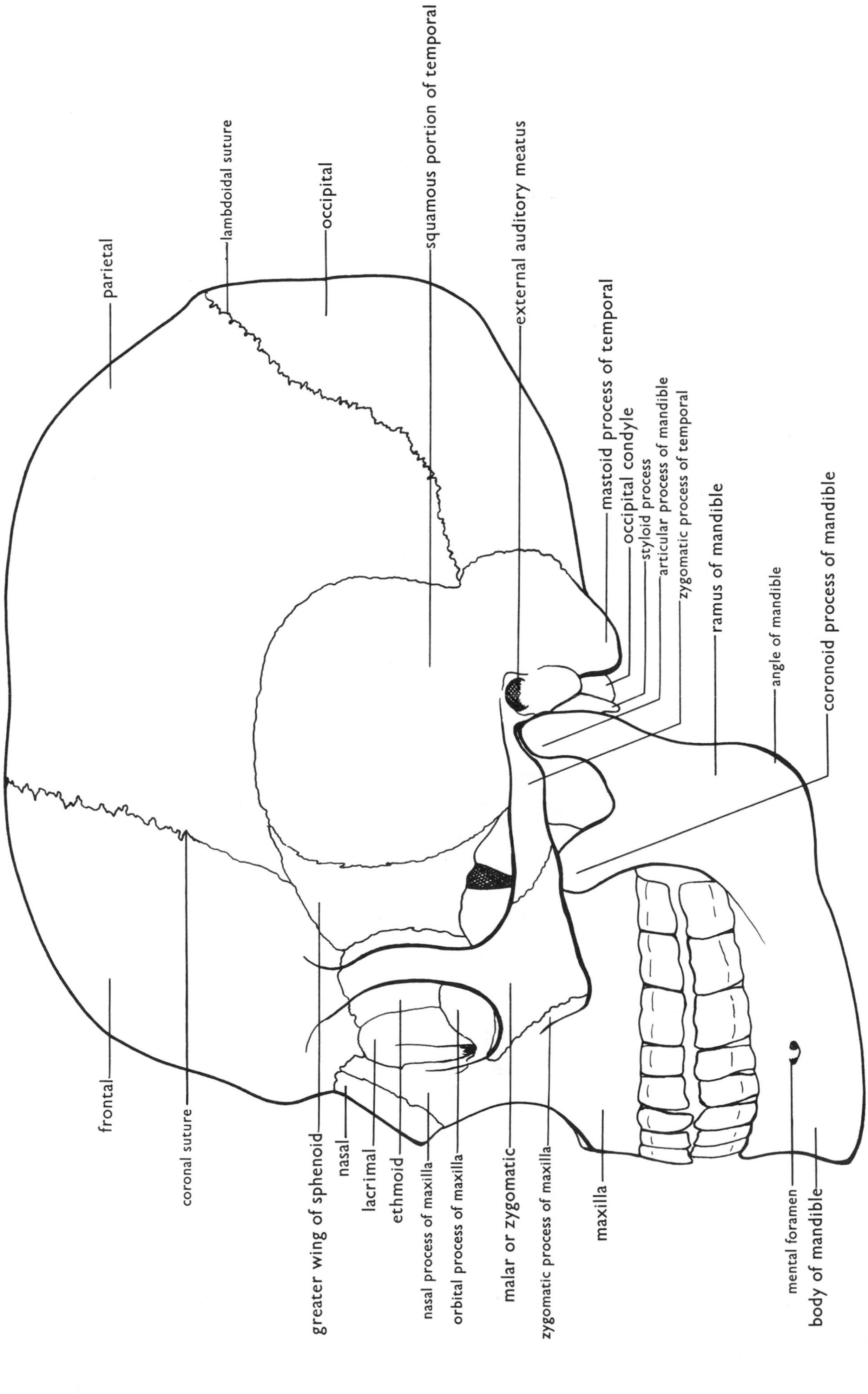

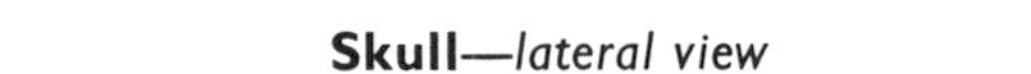

Skull—*lateral view*

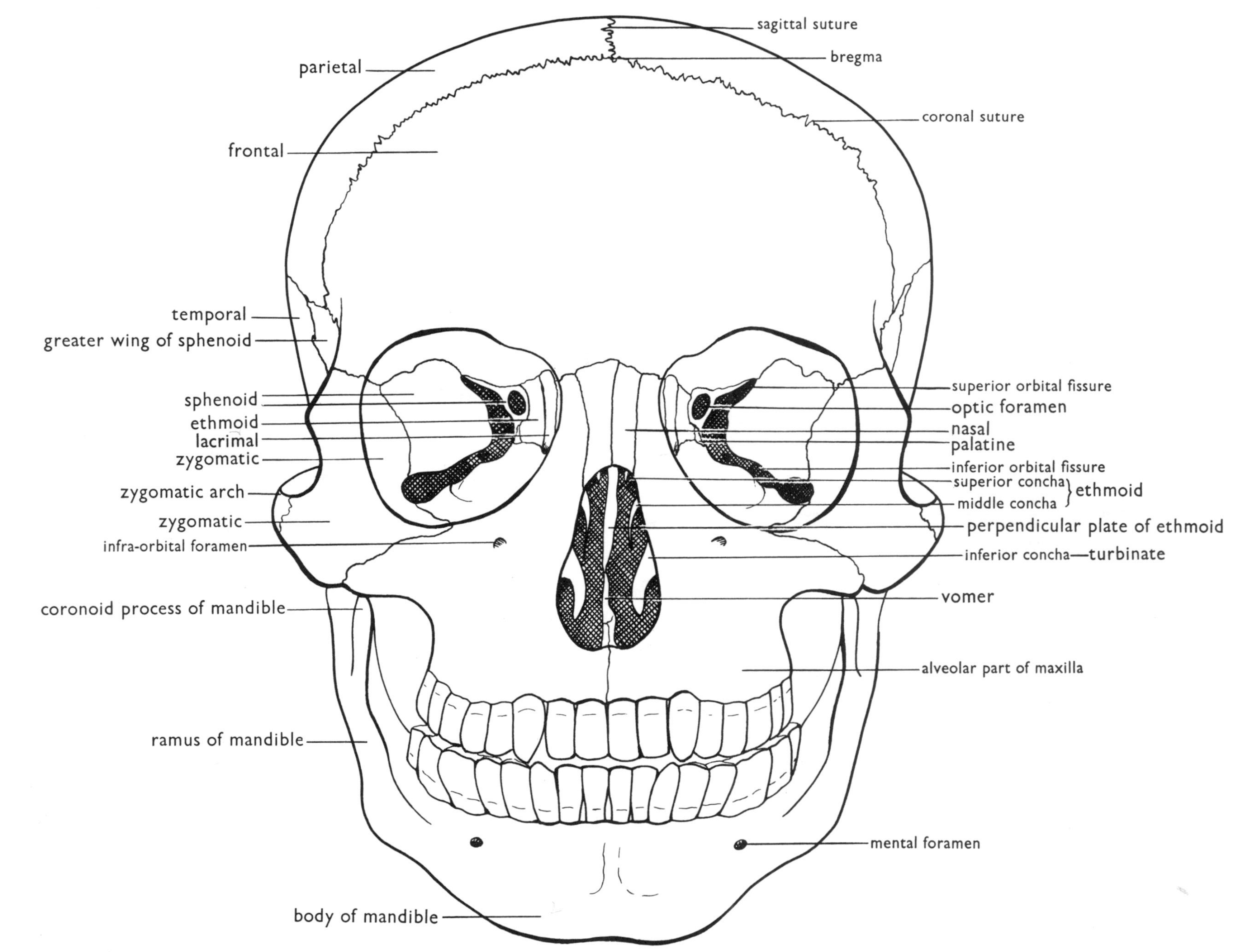

Skull—*frontal view*

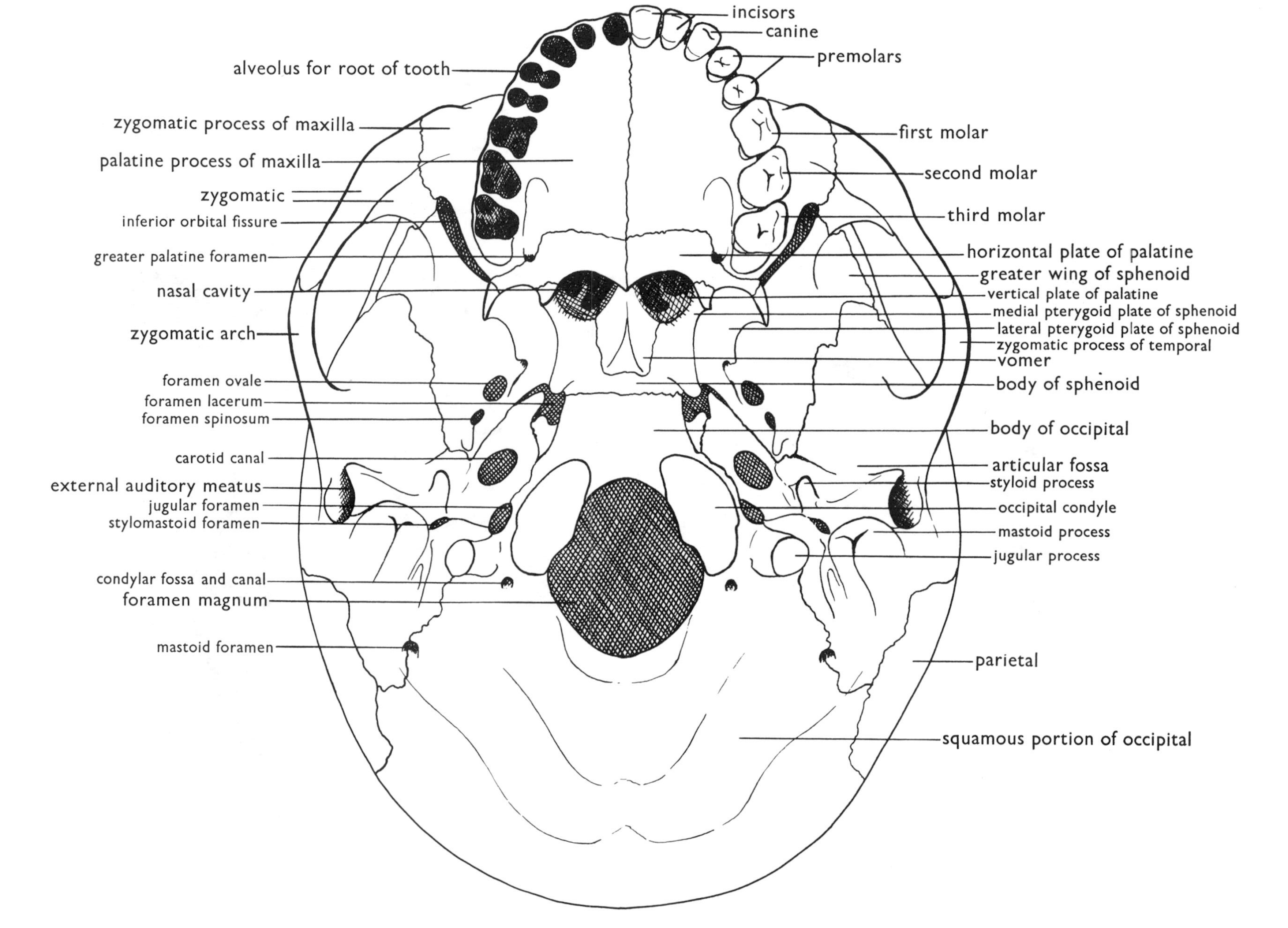

Skull—*external view of the base*

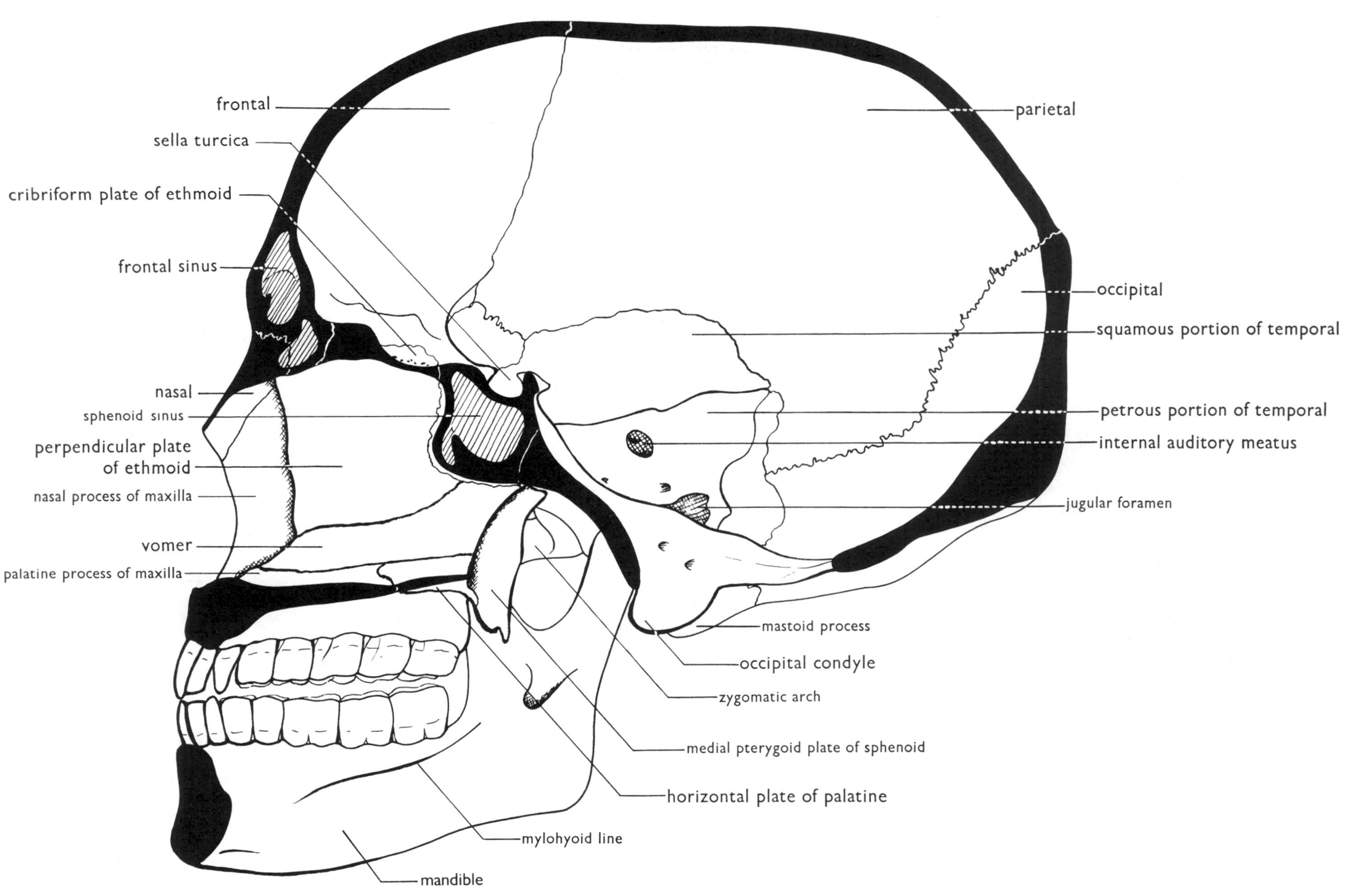

Skull—*vertical section slightly to the left of the mid-line*

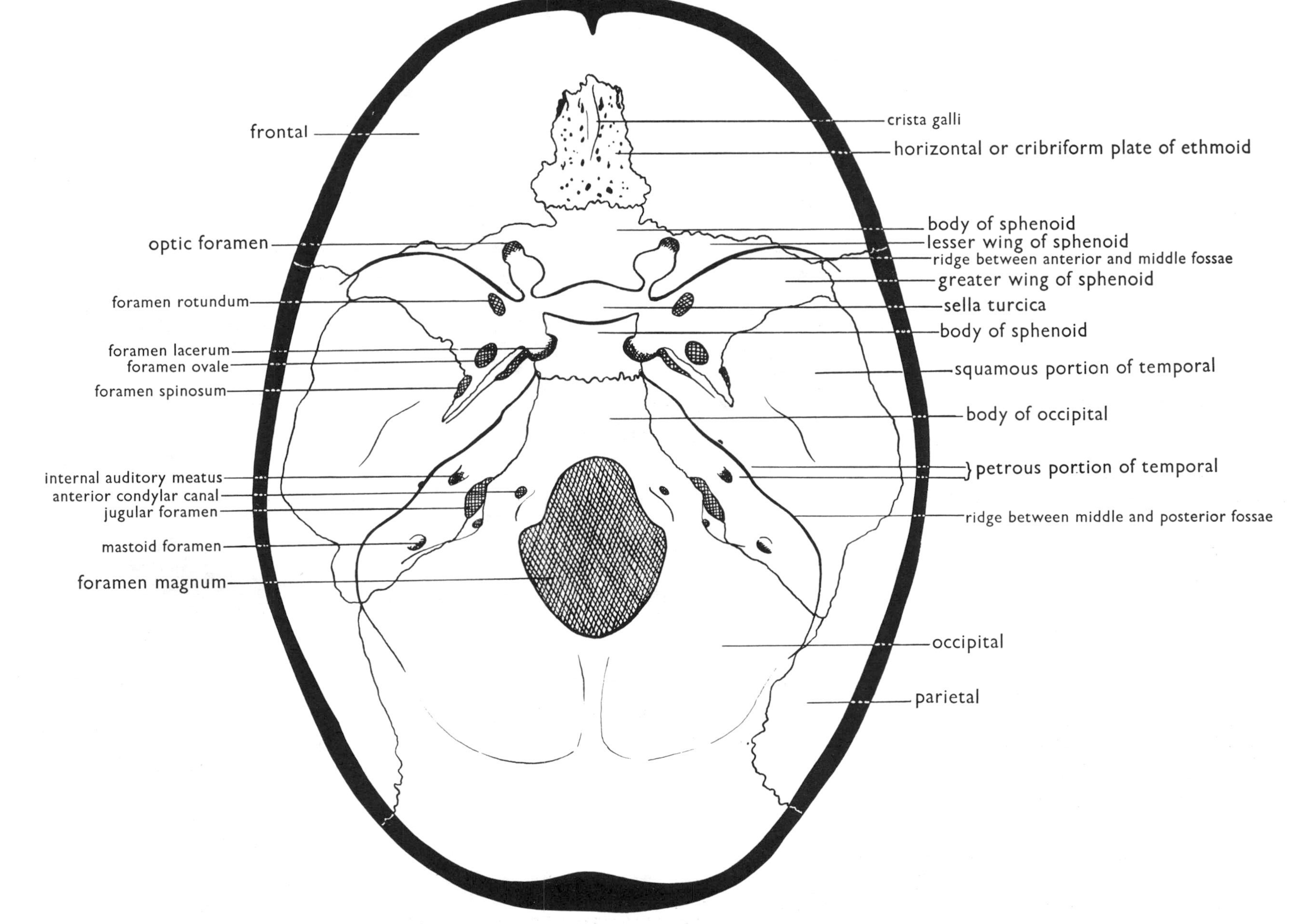

Skull—*view of base with crown removed*

The sinuses of the nasal region and the nasal cavity
—*superior and middle conchae removed*

The sinuses

The **paranasal sinuses** are air-filled spaces in the frontal, sphenoid, ethmoid and maxillary bones. They communicate with the nasal air cavities and are lined with mucous membrane, continuous with the nasal mucosa. The maxillary sinuses are very large and are also known as the **antra of Highmore**.

The **mastoid antra** (**sinuses**) are air-filled spaces in the mastoid portion of each temporal bone. They communicate with the tympanic cavities—see page 74.

The auditory ossicles

The auditory ossicles lie in the middle-ear cavity. They are a chain of three small bones stretching between the tympanic membrane and the fenestra vestibuli (ovalis)—see page 75. They are called the **malleus**, the **incus** and the **stapes**.

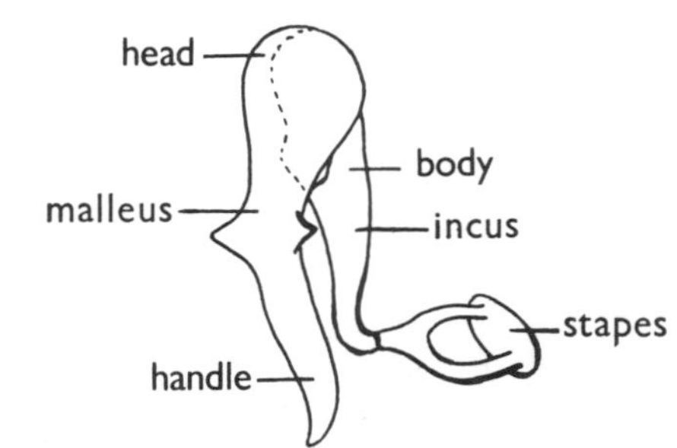

Auditory ossicles

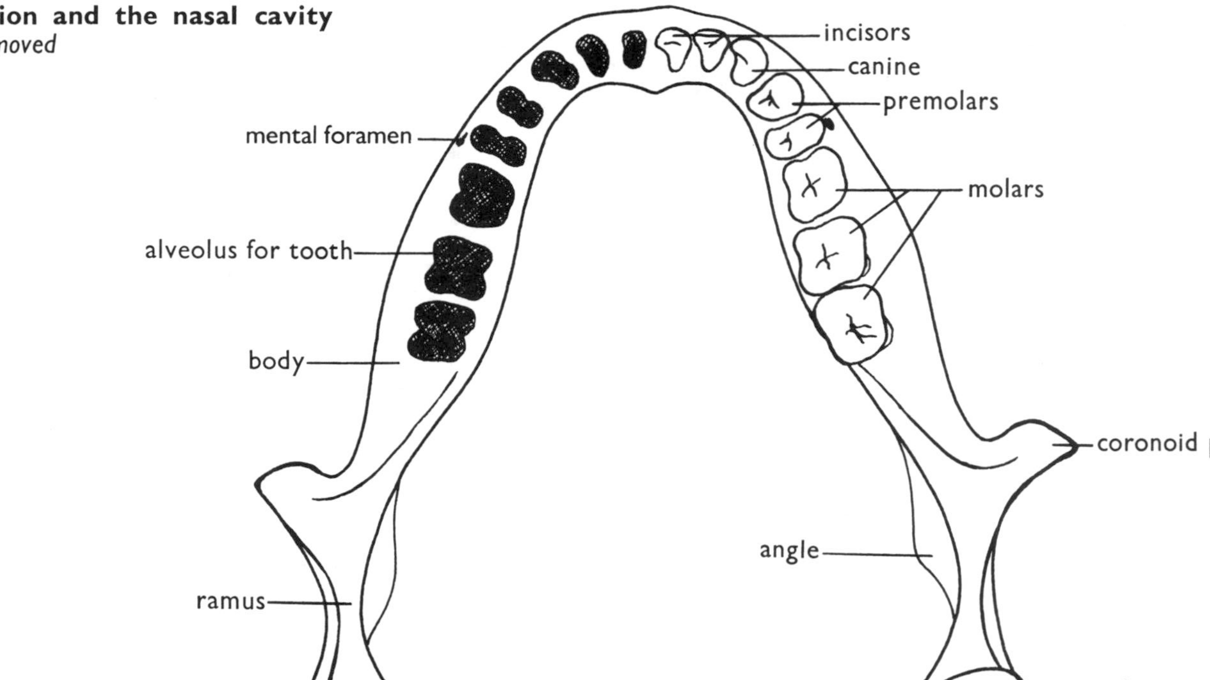

Mandible—*superior view*

The teeth

The teeth are set in sockets in the jaw bones called **alveoli**—see pages 13 and 16. Each tooth has a **crown** above the gum, a **neck** at gum level and a **root** embedded in the bone. The bulk of the tooth is made of a bone-like substance called **dentine**, formed by the activity of **odontoblasts** and hardened by deposits of calcium phosphate. The crown is capped with very hard, non-living **enamel** formed in the tooth-bud by **ameloblasts**. The root is invested in a thin layer of **cement** and joined to the jaw bone by the **periodontal membrane** in which there are blood vessels. The dentine is served by nerves and blood vessels which lie amongst spongy pulp cells in the central **pulp cavity**.

Two sets of teeth are developed during life. The tooth rudiments formed from the **dental lamina** are present in the jaws before birth, but erupt gradually as the jaws grow big enough to accommodate them. The temporary or **milk dentition** begins to appear at about five months and is usually complete by 20 months. This dentition consists of 20 teeth—2 incisors, 1 canine and 2 molars in each half of each jaw. (Though placed in front of the molars, the canines usually erupt later than the first molars.) The **permanent dentition** consists of 32 teeth. Twenty of these—2 incisors, 1 canine and 2 premolars in each half of each jaw—replace the milk teeth when these are shed between the ages of 7 and 11 and are larger than their precursors. In addition to these replacement teeth the permanent dentition has three groups of non-replacement teeth—the 6-year-old, 12-year-old and 18-year-old molars respectively. The last are also known as the wisdom teeth and are often very late in erupting.

incisors
canine
milk molars

Milk dentition

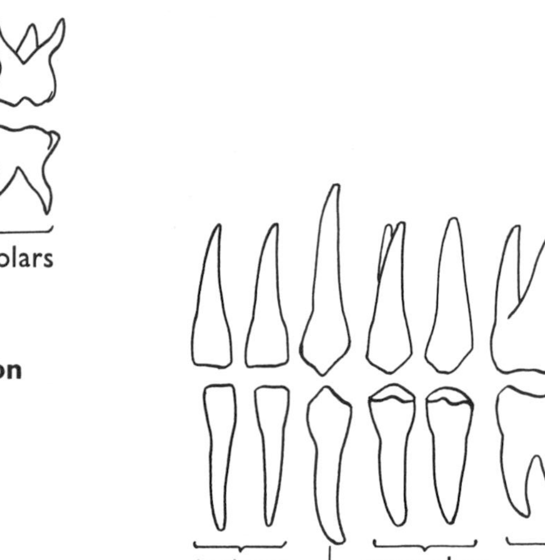

Permanent dentition

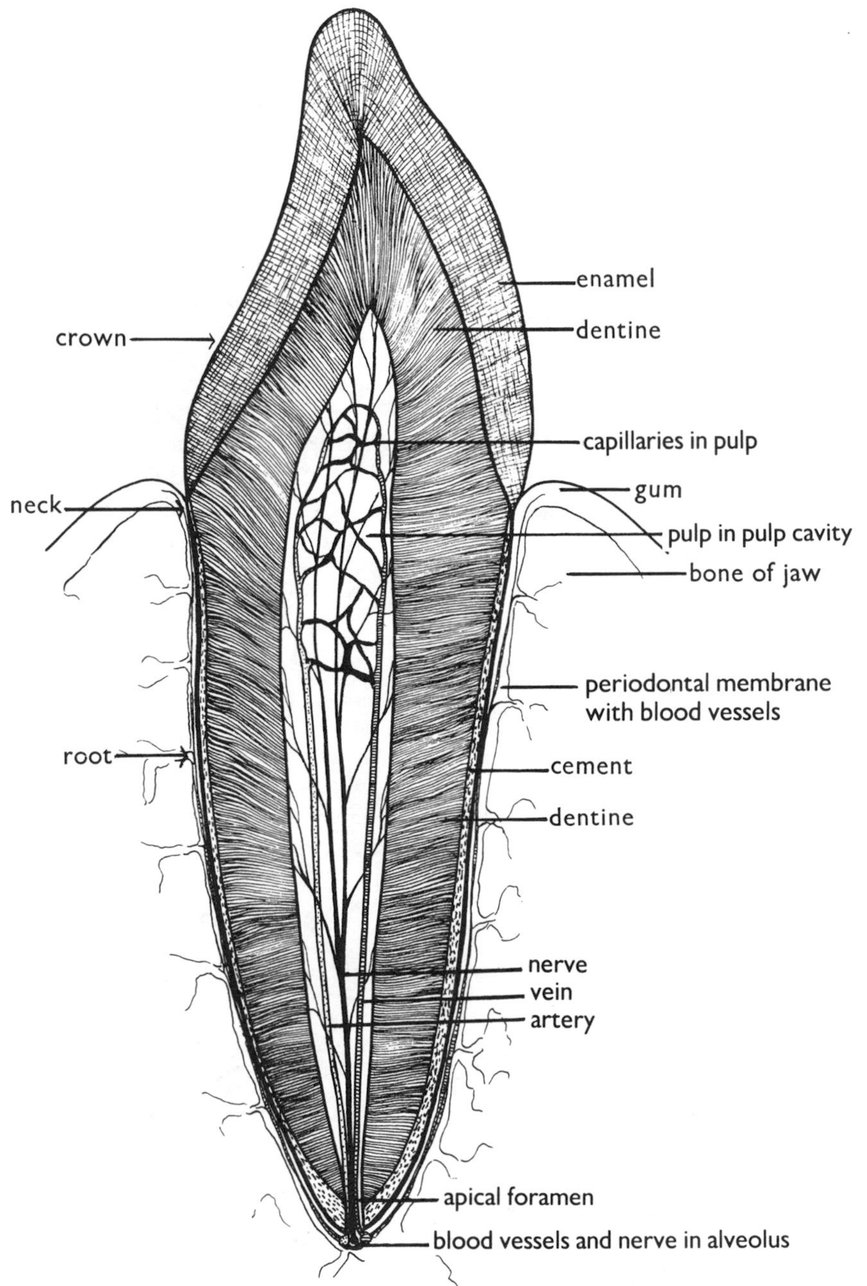

Section of single rooted tooth in alveolus—
distribution of blood vessels and nerves shown diagrammatically

Stages in the development of teeth

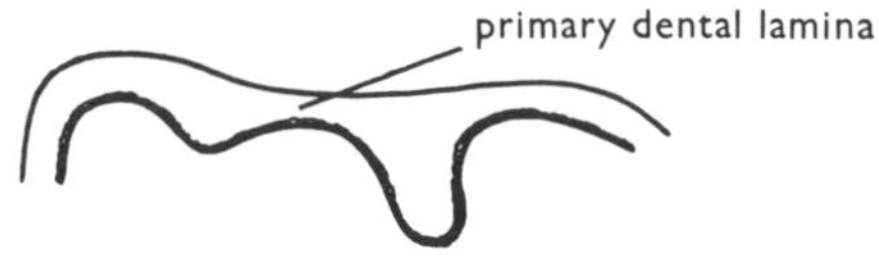

1. The foetal gum forms, continuous with the lip.

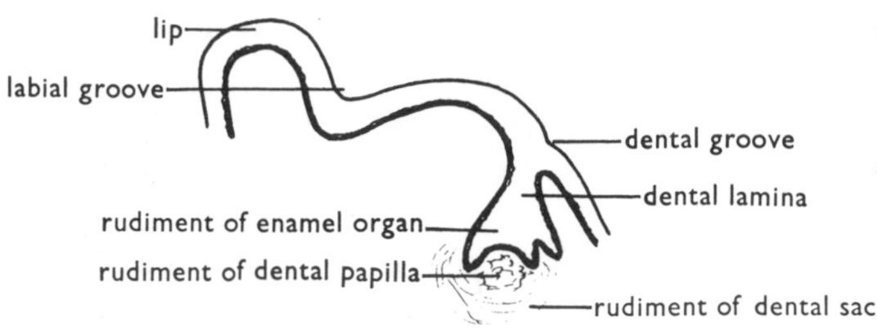

2. The dental lamina forms as a fold of the foetal gum. Buds arise on the outer side of the lamina, each bud becoming an enamel organ. Under the enamel organ connective tissue forms the dental papilla.

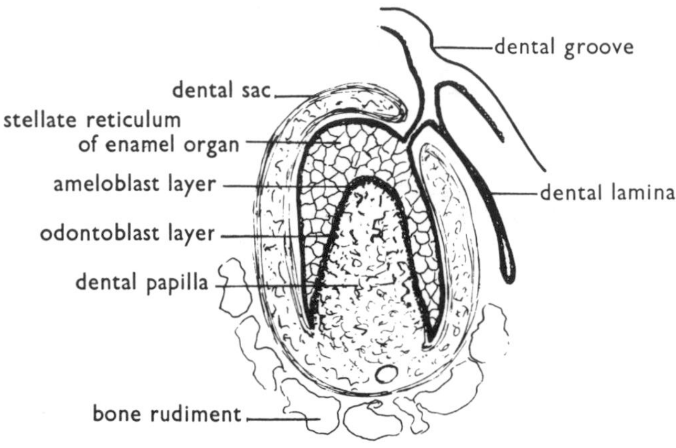

3. The enamel organ develops a distinct ameloblast layer and stellate reticulum. The dental papilla develops an odontoblast layer. Connective tissue around the tooth rudiment becomes the dental sac.

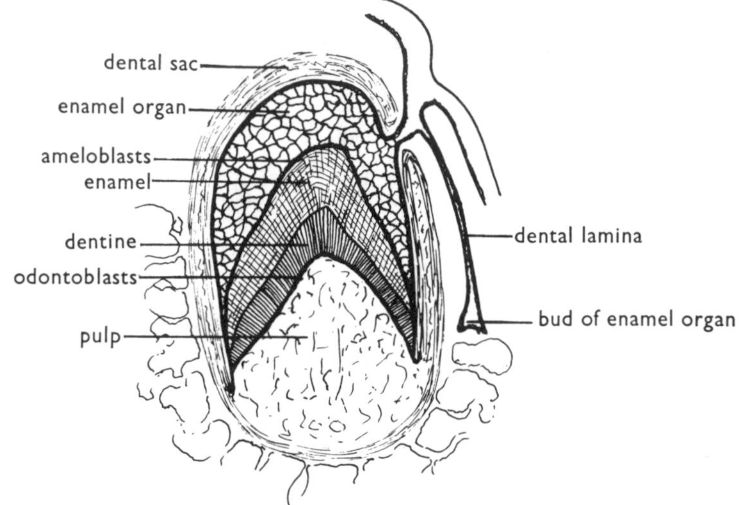

4. The ameloblasts of the enamel organ lay down enamel and the odontoblasts of the dental papilla produce dentine matrix which gradually becomes calcified.

The dental lamina continues to grow and produces a second series of buds from which the second set of teeth develop later.

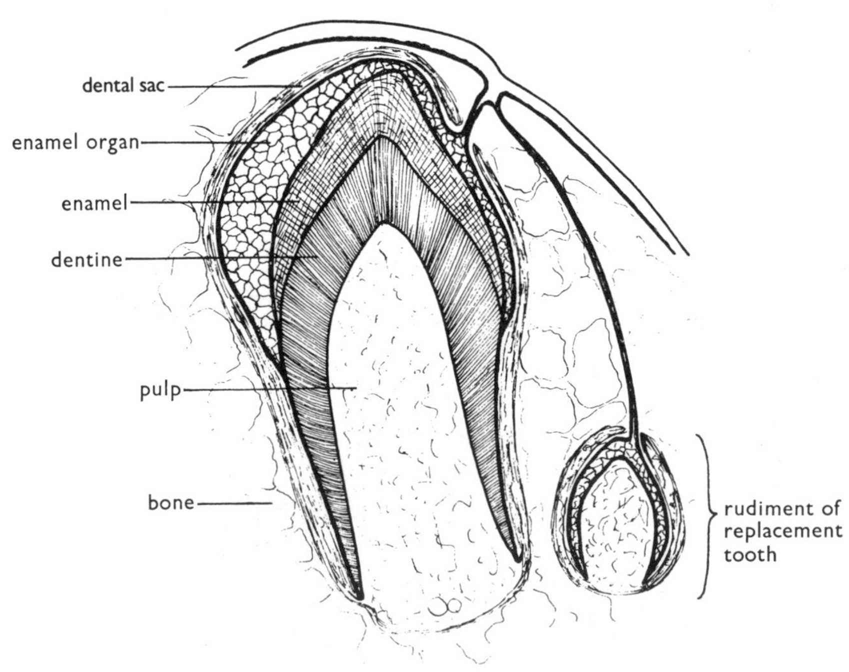

5. The tooth is ready to erupt. The pulp cavity is still widely open at the base. The rudiment of the replacement tooth is formed.

Stages in the development of teeth—*continued*

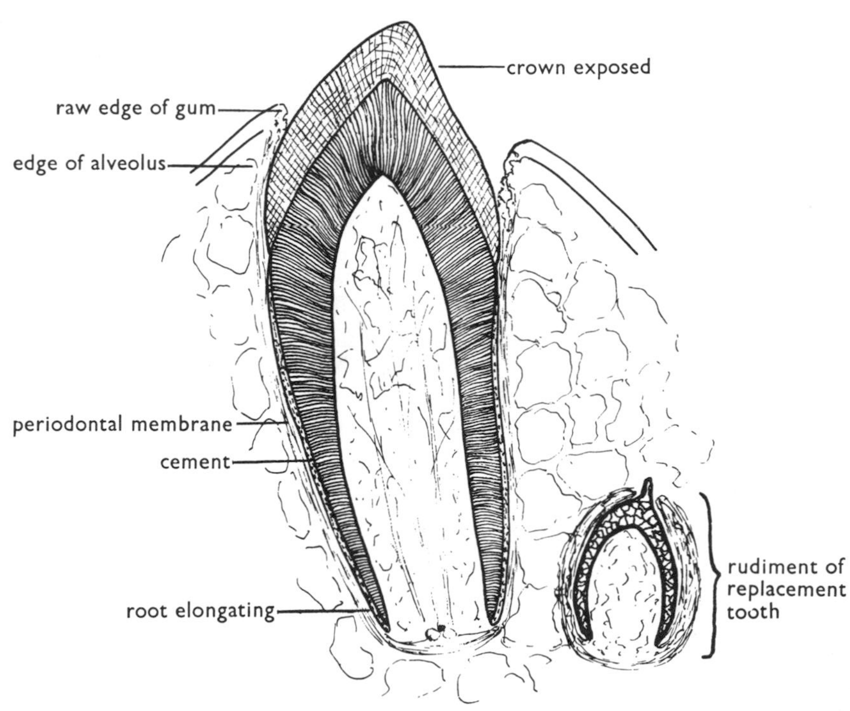

6. The tooth erupts by growth of the root accompanied by rupture of the gum, dental sac and enamel organ and resorption of some bone around the edges of the alveolus. The inner layer of the dental sac forms cement while the outer layer becomes the periodontal membrane.

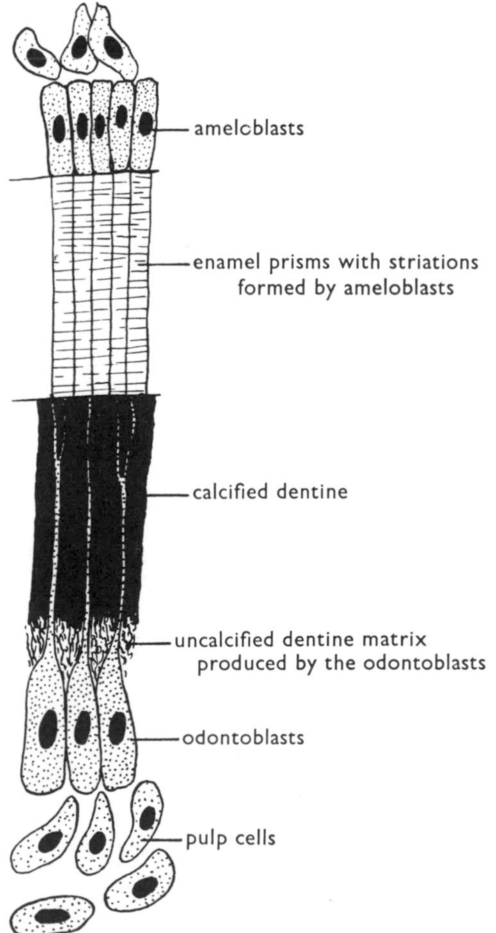

Diagram to show the growth of enamel and dentine

The hyoid bone

The hyoid bone lies at the base of the tongue. It has a **body** and two pairs of horns or **cornua**.

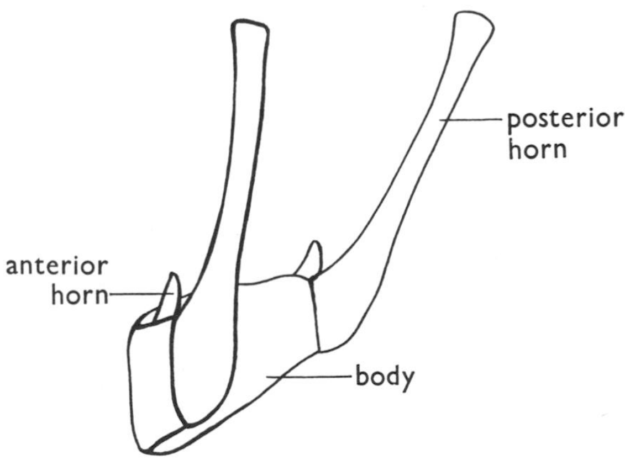

Hyoid bone—*postero-lateral view*

The vertebral column

The vertebral column is composed of 33 **vertebrae**, some of which are fused so that there are only 26 bones. It has the following regions:

cervical region in the neck—7 vertebrae;
thoracic or dorsal region in the thorax—12 vertebrae;
lumbar region in the small of the back—5 vertebrae;
sacral region in the pelvis—5 vertebrae fused to form the sacrum;
coccygeal region below the sacrum—4 vertebrae fused to form the coccyx.

The column shows curvatures in the cervical, thoracic, lumbar and sacral regions. The thoracic and sacral curvatures are primary and are present before a baby is born. The cervical and lumbar curvatures are secondary. The former develops when the child lifts its head and the latter when it starts to walk.

A **typical vertebra** has a **body**, a **neural arch** and **seven processes**.
The body is approximately cylindrical with flattened upper and lower surfaces, which articulate with adjacent vertebrae through **intervertebral discs**—see page 34. The neural arch is formed of two stalk-like **pedicles** and two flattened **laminae**. The processes are attached to the neural arch as follows:

a **spinous process** at the junction of the two laminae;
two **transverse processes**, one on either side, at the junction of each pedicle with the corresponding lamina;
two **superior** and two **inferior articular processes** on the upper and lower edges of the laminae respectively.

The articular processes are slanted to fit the corresponding processes of the adjacent vertebrae and form gliding joints —see page 34.
The neural arch encloses a canal called the **vertebral foramen** through which runs the spinal cord. **Intervertebral notches** between the pedicles of adjacent vertebrae permit the emergence of spinal nerves—see page 66.

The **cervical vertebrae** can be recognised by the possession of a canal called a **foramen transversarium** in each transverse process. The second to sixth vertebrae have cloven spinous processes. The first two cervical vertebrae, the **atlas** and the **axis**, are specialised to support and allow free movement of the head. The atlas has no body, no spinous process and no articular processes. Two large, concave superior articular surfaces fit the occipital condyles of the skull and permit nodding. Two inferior articular surfaces and a facet on the inner surface of the anterior arch articulate with the axis and permit turning movement round the **odontoid process** of the axis. In life the canal of the atlas is traversed by a ligament designed to hold the odontoid process in place and prevent damage to the spinal cord during nodding.
The axis has the large odontoid process attached to the superior surface of the body. On either side of this is a convex facet (superior articular surface) which articulates with the atlas, but there are no superior articular processes on the neural arch. On the anterior surface of the odontoid process is a curved facet which articulates with the socket of the atlas. Nodding movement takes place between the skull and the atlas, while turning of the head takes place between the atlas and the axis.

The **thoracic vertebrae** can be recognised by the facets for articulation of the ribs. The first ten thoracic vertebrae have two pairs of demi-facets on the body because the heads of the second to tenth pairs of ribs overlap on to the vertebrae above those to which they belong, but the first and the last two pairs of ribs articulate directly with the corresponding vertebrae. The transverse processes of the first ten thoracic vertebrae bear facets for the tubercles of the ribs.

The **lumbar vertebrae** are larger and stronger than the other vertebrae.

The **sacrum** is formed of five vertebrae whose bodies and transverse processes are fused. The junctions can be identified and there are foramina between them for the emergence of nerves. The spinous processes are reduced. Only the first sacral vertebra has free articular processes.

The **coccyx** forms a small appendix to the sacrum. It consists of four vertebrae which are usually fused together though the first may remain a separate piece.

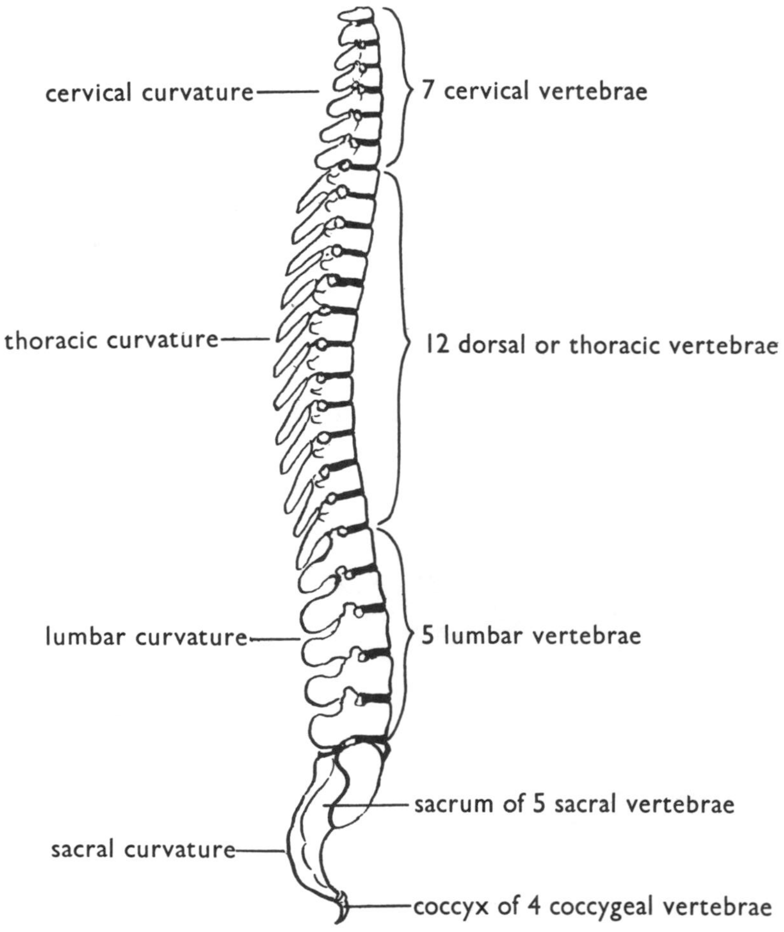

cervical curvature
7 cervical vertebrae
thoracic curvature
12 dorsal or thoracic vertebrae
lumbar curvature
5 lumbar vertebrae
sacrum of 5 sacral vertebrae
sacral curvature
coccyx of 4 coccygeal vertebrae

Vertebral column—*lateral view*

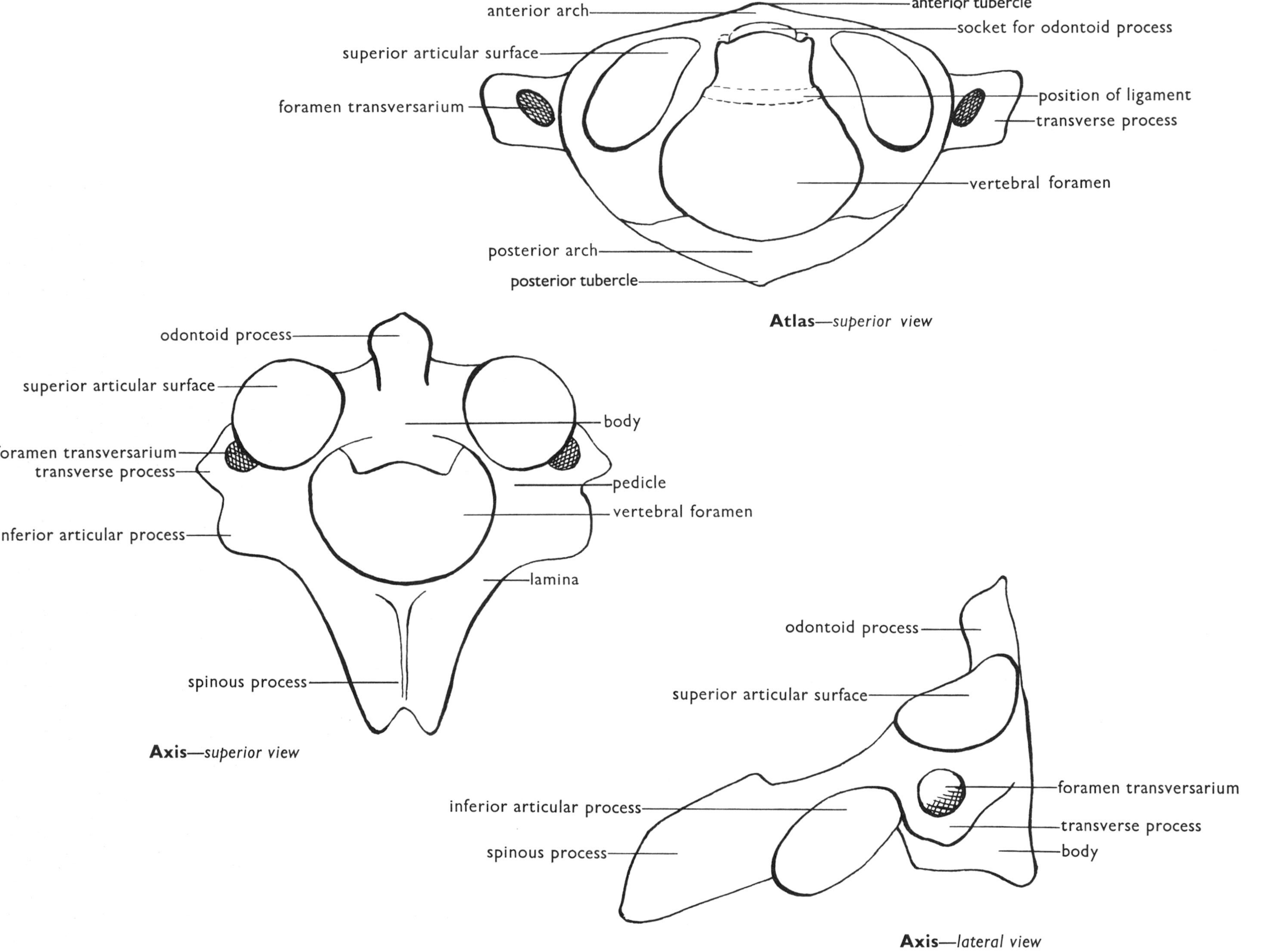

Atlas—*superior view*

Axis—*superior view*

Axis—*lateral view*

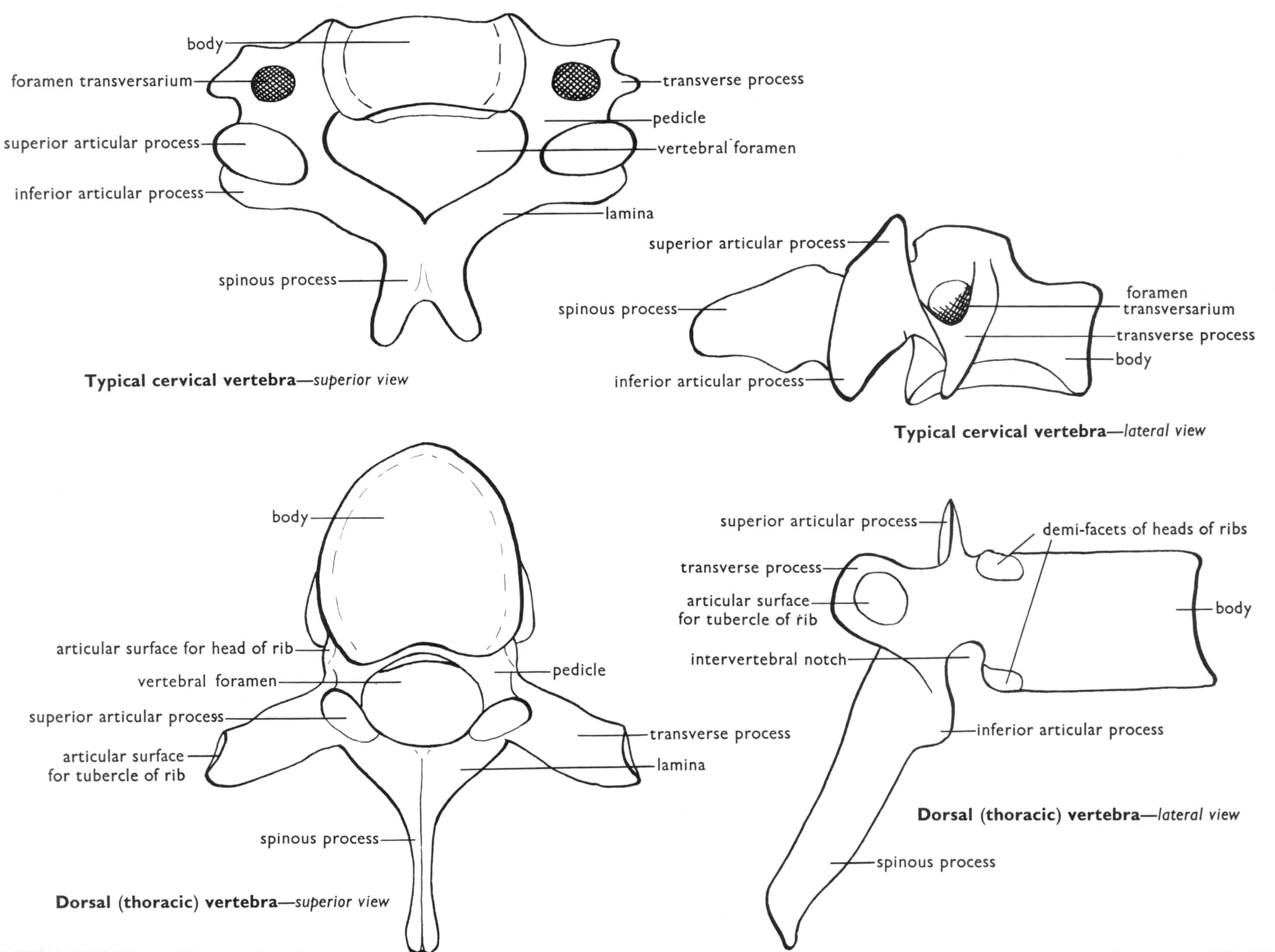

Typical cervical vertebra—*superior view*

Typical cervical vertebra—*lateral view*

Dorsal (thoracic) **vertebra**—*superior view*

Dorsal (thoracic) **vertebra**—*lateral view*

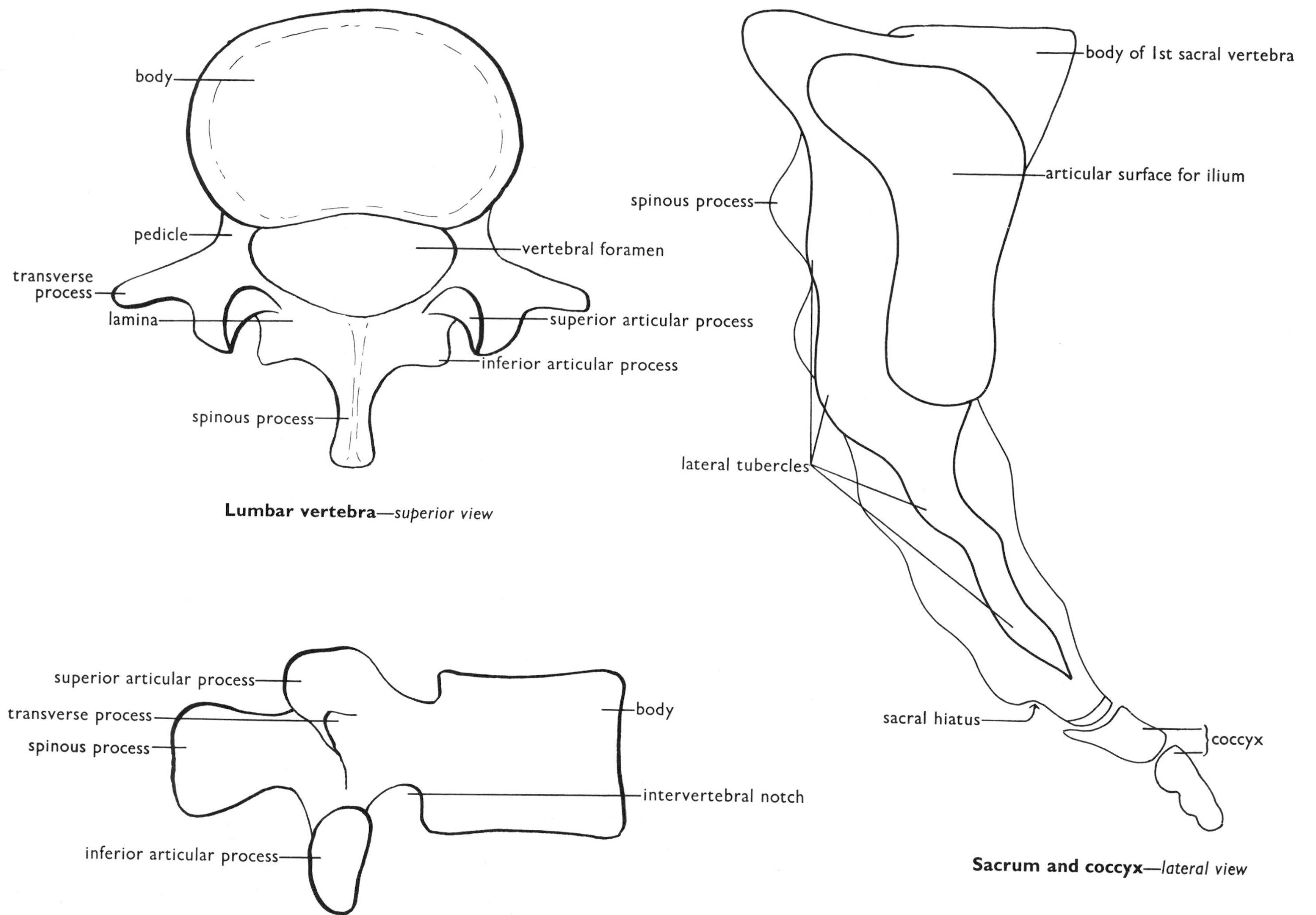

Lumbar vertebra—*superior view*

Lumbar vertebra—*lateral view*

Sacrum and coccyx—*lateral view*

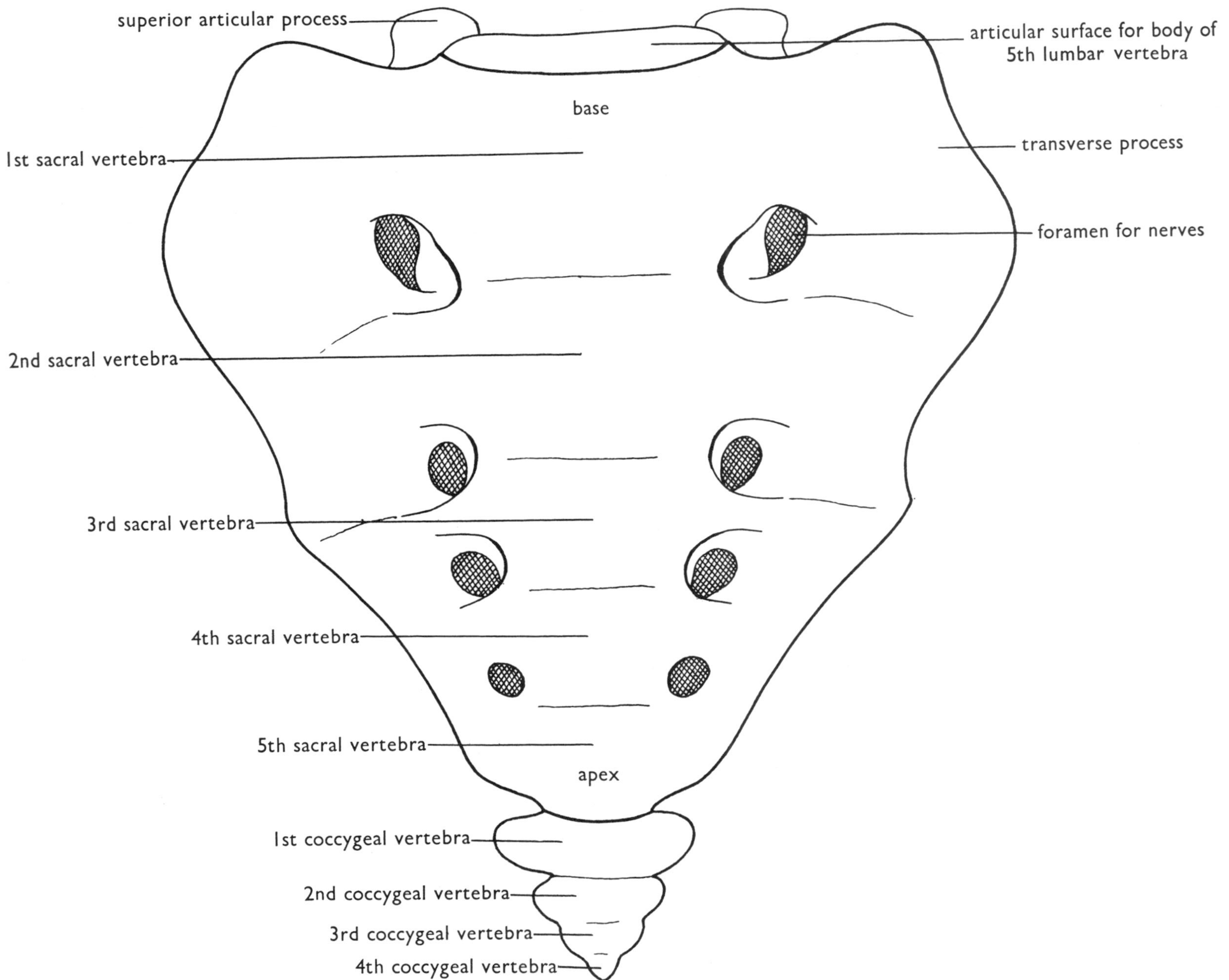

Sacrum and coccyx—*ventral view*

The thoracic cage

The thoracic cage is formed of the twelve thoracic vertebrae posteriorly, the **sternum** anteriorly and the twelve pairs of **ribs** laterally. It is conical in shape.

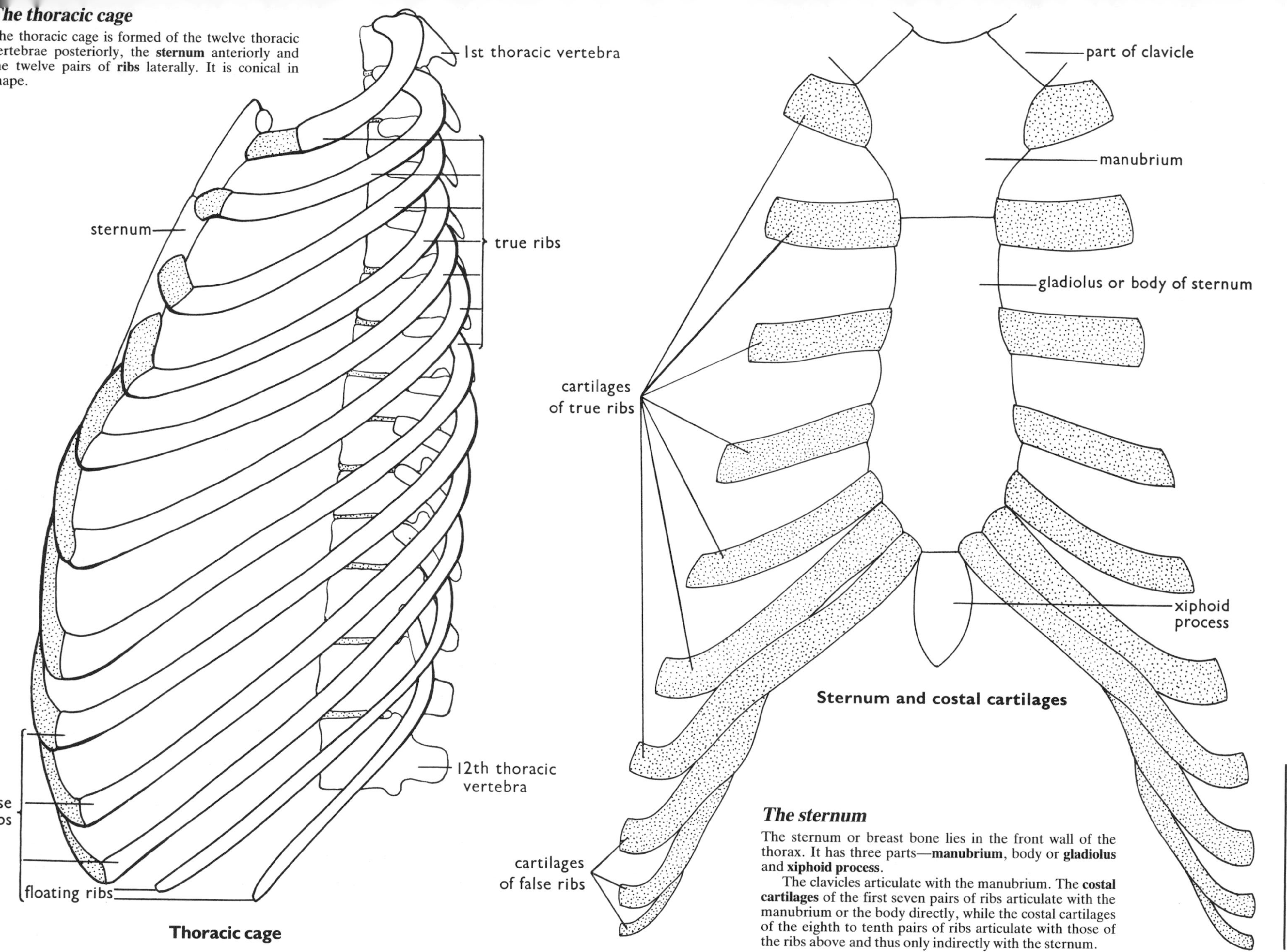

Thoracic cage

Sternum and costal cartilages

The sternum

The sternum or breast bone lies in the front wall of the thorax. It has three parts—**manubrium**, body or **gladiolus** and **xiphoid process**.

The clavicles articulate with the manubrium. The **costal cartilages** of the first seven pairs of ribs articulate with the manubrium or the body directly, while the costal cartilages of the eighth to tenth pairs of ribs articulate with those of the ribs above and thus only indirectly with the sternum.

The ribs

There are twelve pairs of ribs. Each rib typically has a **head**, which articulates with the bodies of two vertebrae, a **tubercle** which articulates with the transverse process of a vertebra, and a **shaft** which curves round the side wall of the thorax. The last two pairs of ribs have no tubercles.

The anterior ends of the first ten pairs of ribs articulate with **costal cartilages**, while the last two pairs of ribs are free or floating ribs. The seven pairs of ribs whose cartilages articulate directly with the sternum are called true ribs, while the five pairs of ribs with indirect or with no attachment to the sternum are called false ribs.

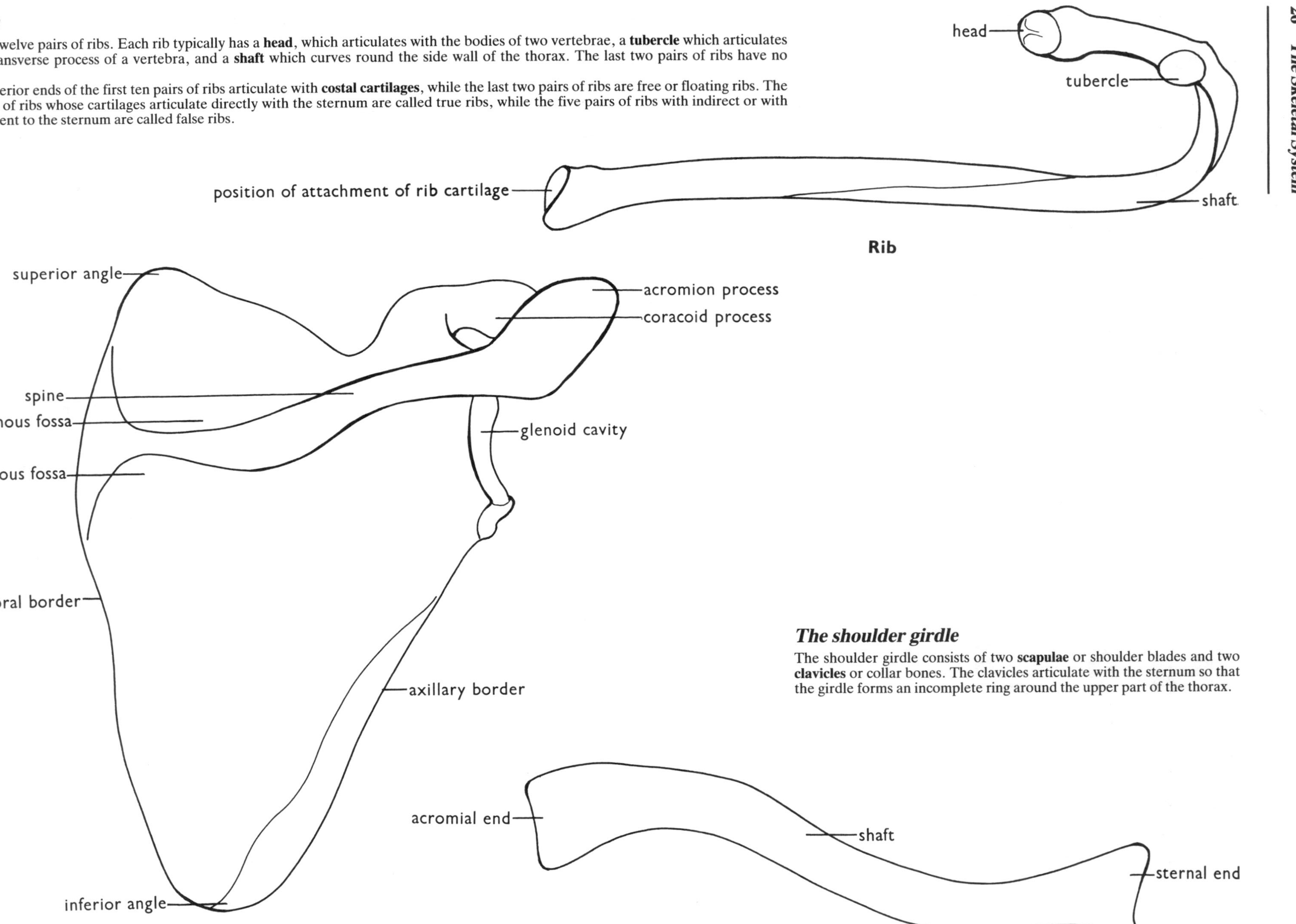

Rib

Right scapula—*dorsal view*

The shoulder girdle

The shoulder girdle consists of two **scapulae** or shoulder blades and two **clavicles** or collar bones. The clavicles articulate with the sternum so that the girdle forms an incomplete ring around the upper part of the thorax.

Right clavicle—*superior view*

The arm

The **humerus** in the upper arm and the **radius** and **ulna** in the forearm are long bones.

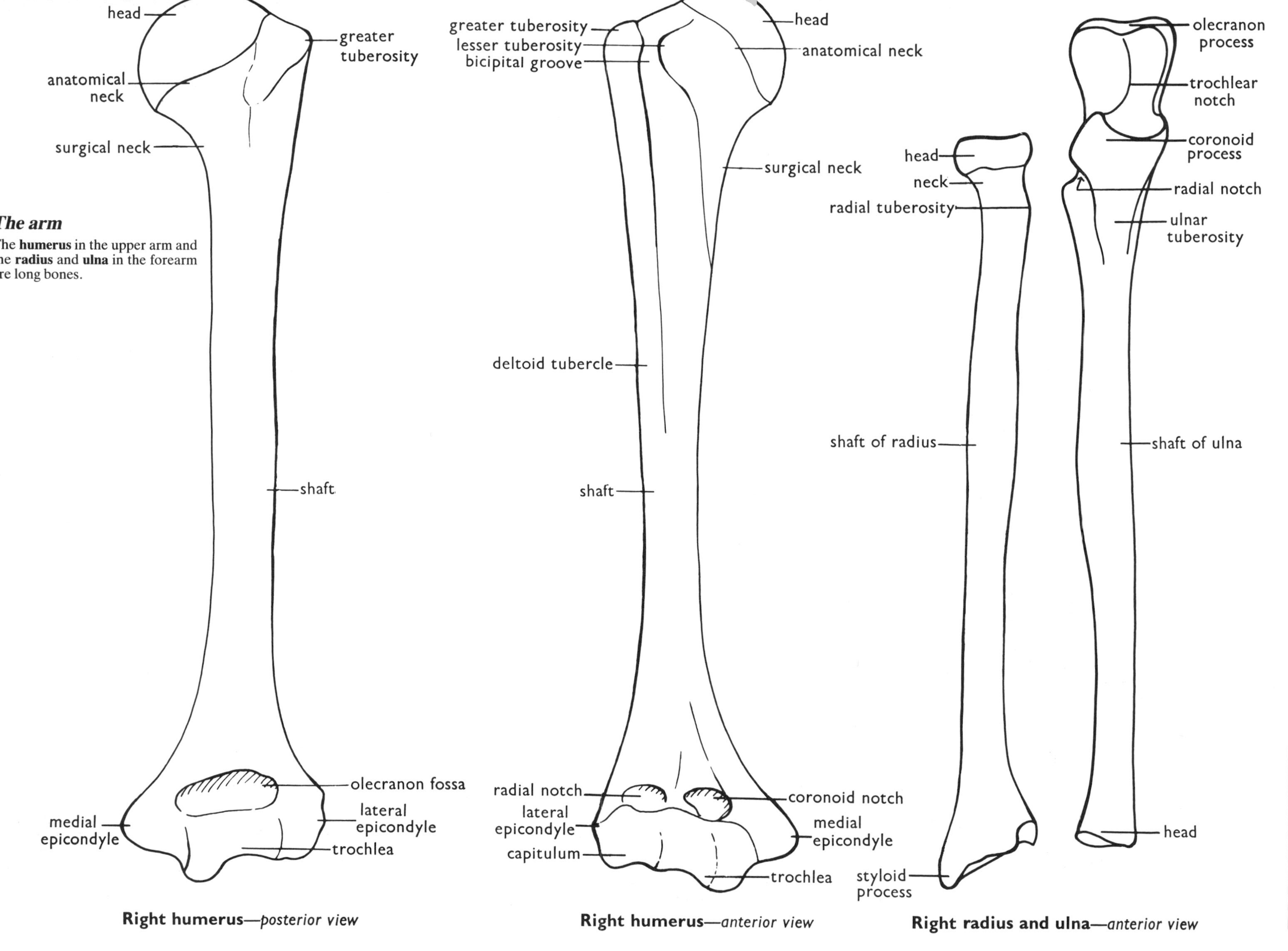

Right humerus—*posterior view*

Right humerus—*anterior view*

Right radius and ulna—*anterior view*

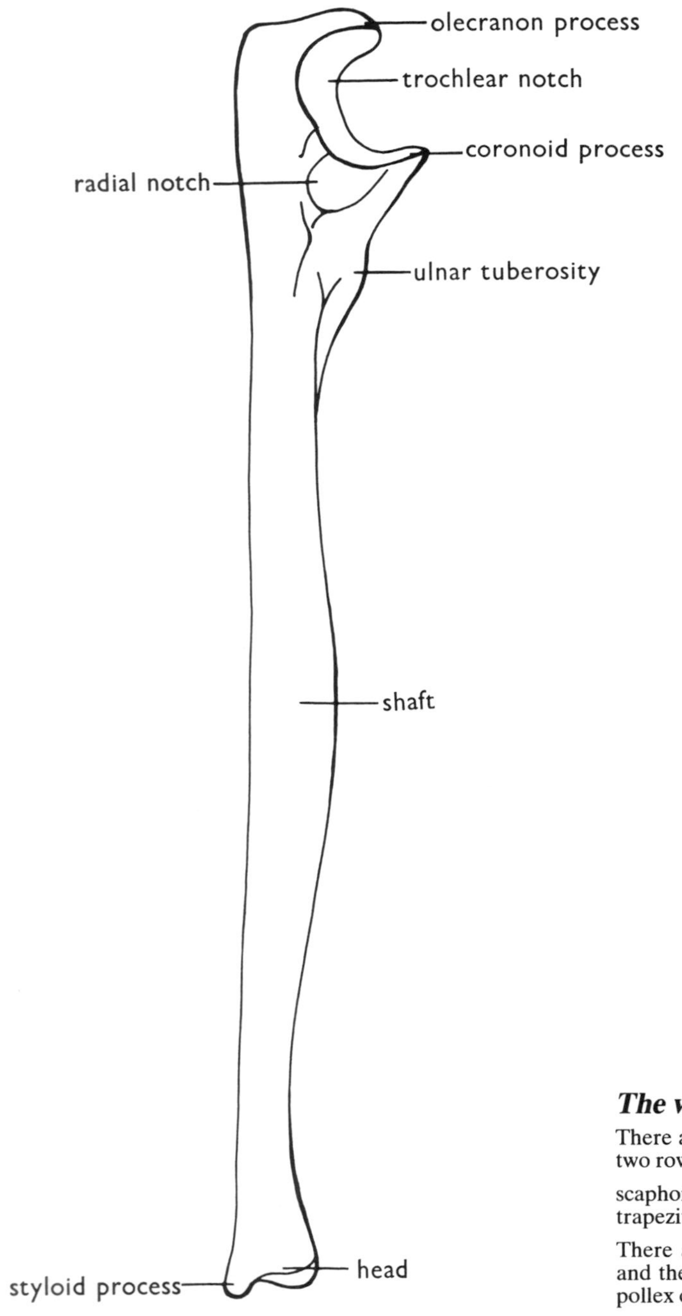

Right ulna—*lateral view*

The wrist and hand

There are eight small **carpal** bones in the wrist arranged in two rows of four:

scaphoid—lunate—triquetral—pisiform
trapezium—trapezoid—capitate—hamate

There are five **metacarpal** bones in the palm of the hand and there are fourteen **phalanges**, two of which are in the pollex or thumb and three in each of the other digits.

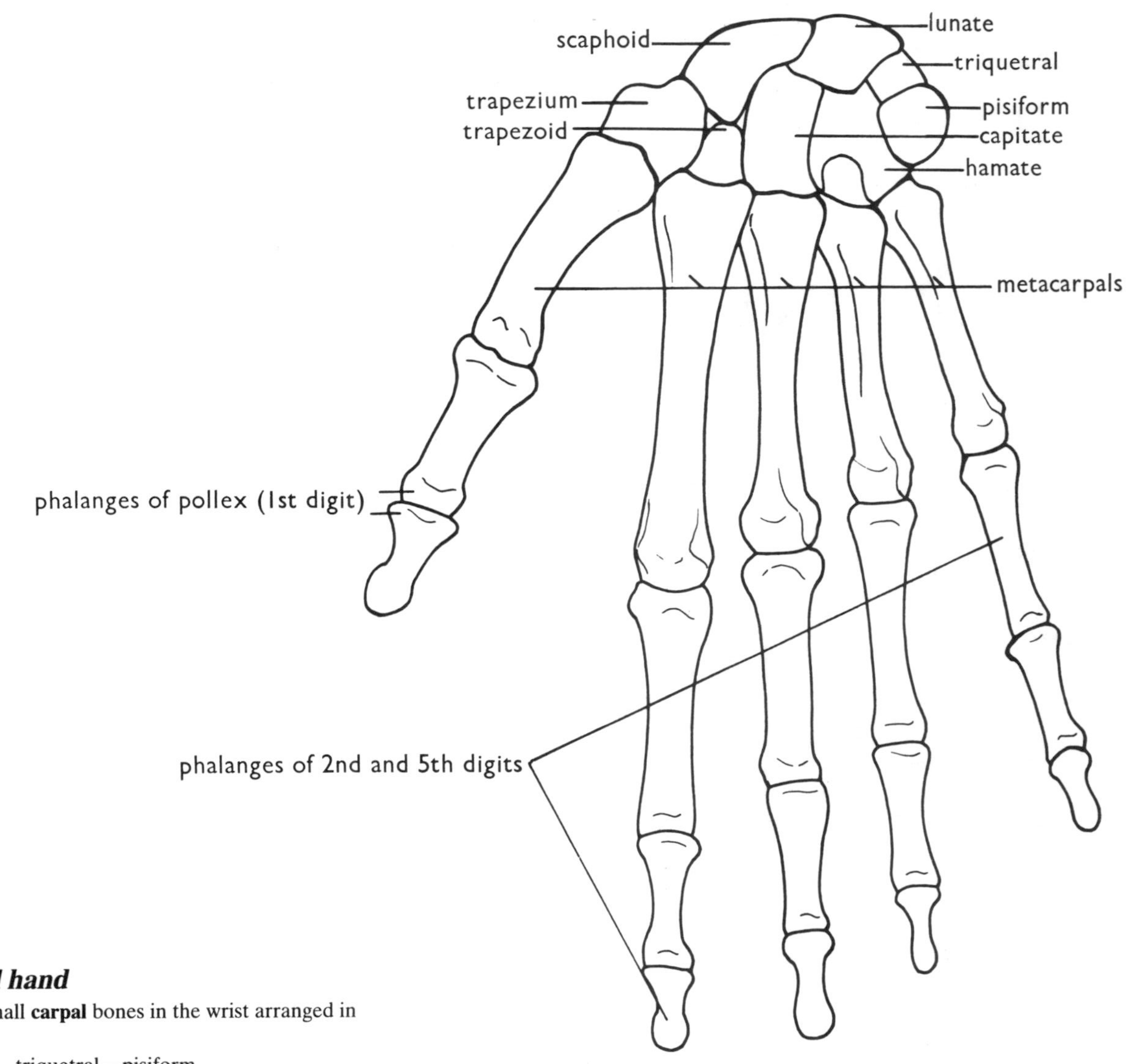

Right wrist and hand—*ventral view*

The hip girdle

The hip or **pelvic girdle** is formed of two **pelvic bones**, also known as **innominate** or **coxal** bones, which are firmly joined to the sacrum and to one another to produce a rigid support for the legs.

Each pelvic bone arises by fusion of three bones, **ilium**, **ischium** and **pubis**. The ilia are large and wing-like and articulate with (meet) the sacrum at strong, immovable **sacro-iliac joints**. The pubes articulate with one another at the **pubic symphysis** where there is a pad of fibro-cartilage, i.e. cartilage with many extra chondrin fibres to give it strength. At the fused junction of the ilium, ischium and pubis there is a socket called the **acetabulum** into which the head of the femur (thigh bone) fits. The lower part of the rim of the acetabulum is completed by ligament—see page 36.

The pelvic bones with the sacrum and coccyx form the **pelvis**, which surrounds the **pelvic cavity** and protects the organs of the lower part of the trunk—see page 82. The pelvis of the female is wider and shallower than that of the male. Its fibro-cartilage pad softens during childbirth to allow the stretching necessary for the passage of the baby's head.

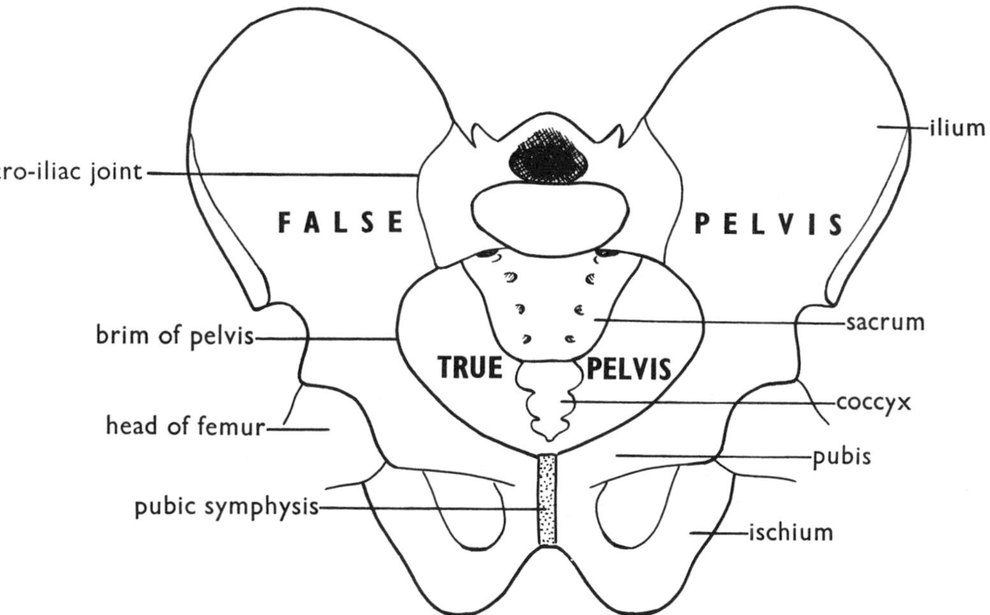

Female pelvis

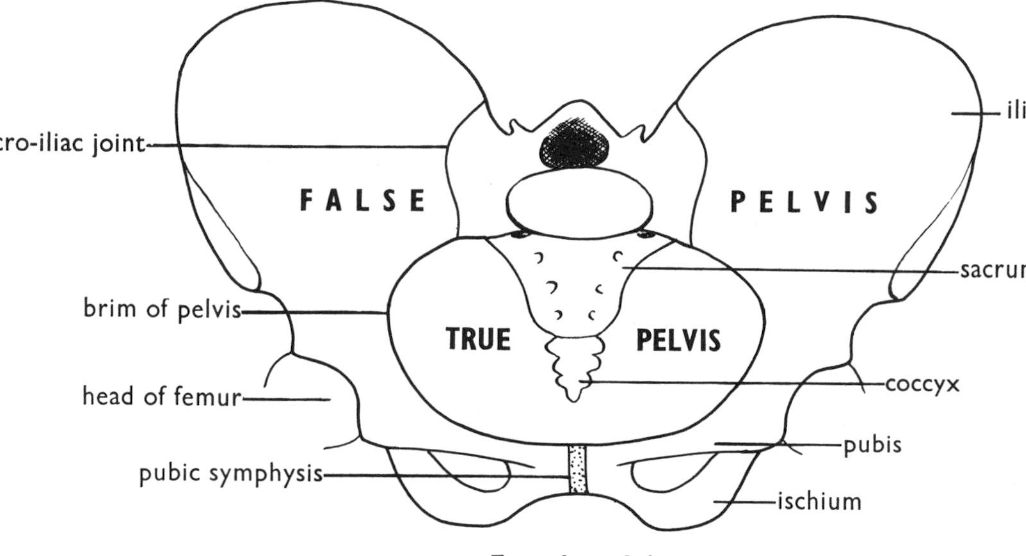

Male pelvis

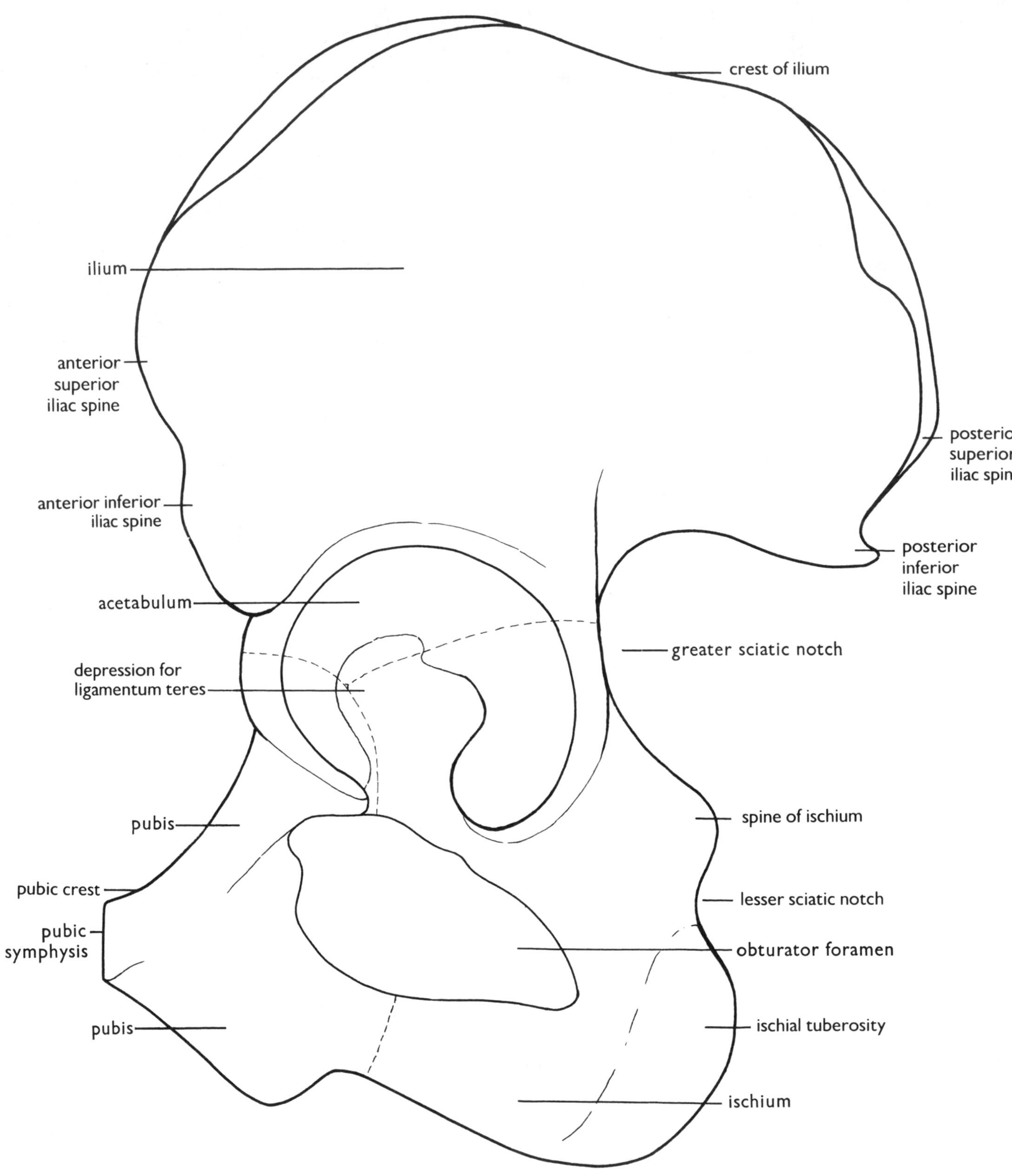

Left pelvic bone—*lateral view*

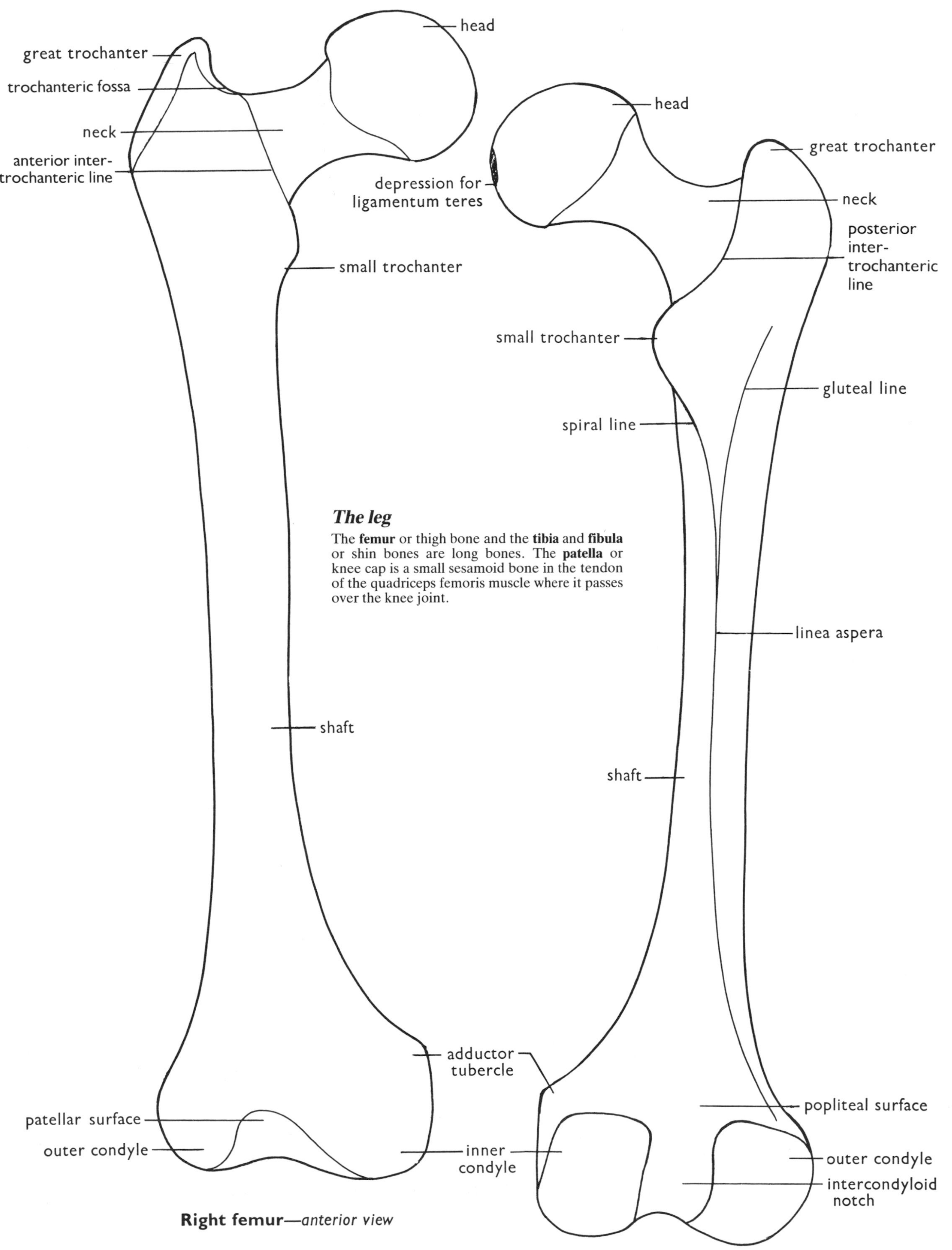

The leg

The **femur** or thigh bone and the **tibia** and **fibula** or shin bones are long bones. The **patella** or knee cap is a small sesamoid bone in the tendon of the quadriceps femoris muscle where it passes over the knee joint.

Right femur—*anterior view*

Right femur—*posterior view*

The ankle and foot

In the foot there are seven **tarsal** bones, five **metatarsal** bones and fourteen **phalanges**. The tarsals are:

1. calcaneum
2. talus
3. navicular
4. cuboid
5. outer cuneiform
6. middle cuneiform
7. inner cuneiform.

The tarsals and metatarsals together support the arches of the foot.

The toes are supported by the phalanges, two of which are in the hallux or great toe and three in each of the other digits. The two distal phalanges of the little toe are often fused.

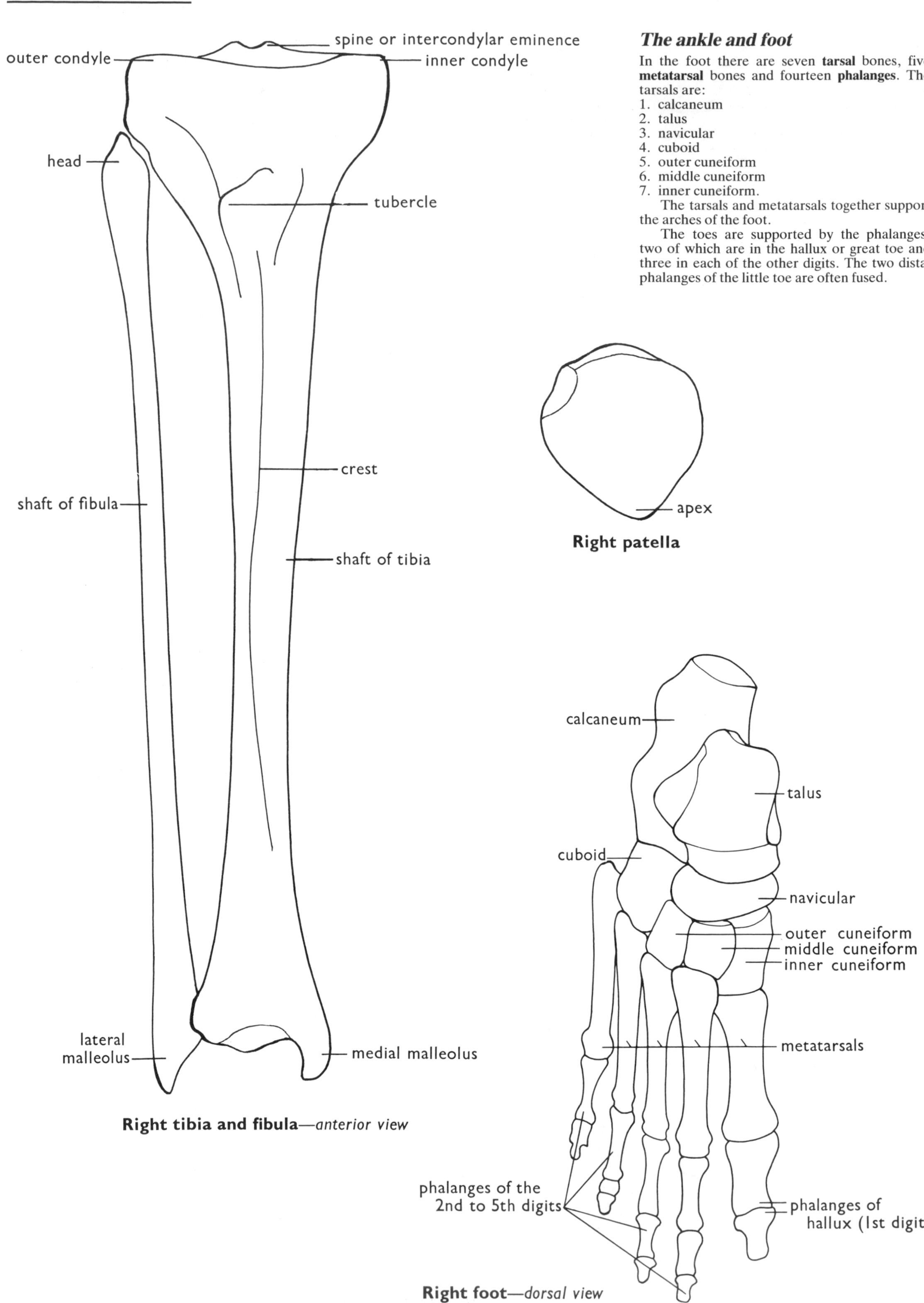

Right tibia and fibula—*anterior view*

Right patella

Right foot—*dorsal view*

JOINTS

Wherever one bone or cartilage meets another there is a joint. There are more joints in a child than in an adult because as growth proceeds some of the bones fuse together—e.g. the ilium, ischium and pubis to form the pelvic bone; the two halves of the infant frontal bone, and of the infant mandible; the five sacral vertebrae and the four coccygeal vertebrae.

Joints are classified according to the amount of movement possible between the articulating surfaces.

Synarthroses are fixed joints at which there is no movement. The articular surfaces are joined by tough fibrous tissue. Often the edges of the bones are dovetailed into one another as in the sutures of the skull.

Amphiarthroses are joints at which slight movement is possible. A pad of cartilage lies between the bone surfaces, and there is a fibrous capsule to hold the bones and cartilage in place. The cartilages of such joints also act as shock absorbers, e.g. the intervertebral discs between the bodies of the vertebrae, where the cartilage is strengthened by extra collagen fibres.

Diarthroses or **synovial joints** are known as freely movable joints, though at some of them the movement is restricted by the shape of the articulating surfaces and by the **ligaments** which hold the bones together. These ligaments are of elastic connective tissue—see diagram on page 39.

A synovial joint has a fluid-filled cavity between articular surfaces which are covered by articular cartilage. The fluid, known as synovial fluid, is a form of lymph produced by the synovial membrane which lines the cavity except for the actual articular surfaces and covers any ligaments or tendons which pass through the joint. Synovial fluid acts as a lubricant.

The **movements** possible at synovial joints are:

angular — flexion: decreasing the angle between two bones;
extension: increasing the angle between two bones;
abduction: moving the part away from the mid-line;
adduction: bringing the part towards the mid-line.

rotary — rotation: turning upon an axis;
circumduction: moving the extremity of the part round in a circle so that the whole part inscribes a cone.

gliding — one part slides on another.

The form of the articulating surfaces controls the type of movement which takes place at any joint. Synovial joints are classified accordingly as:

gliding—gliding movement;
hinge—flexion and extension;
condyloid—flexion and extension, abduction and adduction, and limited circumduction;
saddle—flexion and extension, abduction and adduction, and circumduction;
ball and socket—flexion and extension, abduction and adduction, circumduction, and rotation;
pivot—rotation only.

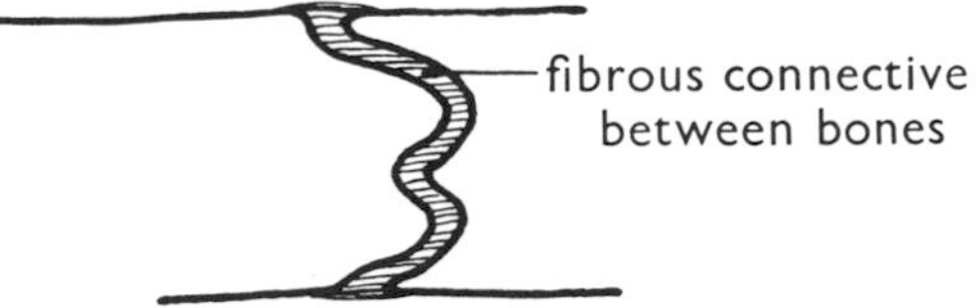

Synarthrosis—*fibrous joint*

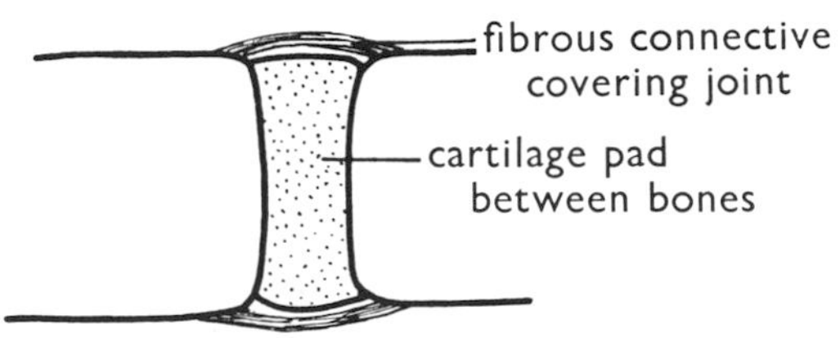

Amphiarthrosis—*cartilaginous joint*

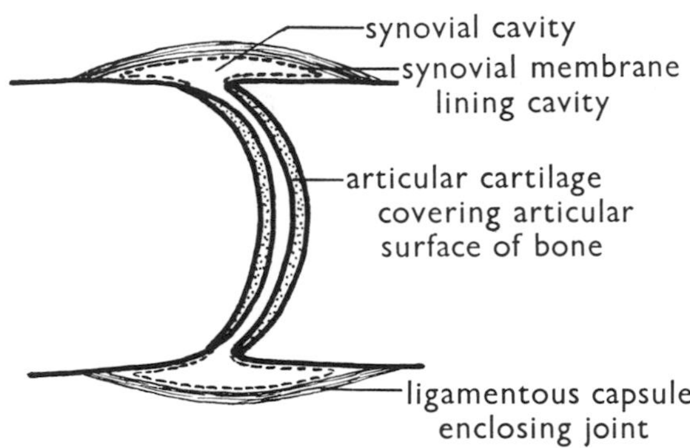

Diarthrosis—*synovial joint*

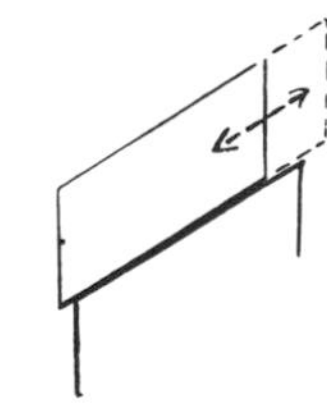

Gliding movement

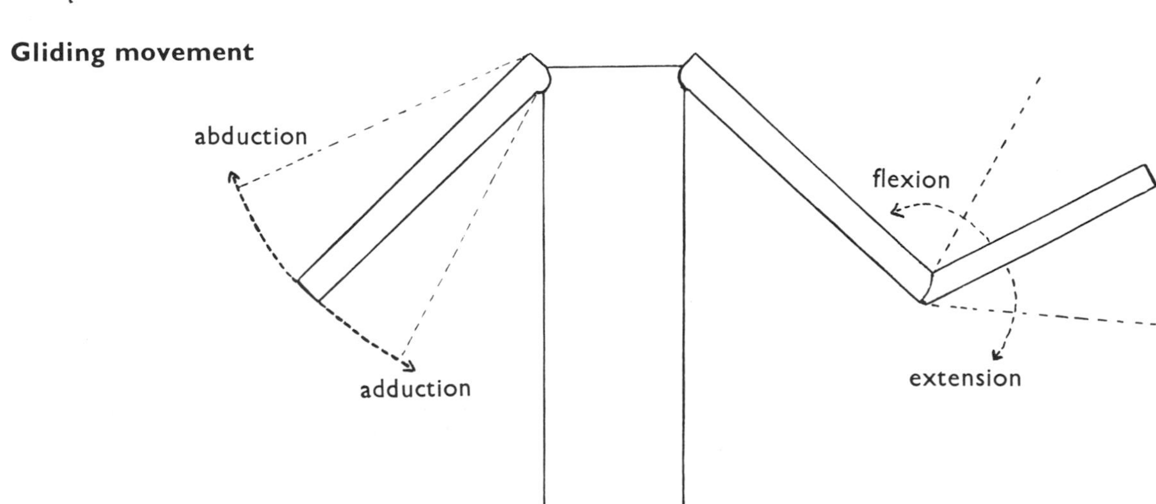

Angular movement

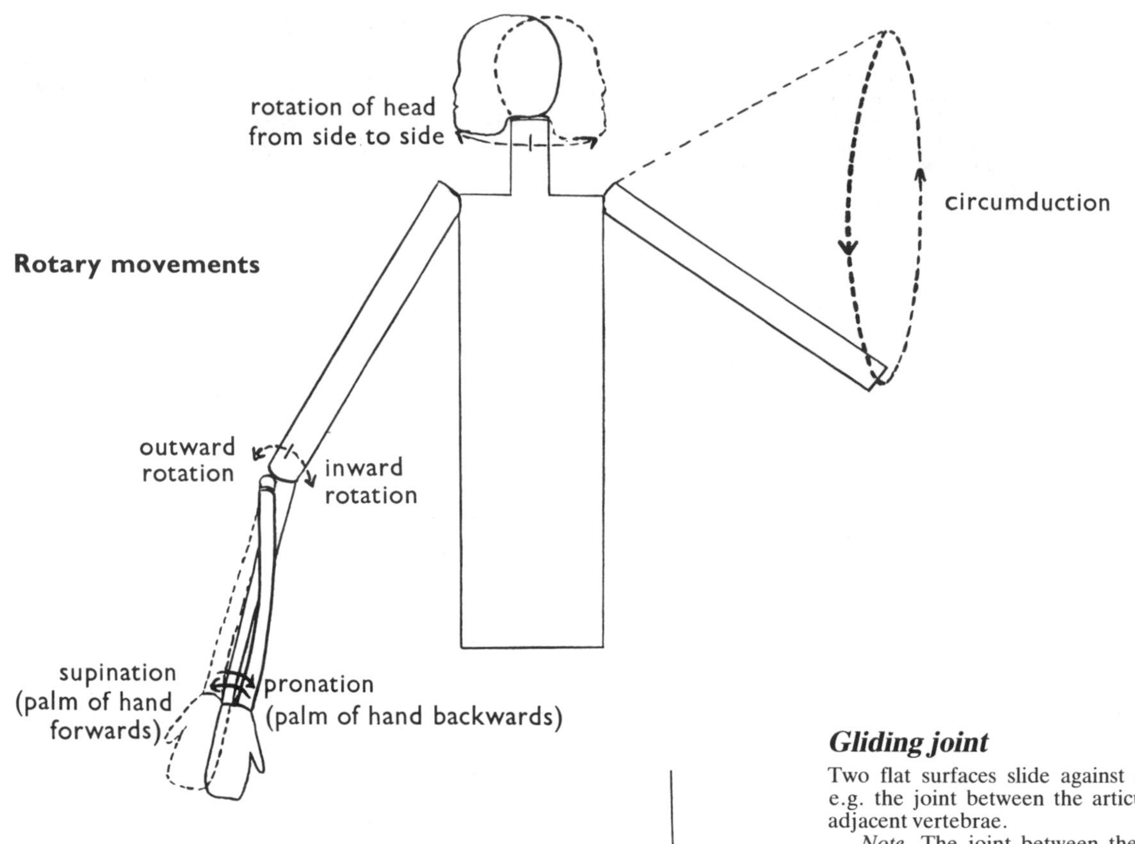

Rotary movements

Gliding joint

Two flat surfaces slide against one another—e.g. the joint between the articular process of adjacent vertebrae.

Note. The joint between the bodies of the vertebrae is cartilaginous and not freely movable.

Gliding joint

Hinge joint

A convex surface fits a concave surface of such form that movement is only possible in one plane—e.g. the elbow joint.

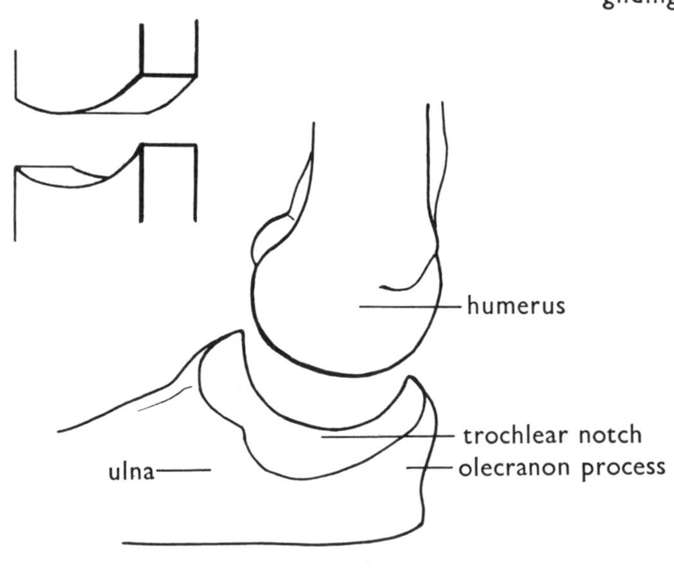

Hinge joint

Condyloid joint

A convex surface fits a concave surface. The radii of curvature are not the same in all planes. Movement is angular in two planes with slight circumduction—e.g. the wrist joint.

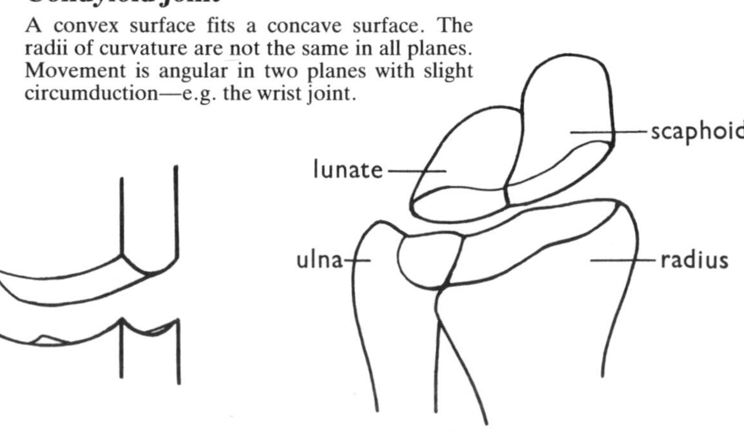

Condyloid joint

Saddle joint

A concavo-convex surface fits into a convexo-concave surface. Angular movements and circumduction can be performed freely, but rotation is impossible—e.g. carpo-metacarpal joint of the thumb.

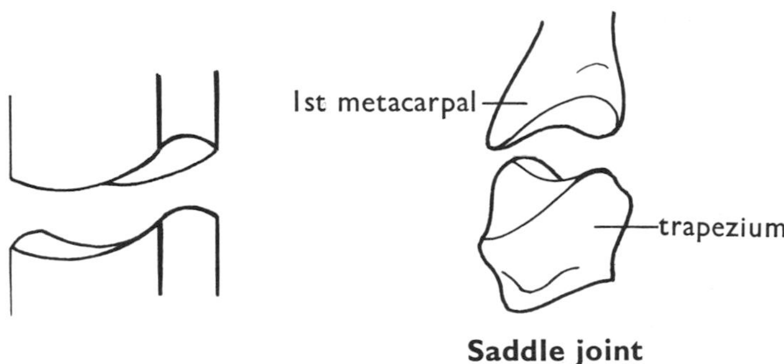

Saddle joint

Ball and socket joint

A rounded head fits in a cup-shaped cavity. Angular movements, circumduction and rotation can be performed freely—e.g. hip joint.

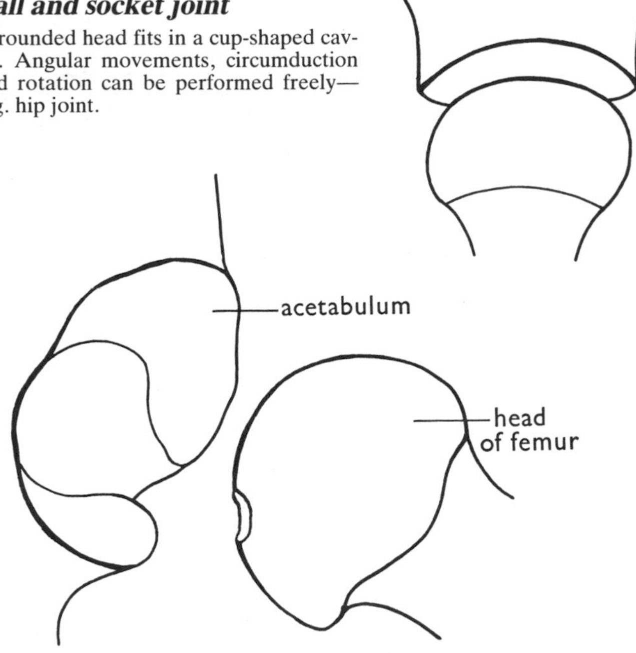

Ball and socket joint

Pivot joint

A process rotates in a socket, e.g.
(a) the odontoid process of the axis within the ring of the atlas allows turning of the head;
(b) the head of the radius in the annular ligament at the superior radio-ulnar joint allows pronation and supination of the hand.

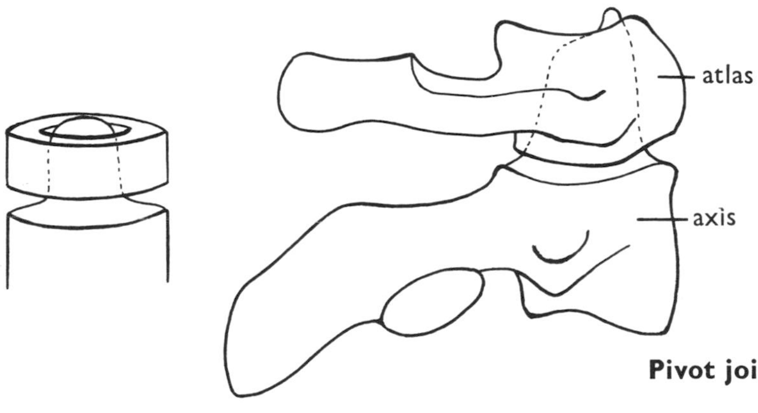

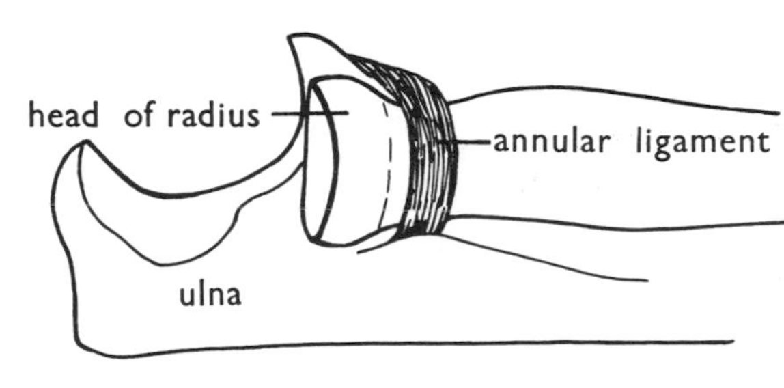

Pivot joint

The principal joints

Fibrous joints or **synarthroses** are found between:

1. most of the **skull** bones—sutures formed by dovetailing;
2. **teeth** and their sockets—gomphoses;
3. **ribs** and **costal cartilages;**
4. lower ends of **tibia** and **fibula**—syndesmosis—slight movement is possible.

Cartilaginous joints or **amphiarthroses** are found between:

1. bodies of the **free vertebrae** except atlas and axis—cartilages are intervertebral discs;
2. **manubrium** and **gladiolus** } both these joints may become synarthroses later in life;
3. **gladiolus** and **xiphoid** }
4. the two **pubes**, i.e. pubic symphysis; cartilage of this joint softens during pregnancy and allows greater movement at childbirth.

Synovial joints or **diarthroses** are found between:

1. **mandible** and **temporal** bones—condyloid;
2. **occipital condyles** and **atlas**—condyloid;
3. **atlas** and **axis**—pivot;
4. **articular processes** of **vertebrae**—gliding;
5. **vertebrae** and **ribs**—gliding;
6. **costal cartilages** and **sternum**—gliding;
7. **sternum** and **clavicle**—gliding with articular disc of cartilage;
8. **clavicle** and **scapula**—gliding with slight rotation;
9. **scapula** and **humerus**—ball and socket;
10. **humerus** and **ulna**—hinge;
11. **humerus** and **radius**—hinge, head of radius can also rotate;
12. head of **radius** and **ulna**—pivot with annular ligament around the radius;
13. lower ends of **radius** and **ulna**—pivot;
14. **fore-arm bones** and **carpals**, i.e. wrist joint—condyloid;
15. **carpals** and one another—gliding, results in various movements of the wrist as a whole;
16. **trapezium** and **metacarpal** of thumb—saddle, allows apposition of the thumb;
17. **carpals** and other **metacarpals**—gliding;
18. **metacarpals** and one another—gliding;
19. **metacarpals** and **phalanges**—condyloid;
20. **phalanges** of each finger—hinge;
21. **ilia** and **sacrum**—gliding but roughened surfaces prevent any actual movement, except at childbirth;
22. **acetabulum** and head of **femur**—ball and socket;
23. **femur** and **tibia**—structurally condyloid but functionally hinge with semilunar cartilages;
24. **femur** and **patella**—gliding;
25. head of **fibula** and **tibia**—gliding;
26. **shin bones** and **talus**, i.e. ankle joint—hinge;
27. **tarsals** and one another—gliding, summation of movements produces inversion and eversion of the foot;
28. **tarsals** and **metatarsals**—gliding;
29. **metatarsals** and one another—gliding;
30. **metatarsals** and **phalanges**—condyloid;
31. **phalanges** of toes—hinge.

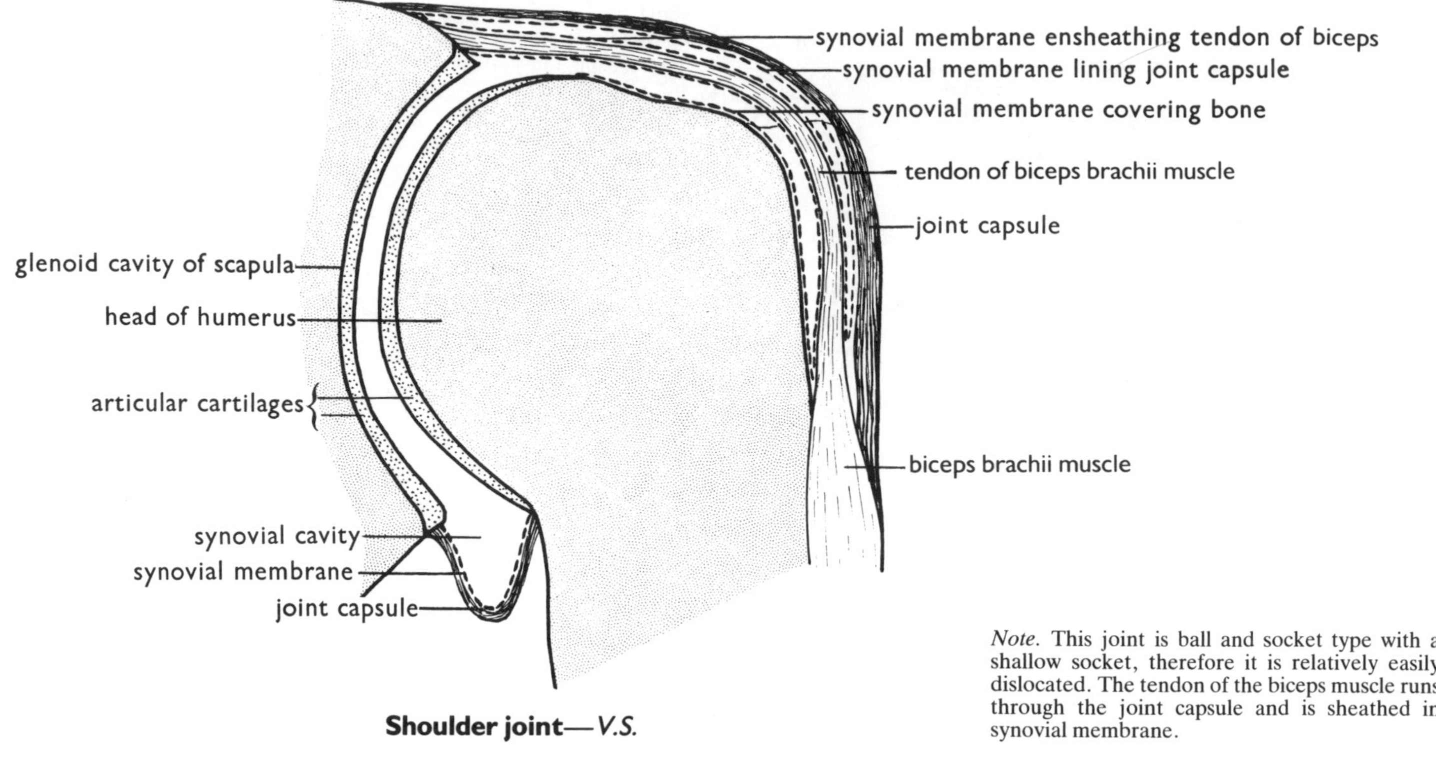

Shoulder joint—*V.S.*

Note. This joint is ball and socket type with a shallow socket, therefore it is relatively easily dislocated. The tendon of the biceps muscle runs through the joint capsule and is sheathed in synovial membrane.

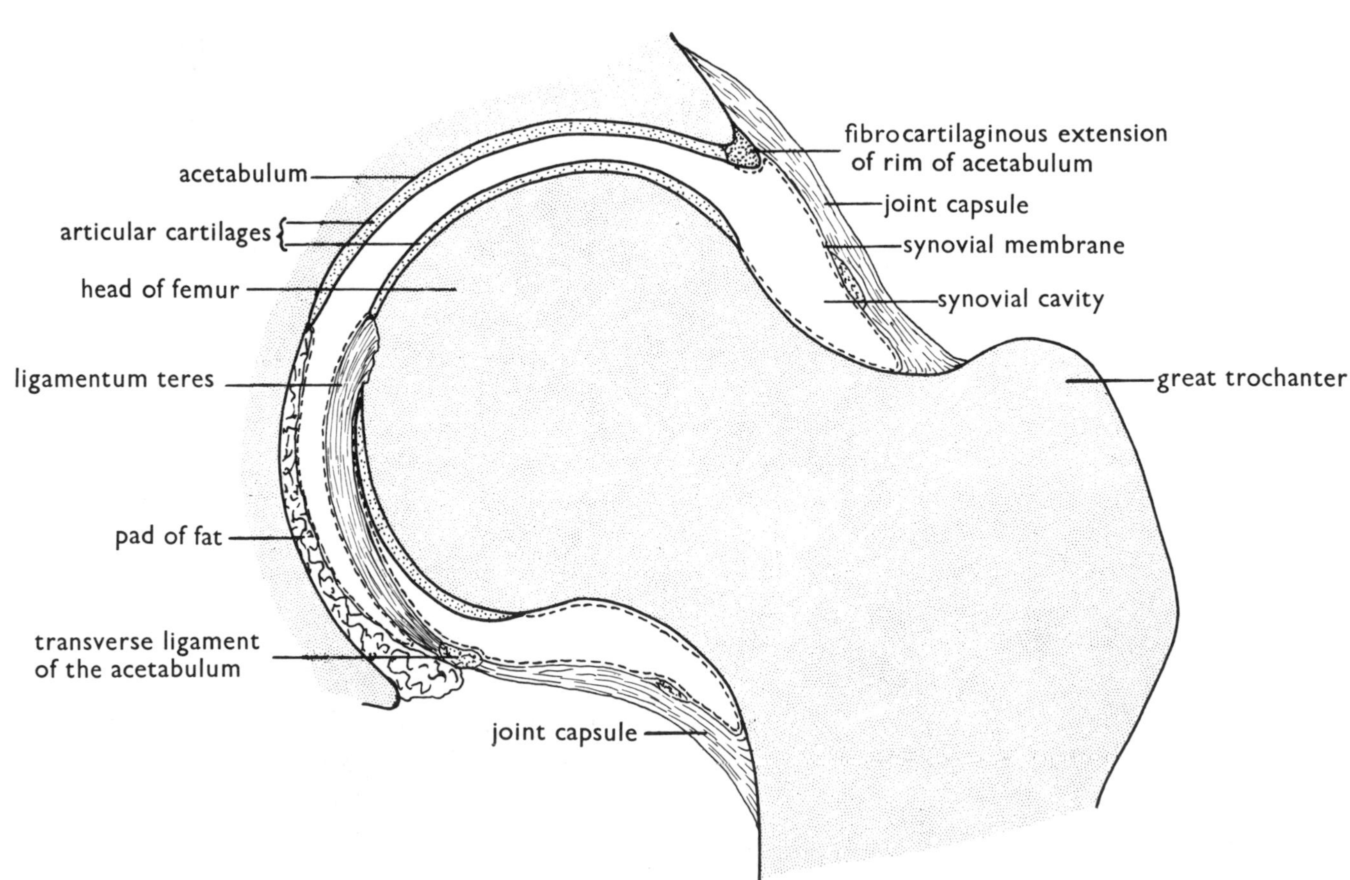

Note. This joint is ball and socket type with a very deep socket increased by a cartilaginous ridge. It has a very tough capsule and a ligament called the ligamentum teres between the transverse ligament of the acetabulum and the head of the femur. These features strengthen the joint but limit its movements.

Hip joint—*V.S.*

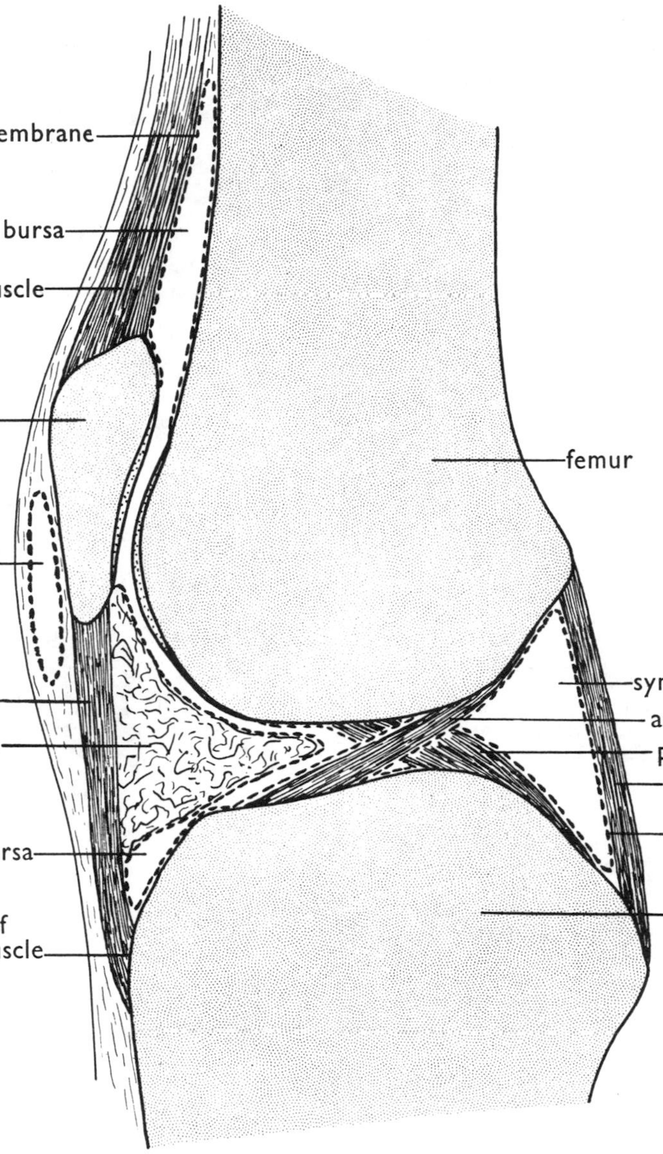

synovial membrane
suprapatellar bursa
tendon of
quadriceps femoris muscle
patella
femur
prepatellar bursa
synovial cavity
patellar ligament
anterior cruciate ligament
pad of fat
posterior cruciate ligament
joint capsule
synovial membrane
infrapatellar bursa
tibia
insertion of tendon of
quadriceps femoris muscle

Knee joint—*sagittal section*

Note. This joint is structurally condyloid but functionally hinge, because the semilunar cartilages and the cruciate ligaments prevent lateral movements.

femur
medial condyle
lateral condyle
synovial membrane
cruciate ligaments
medial semilunar cartilage
lateral semilunar cartilage
joint capsule
tibia
fibula

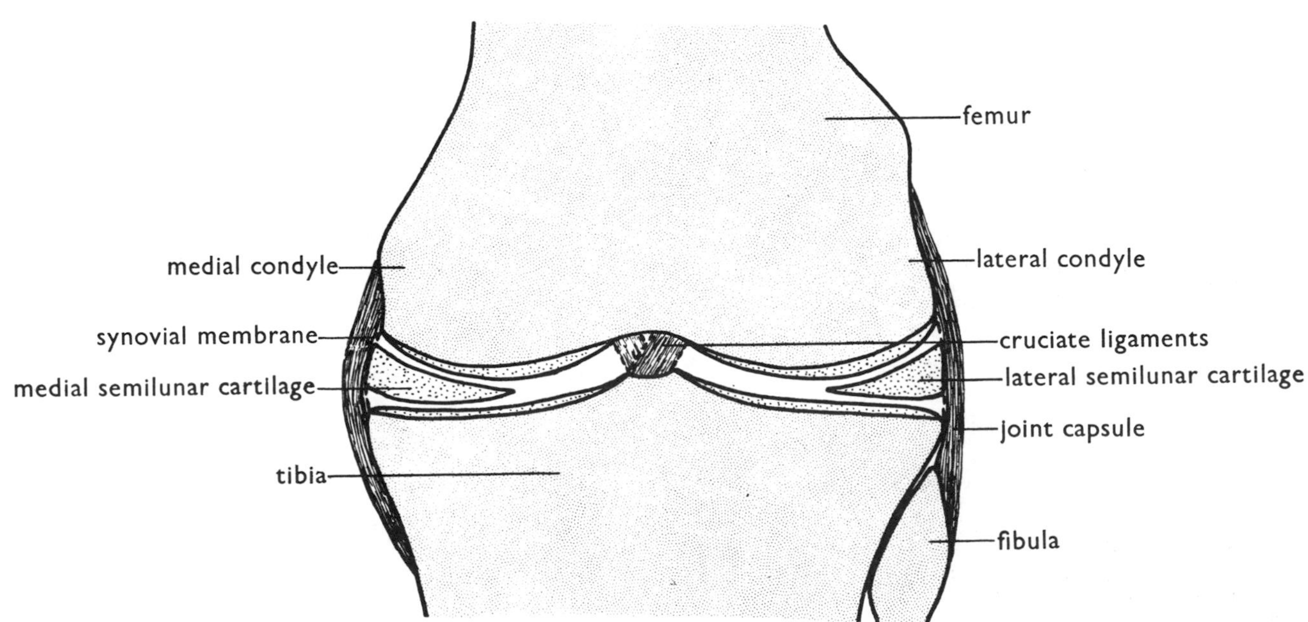

Knee joint—*coronal* (*frontal*) *section*

The Muscular System

The muscular system is concerned with movement. Muscle tissue consists of cells which are capable of contraction. They have no matrix of their own, but are associated with connective tissue. **Skeletal muscles** work in conjunction with the skeleton, while **smooth** and **cardiac** (heart) **muscles** produce movement of the internal organs.

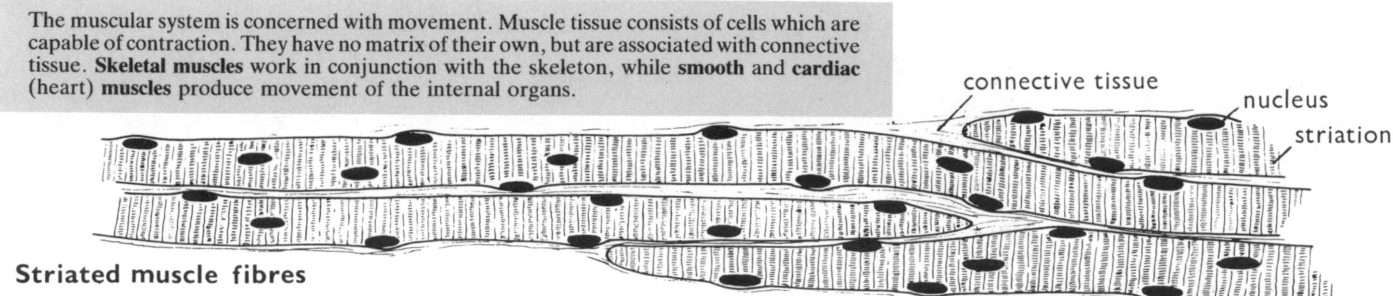

Striated muscle fibres

Note. There are numerous nuclei in each cell and fibrils in the cytoplasm show marked striations. There is connective tissue between the cells.

SKELETAL MUSCLE

Skeletal muscle is controlled by the **voluntary** parts of the nervous system—see page 69. It is composed of large, conspicuously **striated** (banded) cells, bound together into bundles or sheets by **areolar connective tissue**, which contains both tough **collagen fibres** and **yellow elastic fibres**, gelatinous **ground substance** and some **fibroblasts** (fibre-building cells) and **mast** cells which have come from the blood—see page 100.

Individual muscle cells vary in size from 10 to 100 μm in diameter and from 2 to 300 mm in length. Each cell has many nuclei which control the metabolism of the large unit. The outer surface membrane, called the **sarcolemma**, is a combination of plasma membrane and a polysaccharide layer containing white collagen fibres, which link it with the surrounding connective tissue. At intervals the plasma membrane component is extended into the cell where it lines **transverse tubules**. The cytoplasm within the muscle cell is called the **sarcoplasm**. Besides the numerous nuclei, which lie just under the sarcolemma, there are bundles of **myofibrils**, numerous **mitochondria** and an extensive **sarcoplasmic reticulum**, swollen into cisternae near the transverse tubules, cf. the parts of a generalised cell—see pages 5–7.

The myofibrils are arranged in a specific pattern, which gives rise to the striated appearance. The bands are identified by letters as shown in the diagram. Each repeated unit, or **sarcomere**, has thick longitudinal myofibrils, formed mainly of **myosin**, overlapping thin myofibrils, formed mainly of **actin**. Each myosin fibril is made of many rod-like myosin molecules, whose heads protrude on flexible necks at regular intervals. Each head is capable of forming a **cross-bridge** with one of the adjacent actin molecules. The actin molecules form double-stranded coils around fine double strands of **tropomyosin**, with associated molecules of a globular protein called **troponin**.

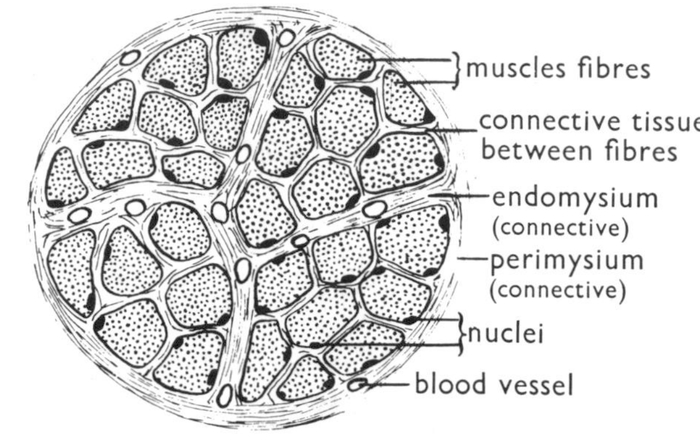

Skeletal muscle—*T.S.*

Note. The fibres are bound in bundles called fasciculi.

Functioning of skeletal muscle

Skeletal muscles contract and relax promptly. Each muscle fibre is served by at least one nerve fibre which ends in a **neuromuscular junction**—see page 69, through which the stimulus to contract is passed. This stimulus spreads rapidly across the sarcolemma and via the transverse tubules to the sarcoplasmic reticulum, bringing about release of **calcium ions** stored therein. Ca^{2+} binds to sites on the troponin causing change of shape which pulls the tropomyosin strands away from the binding sites on the actin. At the same time the Ca^{2+} removes inhibition of the enzyme **ATP-ase**, which then stimulates breakdown of ATP associated with the myosin heads. The latter become activated and capable of combining with actin to complete the cross-bridges and form **actino-myosin**. The necks shorten and pull the actin fibrils along relative to the myosin fibrils. The bridges then break and the sequence is repeated. As each cross-bridge is released, the neck moves to its original position and length and the head establishes a new attachment. The result is a hand-over-hand type pull which shortens the whole sarcomere. For relaxation the calcium must be removed. It is actively pumped back into the sarcoplasmic reticulum.

After a brief pause, the **refractory period**, the muscle fibre may be ready to contract again, but continuance of the process ultimately depends on availability of ATP, which in turn depends on its regeneration from ADP during oxidative processes in the mitochondria—see pages 4 and 7. Resting muscle accumulates stores of glycogen from which glucose can be made available as required. Availability of oxygen is also important. Without glucose and oxygen muscle becomes fatigued. **Heat** is released both during ATP breakdown—initial heat—and its restoration—recovery heat. Only about 25% of the energy made available goes into the muscular contraction.

Muscle fibres are of two types:

1. **Red muscle fibres**. These contain myoglobin, which holds oxygen reserves, cf. haemoglobin—see page 97. They are slow to react because they have little ATP-ase, but resist fatigue because the ATP can be regenerated synchronously with need.
2. **White muscle fibres**. These have less myoglobin, but more ATP-ase. They react more quickly, but fatigue more easily unless there is high vascularity (much blood supply).

Most muscles contain a mixture of fibres but in different proportions, e.g. postural muscles have many red fibres, leg muscles have red and white fibres and much vascularity, arm muscles have red and white fibres but less vascularity. The ratios of red and white fibres can be changed by training.

The fibres are bound into bundles by connective tissue collectively called **fasciae**, which merge with **tendons** or with tough flat sheets called **aponeuroses**.

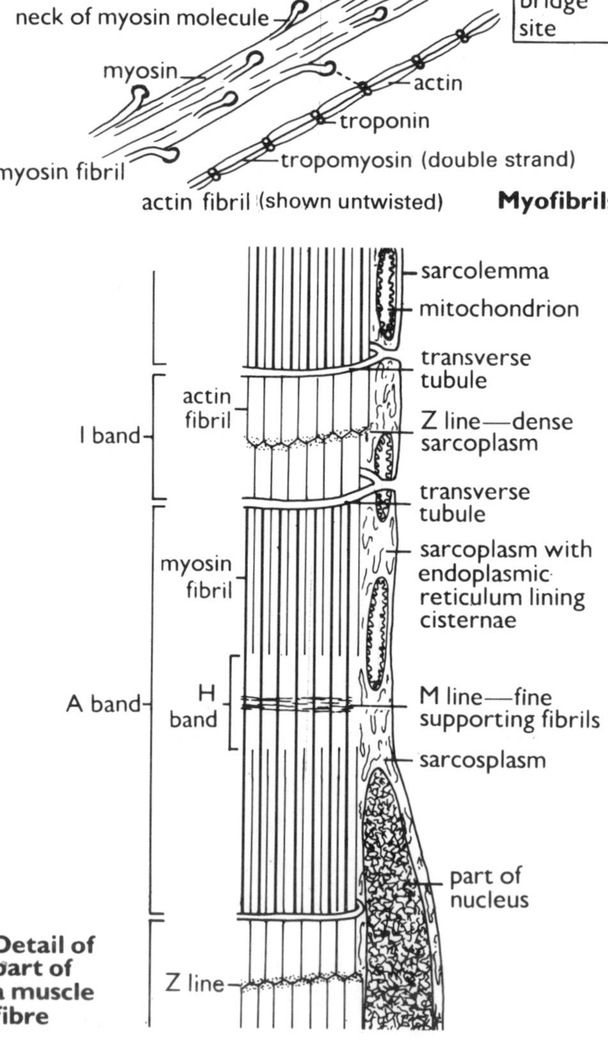

Myofibrils

Detail of part of a muscle fibre

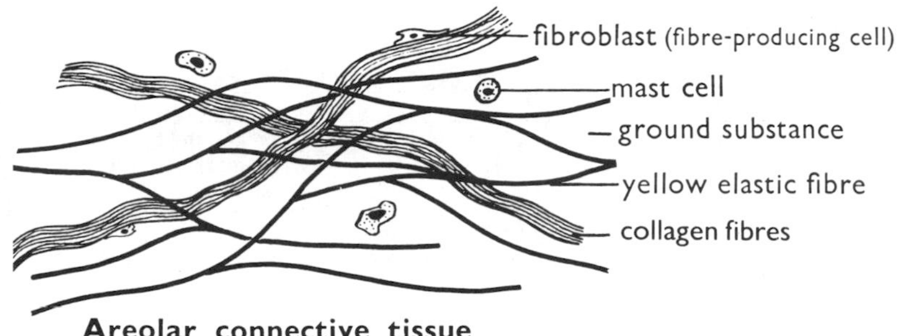

Areolar connective tissue

Note. The mixture of white collagen fibres and yellow elastic fibres and ground substance varies.

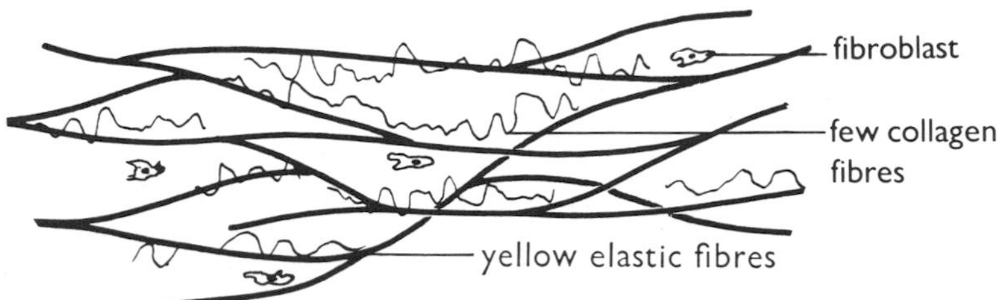

Yellow elastic connective tissue—*teased ligament*

Note. The yellow elastic fibres anastomose (form a network). The few collagen fibres are not in bundles.

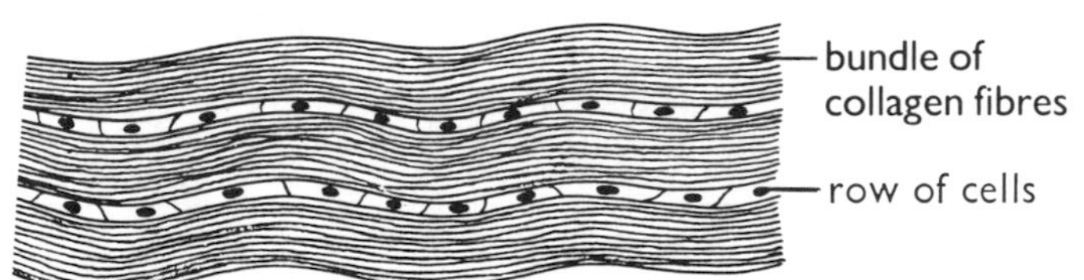

White fibrous connective tissue—*L.S. tendon*

Note. The fibres form parallel bundles between which the cells are in rows.

Action of skeletal muscles

Each individual muscle fibre acts in an **all-or-none** manner, therefore the strength of the overall contraction is related to the number of fibres activated. Normally a few fibres are stimulated in turn, maintaining **tone**, even when the muscle as a whole is at rest.

A sudden rapid stimulation of many fibres causes a **twitch**, while too frequent stimulation may cause a sustained contraction known as a **tetanus**. Most commonly, however, many fibres are stimulated in turn, in a controlled manner which produces either shortening of the whole muscle and thus movement (**isotonic contraction**) or increase in tension between the two ends without movement (**isometric contraction**). The end which usually remains fixed during isotonic contraction is the **origin** and the end which moves is the **insertion**.

In many cases the bones act as **levers**. The joint is usually the **fulcrum** (balance point of the lever), the skeletal muscle is the **effort** and the part supported or moved is the **load**. There are three orders of levers depending on the relative positions of fulcrum, effort and load. In all cases the force (effort or load) × distance from the fulcrum is known as the **moment** of the force. For equilibrium the moments of effort and load are equal.

1st order levers. The effort and the load are on opposite sides of the fulcrum, and movement of the effort results in movement of the load in the opposite direction.

When the fulcrum is central the effort must equal the load. When the fulcrum is nearer to the effort, the effort must be greater than the load but any distance it moves will be less than the corresponding movement of the load. This type of lever produces movements of the trunk and head with very little actual shortening of the muscles concerned.

Note. The head is very nearly balanced on the neck, so that to maintain upright posture the effort from the muscles at the back of the neck is very slight; but when the head is tilted forwards and more of its weight is in front of the fulcrum, a much stronger pull is needed to hold it steady or bring it upright again.

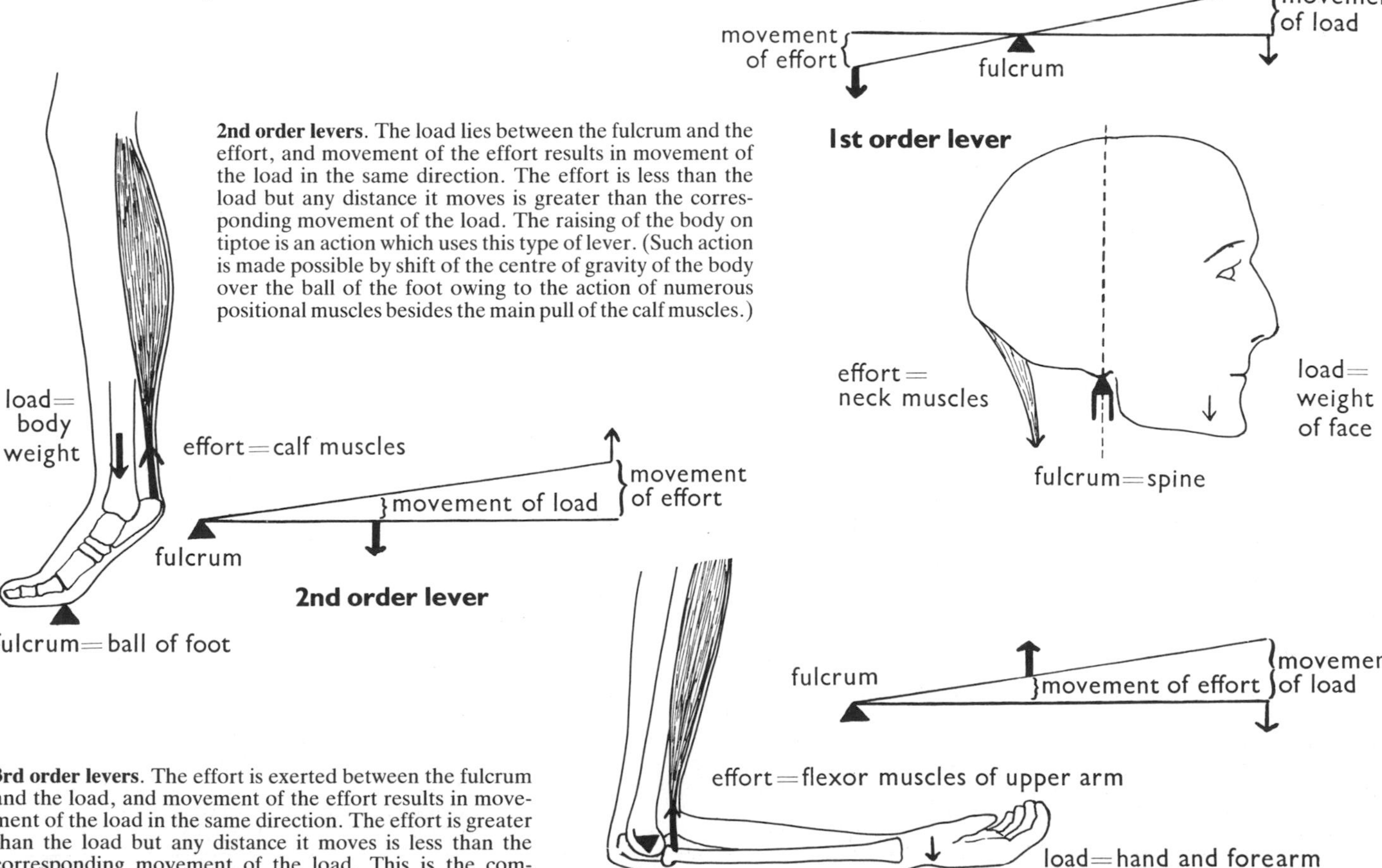

2nd order levers. The load lies between the fulcrum and the effort, and movement of the effort results in movement of the load in the same direction. The effort is less than the load but any distance it moves is greater than the corresponding movement of the load. The raising of the body on tiptoe is an action which uses this type of lever. (Such action is made possible by shift of the centre of gravity of the body over the ball of the foot owing to the action of numerous positional muscles besides the main pull of the calf muscles.)

3rd order levers. The effort is exerted between the fulcrum and the load, and movement of the effort results in movement of the load in the same direction. The effort is greater than the load but any distance it moves is less than the corresponding movement of the load. This is the commonest type of lever in the body and by means of it a variety of large movements can be made with very little shortening of the muscles concerned.

The chief muscles which move the head and trunk

Muscle	Origin	Insertion	Action of two sides together	Action of one side only	Notes
Sternomastoid	Sternum and clavicle	Mastoid process	Flex neck	Flexes neck laterally and rotates it	When the head is fixed this muscle raises sternum.
Splenius	Lower half of ligamentum nuchae and 1st six thoracic vertebrae	Mastoid process and occipital bone	Extend neck	Flexes neck laterally and rotates it	
Semispinalis capitis	4th cervical to 5th thoracic vertebrae	Occipital bone	Extend neck	Rotates neck	
Sacro-spinalis (Erector spinae)	Sacrum and iliac crest with additional fibres from ribs and lower vertebrae	Ribs, vertebrae and mastoid process	Extend trunk	Flexes trunk laterally	This muscle divides into three columns, ilio-costo-cervicalis longissimus and spinalis.
Semispinalis cervicis and semispinalis thoracis	Transverse processes of thoracic vertebrae	Spinous processes of vertebrae 6 or 7 above	Extend trunk and straighten thoracic curvature		These muscles are not shown in the diagrams because they lie under the sacrospinalis. They are chiefly postural in function, aiding rather than causing the larger movements.
Multifidus	Ilium, sacrum and transverse or articular processes of vertebrae	Spinous processes of vertebrae 1 to 4 above	Extend trunk and neck	Flex trunk and neck laterally and rotate them	
Rotatores	Transverse processes of thoracic vertebrae	Laminae of thoracic vertebrae next above		Rotate trunk	
Levatores costarum (12 pairs)	Transverse processes of last cervical and all but last thoracic vertebrae	Ribs next below	Rotate heads of ribs	Flex trunk laterally and rotate it	
Quadratus lumborum	Iliac crest and ilio-lumbar ligament	12th rib and transverse processes of upper four lumbar vertebrae	Extend trunk	Flexes trunk laterally and steadies lowest rib	See also page 50.
Rectus abdominis	Pubis	5th to 7th costal cartilages	Flex trunk ventrally	Flexes trunk laterally	See also page 55.
External oblique and internal oblique	For details see the abdominal wall, page 55		Flex trunk ventrally	Flex trunk laterally and rotate it	
Psoas minor	12th thoracic and 1st lumbar vertebrae	Pubis and iliac fascia	Help to flex trunk ventrally		These muscles are weak and often missing—see diagram page 50.
Psoas major	For details see hip muscles, page 45		Flex trunk ventrally	Flexes trunk laterally and flexes hip	

The chief muscles which move the ribs—see also muscles of respiration, page 58

Muscle	Origin	Insertion	Action	Notes
External intercostals (11 pairs)	Lower borders of upper 11 pairs of ribs	Further forward on upper border of rib next below in each case	Help to raise the ribs	The chief function of these muscles is to maintain the shape of the wall of the thorax.
Internal intercostals (11 pairs)	Inner surfaces of upper 11 pairs of ribs and the costal cartilages	Further back on upper border of rib or costal cartilage next below	Help to raise the ribs	
Scalenus muscles	Transverse processes of cervical vertebrae	1st and 2nd ribs	Raise upper ribs	Acting from below these muscles help to flex the neck.
Serratus posterior superior	Lower part of the ligamentum nuchae and spines of first three thoracic vertebrae	2nd to 5th ribs	Raises ribs	
Serratus posterior inferior	Spines of 11th thoracic to 2nd or 3rd lumbar vertebrae	9th to 12th ribs	Draws lower ribs downward and backwards	This muscle helps to steady the origin of the diaphragm.
Sternocostalis	Sternum	Costal cartilages	Draws down costal cartilages	This muscle is inside the thoracic cage and is not shown in diagram.

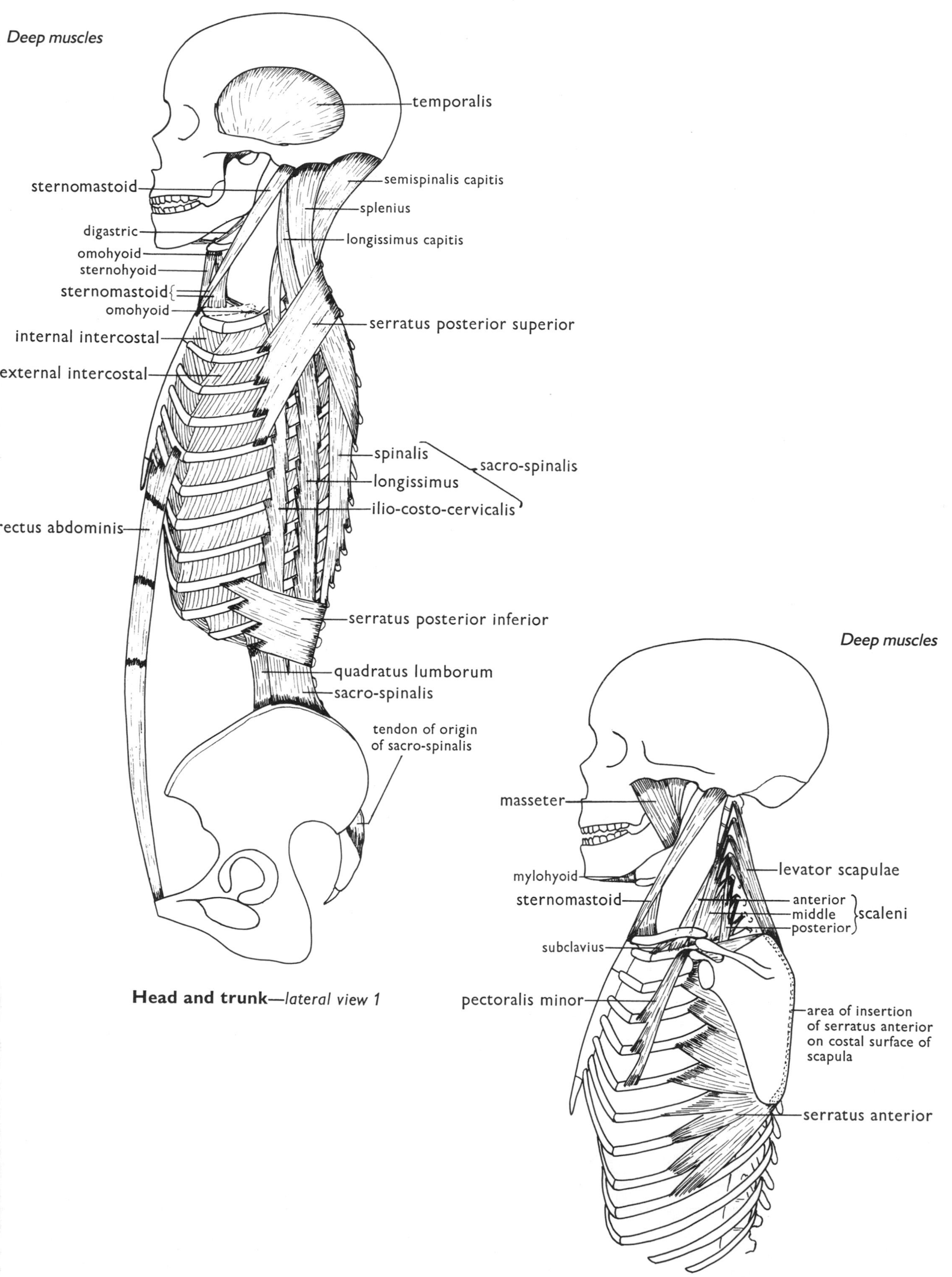

Head and trunk—*lateral view 1*

Head and trunk—*lateral view 2*

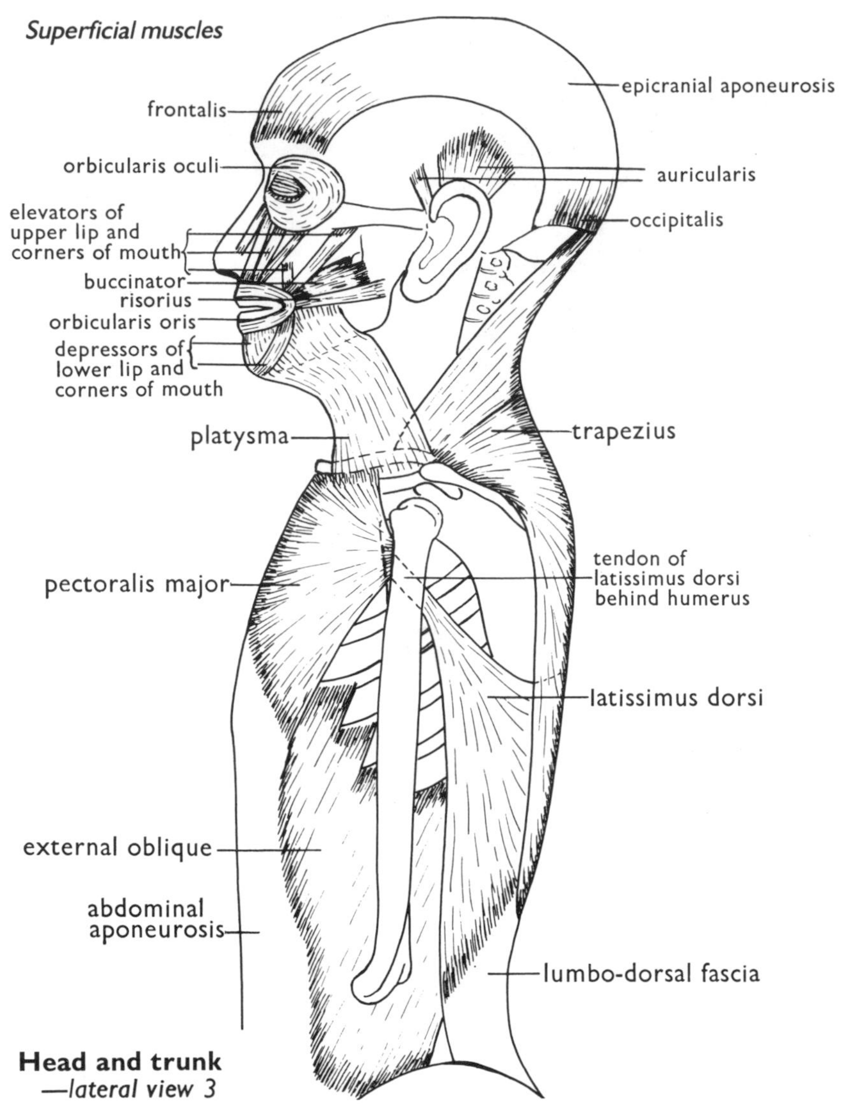

Head and trunk
—lateral view 3

The chief muscles of facial expression

Muscle	Origin	Insertion	Action	Notes
Occipito-frontalis (a) Occipital part (b) Frontal part	 Occipital bone Epicranial aponeurosis	 Epicranial aponeurosis Skin of eyebrow region	Moves scalp, raises eyebrows and wrinkles forehead	
Auricularis	Epicranial aponeurosis and temporal bone	Cartilage of the pinna	Moves pinna slightly	There are 3 of these small muscles, vestiges of those which cock the ears of other mammals.
Orbicularis oculi	Medial parts of rim of orbit	Forms a sphincter round the eye and across the eyelid	Closes lids	
Levator palpebrae	Back of orbit	Upper lid	Opens upper lid	See diagram, page 77.
Orbicularis oris	Sphincter muscle round the mouth		Closes mouth	
Buccinator	Maxilla and mandible	Angle of mouth	Compresses cheeks	
Risorius	Parotid fascia	Angle of mouth	Retracts angle of mouth	
Elevators of corners of mouth	Maxillae	Angles of mouth	Produce cheerful expression	
Depressors of corners of mouth	Mandible	Angles of mouth	Produce dismal expression	
Elevators of upper lip	Maxillae and zygomatic bones	Upper lip	Open mouth	
Depressors of lower lip	Mandible	Lower lip	Open mouth	
Platysma	Fascia over pectoralis major and deltoid muscles	Mandible and skin and muscles of lower part of face	Helps to draw down mandible and lower lip and wrinkles skin of neck	Used during yawning.
Nasal muscles	Maxillae	Nose	Compress and dilate the nasal openings	

The muscles of mastication

Muscle	Origin	Insertion	Action	Notes
Masseter	Zygomatic arch	Mandible	Raises lower jaw	
Temporalis	Temporal fossa	Coronoid process of mandible	Raises and retracts lower jaw	
Medial pterygoid	Pterygoid plate of sphenoid, palatine and maxilla	Angle of mandible	Raises lower jaw	These muscles are internal to the temporalis muscle and coronoid process and are not shown in diagrams. The right lateral pterygoid acting with the left elevators of the jaw and alternating with the left lateral pterygoid and right elevators produces chewing.
Lateral pterygoid	Pterygoid plate and great wing of sphenoid	Neck of mandible	Opens mouth and protrudes lower jaw	

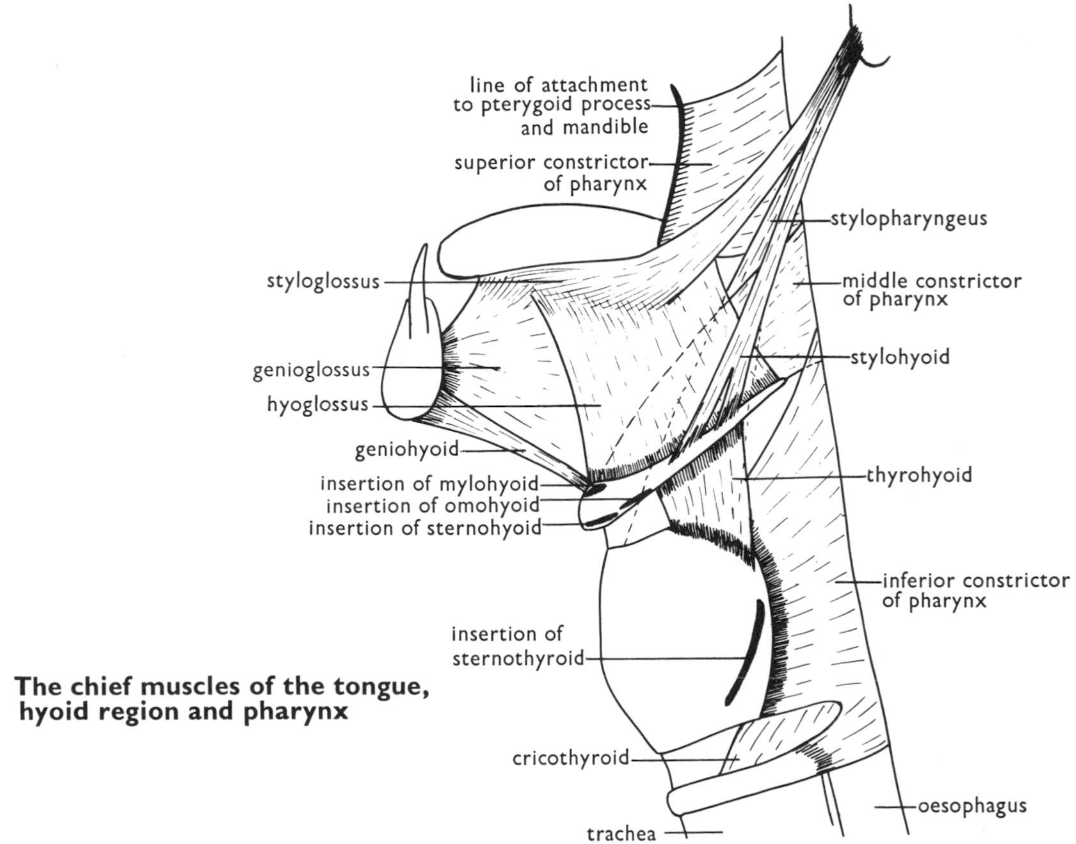

The chief muscles of the tongue, hyoid region and pharynx

The chief muscles which move the tongue, hyoid bone, larynx, pharynx and palate

Muscle	Origin	Insertion	Action	Notes
Genioglossus	Mandible	Tongue	Pulls tongue forwards	These are the extrinsic muscles which move the tongue as a whole.
Hyoglossus	Hyoid bone	Tongue	Depresses tongue	
Styloglossus	Styloid process of temporal bone	Tongue	Pulls tongue upwards and backwards	
Longitudinal tongue muscles	Lie completely inside the tongue		Shorten tongue and curve it up or down	These are the intrinsic muscles of the tongue which change its shape.
Transverse tongue muscles	Lie completely inside the tongue		Make the tongue long and narrow	
Vertical tongue muscles	Lie completely inside the tongue		Make the tongue flat and round	
Mylohyoid	Mandible	Hyoid bone and muscle of opposite side	Supports tongue and raises hyoid	These muscles form the floor of the mouth under the tongue.
Geniohyoid	Mandible	Hyoid bone	Raises hyoid	These muscles move and steady the hyoid bone during movements of the tongue and swallowing. The intermediate tendon of the omohyoid is connected through fascia to the first rib and clavicle.
Digastric	Mandible and temporal bone	Tendon between two halves of muscle connected to hyoid bone by a loop	Raises hyoid	
Stylohyoid	Styloid process	Hyoid bone	Raises hyoid	
Omohyoid	Scapula	Hyoid bone through an intermediate tendon	Depresses hyoid	
Sternohyoid	Sternum and clavicle	Hyoid bone	Depresses hyoid	

continued on page 44

The chief muscles which move the tongue, hyoid bone, larynx, pharynx and palate—continued

Muscle	Origin	Insertion	Action	Notes
Sternothyroid	Sternum and 1st rib	Thyroid cartilage	Depresses larynx	
Thyrohyoid	Thyroid cartilage	Hyoid bone	Depresses hyoid and raises larynx	These muscles are used during swallowing and ensure that food goes the right way and does not enter the larynx or naso-pharynx. The palatopharyngeus and levator palati muscles are not visible in the diagrams.
Stylopharyngeus	Styloid process	Thyroid cartilage and wall of pharynx	Raises larynx	
Palatopharyngeus	Palate	Thyroid cartilage and wall of pharynx	Raises larynx	
Levator palati	Temporal bone	Palate	Raises soft palate	
Constrictors of pharynx	Pterygoid process, mandible, hyoid bone, thyroid and cricoid cartilages	The corresponding muscles of the other side	Constrict pharynx	

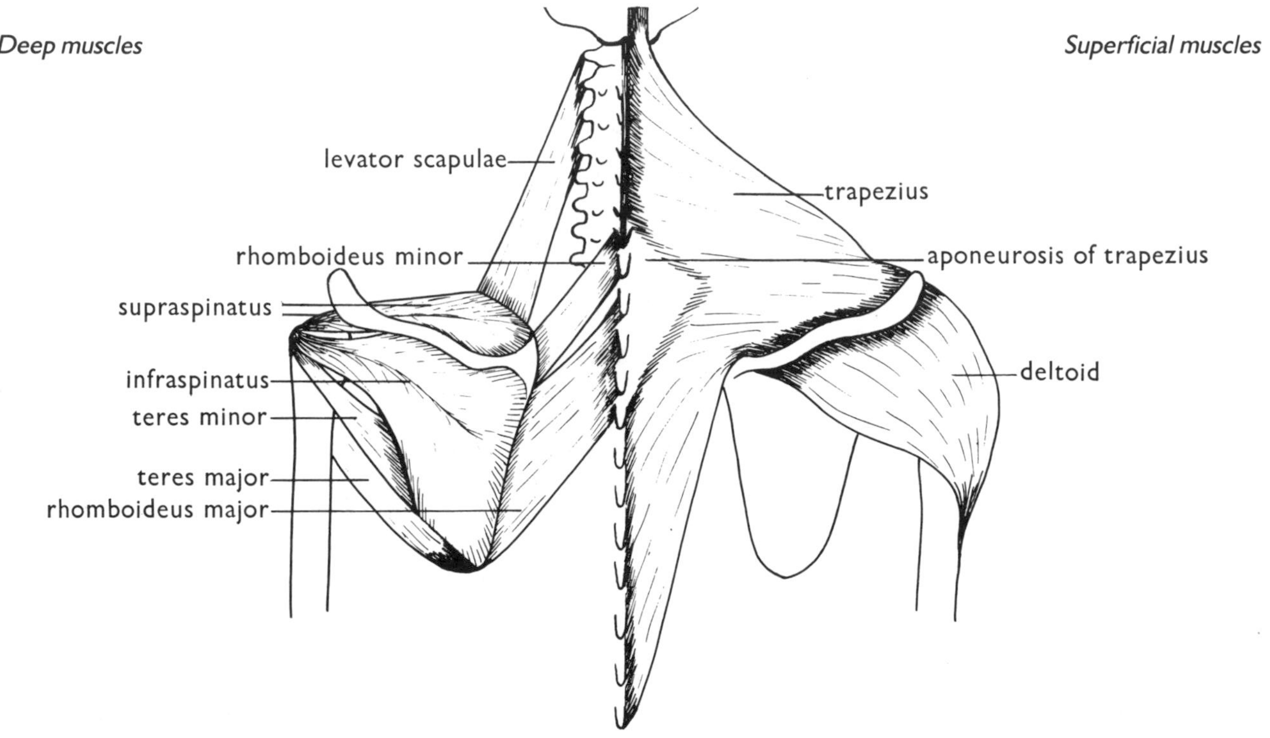

Shoulder region—*posterior view*

The chief muscles which move the shoulder

Muscle	Origin	Insertion	Action	Notes
Sternomastoid	See trunk muscles—page 40			All the muscles which move the shoulder are also used in combination with one another to steady the shoulder when the arm is moved and to adjust the angle of the glenoid cavity so that there is great freedom of movement.
Subclavius	1st rib	Clavicle	Draws shoulder downwards and forwards	
Pectoralis minor	3rd–5th ribs	Coracoid process of scapula	Draws shoulder downwards and forwards	
Serratus anterior	Upper eight or nine ribs	Costal surface of vertebral border of scapula	Draws shoulder forwards and rotates scapula	
Levator scapulae	Upper four cervical vertebrae	Upper part of vertebral border of scapula	Elevates shoulder and rotates scapula	
Rhomboideus minor	Ligamentum nuchae and 1st thoracic vertebrae	Vertebral border of scapula	Braces shoulder and rotates scapula	
Rhomboideus major	2nd–5th thoracic vertebrae	Vertebral border of scapula	Braces shoulder and rotates scapula	
Trapezius	Occipital bone, ligamentum nuchae and all thoracic vertebrae	Clavicle and spine of scapula	Elevates and braces shoulder and rotates scapula	When the scapula is fixed this muscle pulls the head back.

The chief muscles which move the whole arm

Muscle	Origin	Insertion	Action	Notes
Coraco-brachialis	Coracoid process of scapula	Shaft of humerus	Draws arm forwards and medially	When arm is raised it prevents side slip.
Pectoralis major	Clavicle, sternum, cartilages of true ribs and abdominal aponeurosis	Lateral lip of bicipital groove of humerus	Draws arm forwards and medially, adducts and rotates it inwards	These two muscles also cause depression of the shoulder against force and elevation of the body on the arms, e.g. when climbing.
Latissimus dorsi	Lower six thoracic vertebrae and through lumbodorsal fascia from lumbar vertebrae and iliac crest	Bicipital groove of humerus	Draws arm backwards, adducts and rotates it inwards	
Teres major	Inferior angle of scapula	Medial lip of bicipital groove of humerus	Draws arm backwards and medially, adducts and rotates it inwards	
Subscapularis	Subscapular fossa of scapula	Lesser tuberosity of humerus	Rotates arm inwards	These three muscles counteract slip of the head of the humerus when the deltoid is acting.
Teres minor	Axillary border of scapula	Greater tuberosity of humerus	Rotates arm outwards	
Infraspinatus	Infraspinous fossa of scapula	Greater tuberosity of humerus	Rotates arm outwards	
Supraspinatus	Supraspinous fossa of scapula	Greater tuberosity of humerus	Abducts arm	Acting with the three muscles above it helps to steady the head of the humerus.
Deltoid	Clavicle, acromion process and spine of scapula	Deltoid tuberosity of humerus	(a) Front draws arm forwards (b) Back draws arm backwards (c) Middle part abducts arm	Action (c) requires simultaneous rotation of the scapula.

The chief muscles which move the forearm, hand and fingers

Note. When the arm is in a relaxed position hanging by the side of the body, the palm of the hand faces medially, i.e. it is half-way between the prone and the supine positions. In this position the flexors of the hand and wrist are medial and the extensors are lateral, while the flexors of the elbow are anterior and the extensors are posterior. (Most books show the forearm twisted into the supine position with the flexors of the hand and wrist and the brachioradialis anterior, and with the ulnar flexor and the extensors posterior.)

Muscle	Origin	Insertion	Action	Notes
Triceps (a) long head (b) medial head (c) lateral head	 Scapula Humerus Humerus	By strong tendon on olecranon process of ulna	Extends elbow	This is the only muscle at the back of the upper arm.
Anconeus	Lateral epicondyle of humerus	Lateral part of olecranon process of ulna	Extends elbow	
Brachialis	Shaft of humerus	Coronoid process and tuberosity of ulna	Flexes elbow	The main part of this muscle crosses the elbow.
Brachioradialis	Lateral supracondylar ridge of humerus	Distal part of radius	Flexes elbow	The main part of this muscle lies below the elbow.
Biceps (brachii) (a) long head (b) short head	 Scapula (passes through shoulder joint capsule) Coracoid process of scapula	By a strong tendon on tuberosity of radius	Flexes elbow and supinates forearm and hand	This is the main muscle of the front of the upper arm. Because it originates from the scapula it helps to flex and steady the shoulder joint.
Supinator	Lateral epicondyle of humerus	Lateral surface of radius	Supinates forearm and hand	
Pronator teres	Above medial epicondyle of humerus and coronoid process of ulna	Middle of shaft of radius	Pronates forearm and hand	
Pronator quadratus	Distal part of ulna	Distal part of shaft of radius	Keeps radius and ulna together	The pronating action of this muscle is very weak.
Superficial extensors of wrist and fingers	Lateral epicondyle of humerus	Metacarpals and phalanges	Extend wrist and fingers	The radial flexor and extensor of the wrist when acting together abduct the hand while the ulnar flexor and extensor of the wrist adduct the hand. The tendons of all these muscles are held in place at the wrist by fibrous bands called the extensor retinaculum and the flexor retinaculum respectively.
Deep extensors of fingers	Ulna and radius	Phalanges of thumb and forefinger	Extend thumb and forefinger	
Superficial flexors of wrist and fingers	Medial epicondyle of humerus	Metacarpals, palmar aponeurosis and phalanges	Flex wrist and fingers	
Deep flexors of fingers	Ulna and radius	Phalanges	Flex thumb and fingers	

Note. In the palm of the hand there are additional small muscles causing flexion, abduction and adduction of the fingers and apposition of the thumb and little finger.

subscapularis
teres major
coraco-brachialis
Deep muscles
deltoid
Superficial muscle

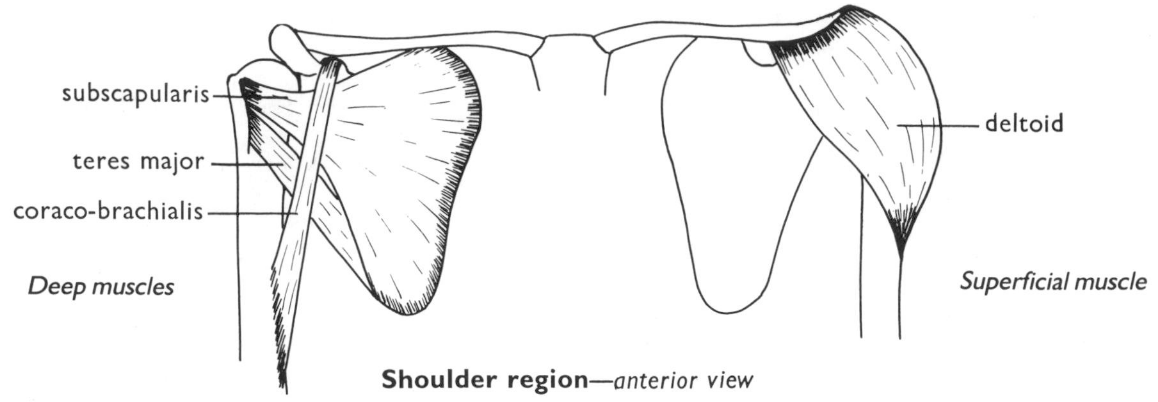

Shoulder region—*anterior view*

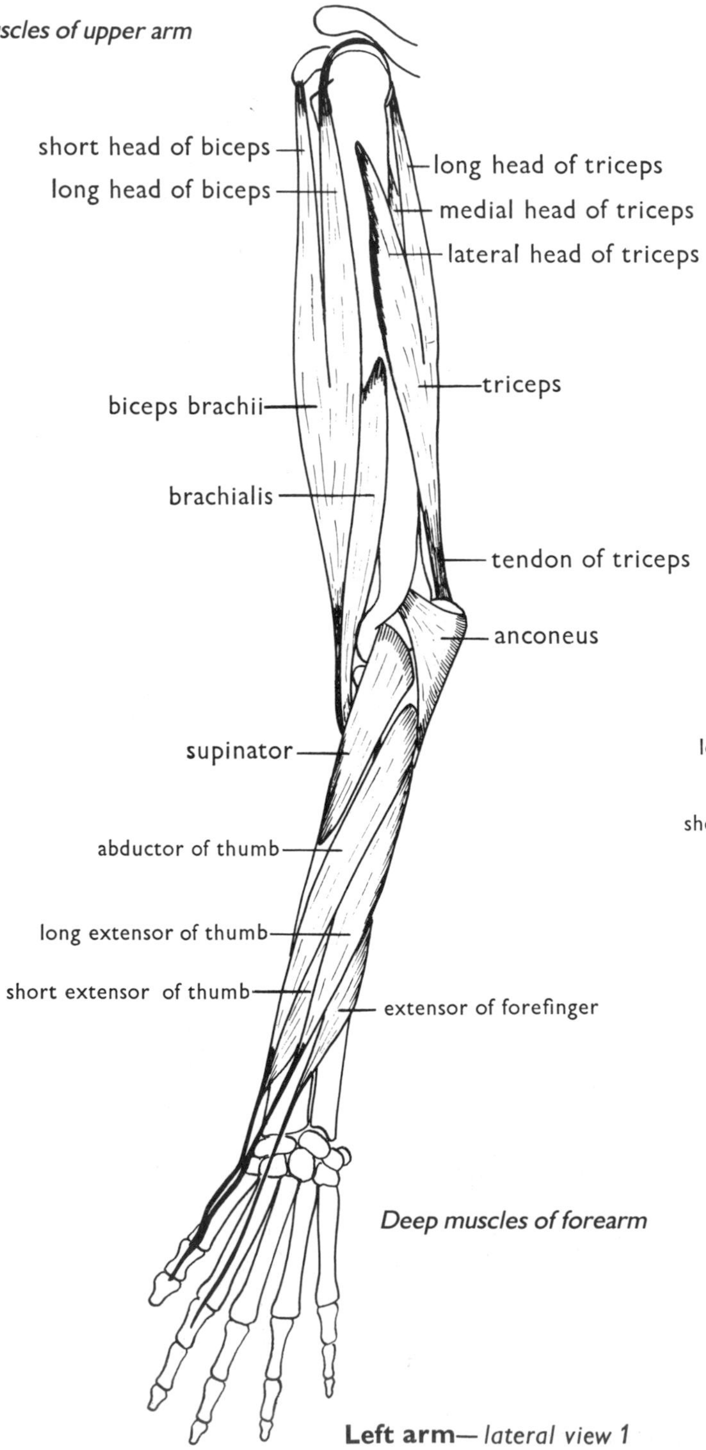

Muscles of upper arm
short head of biceps
long head of biceps
long head of triceps
medial head of triceps
lateral head of triceps
triceps
biceps brachii
brachialis
tendon of triceps
anconeus
supinator
abductor of thumb
long extensor of thumb
short extensor of thumb
extensor of forefinger
Deep muscles of forearm

Left arm—*lateral view 1*

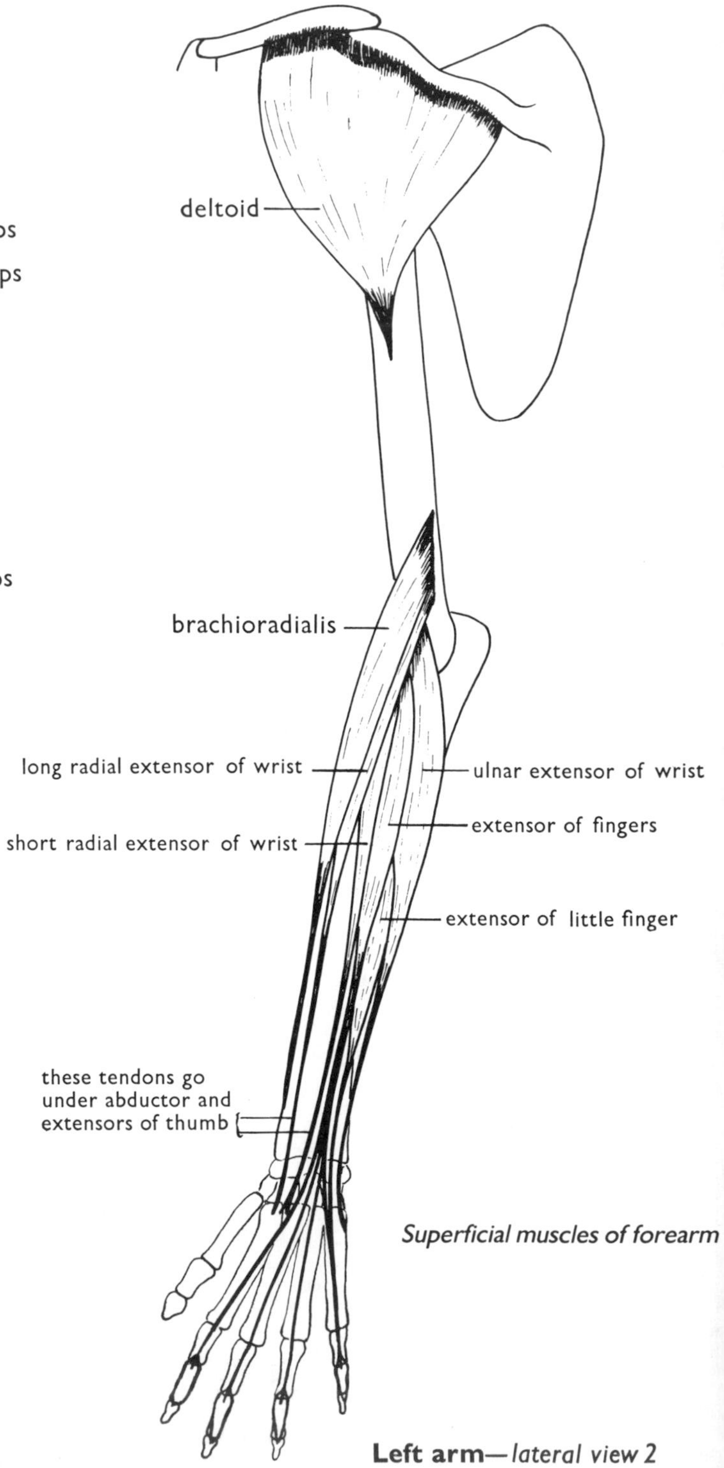

deltoid
brachioradialis
long radial extensor of wrist
ulnar extensor of wrist
short radial extensor of wrist
extensor of fingers
extensor of little finger
these tendons go
under abductor and
extensors of thumb
Superficial muscles of forearm

Left arm—*lateral view 2*

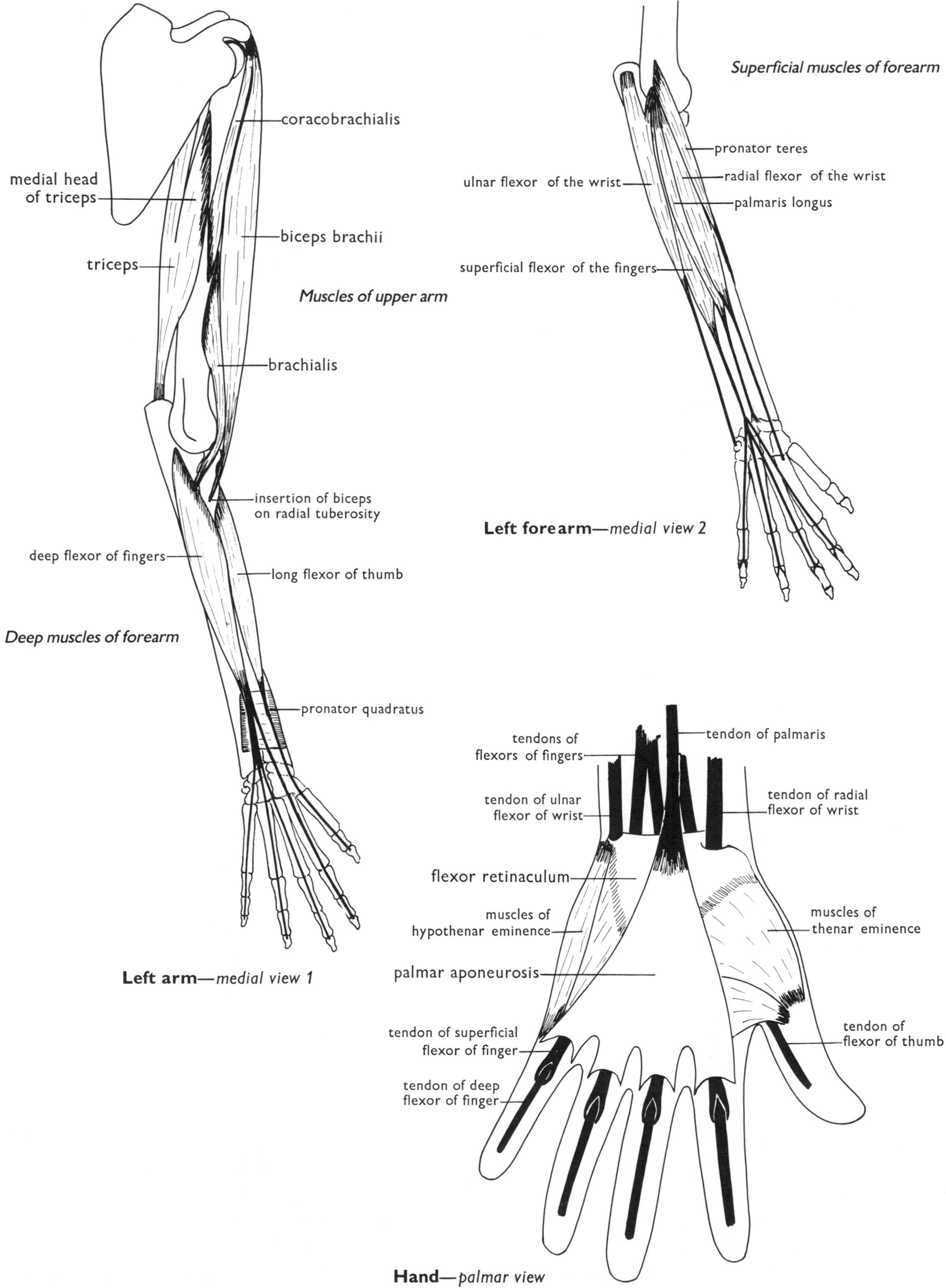

Left arm—*medial view 1*

Left forearm—*medial view 2*

Hand—*palmar view*

The chief muscles which move the lower limb

Note. There is so much overlap of function with muscles acting on hip and knee, and on knee and ankle, that it is impossible to separate the different groups distinctly. Also, because the knee bends in the opposite direction from the hip and ankle some muscles may be both flexors and extensors.

Muscle	Origin	Insertion	Action	Notes
Piriformis	Front of sacrum	Greater trochanter of femur	Rotates femur laterally	This group of short muscles limits the medial rotation inevitable with the action of the large flexor muscles and resists strain on the hip joint so that the capsular ligaments can be thin enough not to restrict movement.
Obturator internus	Inner surface of pelvis	Greater trochanter of femur	Rotates femur laterally	
Gemellus superior and inferior	Spine and tuberosity of ischium	Tendon of obturator internus	Rotate femur laterally	
Quadratus femoris	Tuberosity of ischium	Trochanteric crest of femur	Rotates femur laterally	
Obturator externus	Outer surface of pubis, ischium and obturator membrane	Trochanteric fossa of femur	Rotates femur laterally	The actual movements produced by these muscles are very slight.
Pectineus	Pubis	Near lesser trochanter of femur	Adducts femur and flexes hip	
Adductors	Pubis and ischium	Linea aspera and supra-condylar line	Adduct and rotate femur laterally	These muscles can also draw the abducted leg medially.
Ilio-psoas (a) Psoas major (b) Iliacus	 12th thoracic and all lumbar vertebrae Iliac fossa and front of sacrum	Lesser trochanter of femur	Rotate femur medially and flex hip	These muscles are very important in maintaining the upright posture of the trunk on the legs, standing, walking, etc.
Tensor fasciae latae	Outer part of iliac crest	Fascia lata	Keeps the fascial sheath of the thigh tensed and therefore helps to abduct and rotate femur medially and to extend knee	
Gluteus maximus	Posterior gluteal line of ilium and crest above and also sacrum and coccyx	Fascia lata and gluteal tuberosity of femur	Tenses fascia lata and extends hip, therefore raises trunk after stooping	
Gluteus medius	Between posterior and middle gluteal lines of ilium	Greater trochanter of femur	Abducts femur and anterior fibres rotate it medially	
Gluteus minimus	Between middle and inferior gluteal lines of ilium	Greater trochanter of femur	Abducts femur and anterior fibres rotate it medially	
Gracilis	Pubis and ischium	Below medial condyle of tibia	Adducts and rotates femur medially, flexes knee	
Semitendinosus	Tuberosity of ischium	Below medial condyle of tibia	Extend hip and flex knee and when knee semiflexed rotate femur medially	These muscles are collectively known as the 'hamstrings'.
Semimembranosus	Tuberosity of ischium	Medial condyle of tibia		
Biceps femoris (a) long head (b) short head	 Tuberosity of ischium Linea aspera	Head of fibula and lateral condyle of tibia	Extends hip and flexes knee and when knee semi-flexed rotates femur laterally	
Sartorius	Anterior superior iliac spine	Below medial condyle of tibia	Flexes hip and knee, abducts and rotates femur laterally	
Quadriceps femoris (a) Rectus femoris (b) Vastus lateralis (c) Vastus intermedius (d) Vastus medialis	 Above acetabulum Greater trochanter and linea aspera Shaft of femur Trochanteric and spiral lines and linea aspera	Through the patella and patellar ligament on to tubercle of tibia	Together extend knee and the rectus femoris part helps to flex hip	The tendon of this muscle forms a large part of the capsule of the knee joint.

continued on page 53

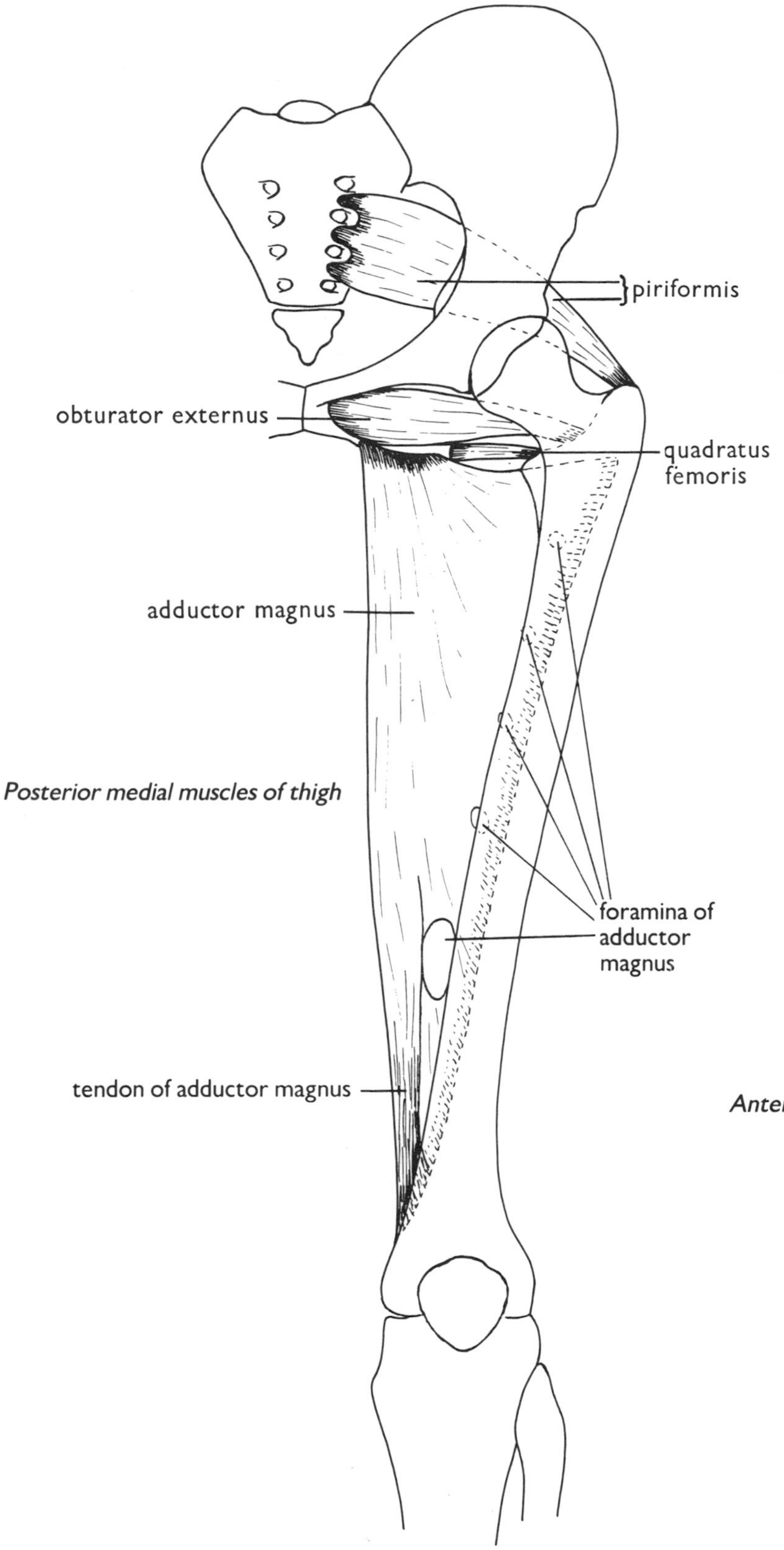

piriformis
obturator externus
quadratus femoris
adductor magnus
Posterior medial muscles of thigh
foramina of adductor magnus
tendon of adductor magnus

Hip and thigh—*anterior view 1*

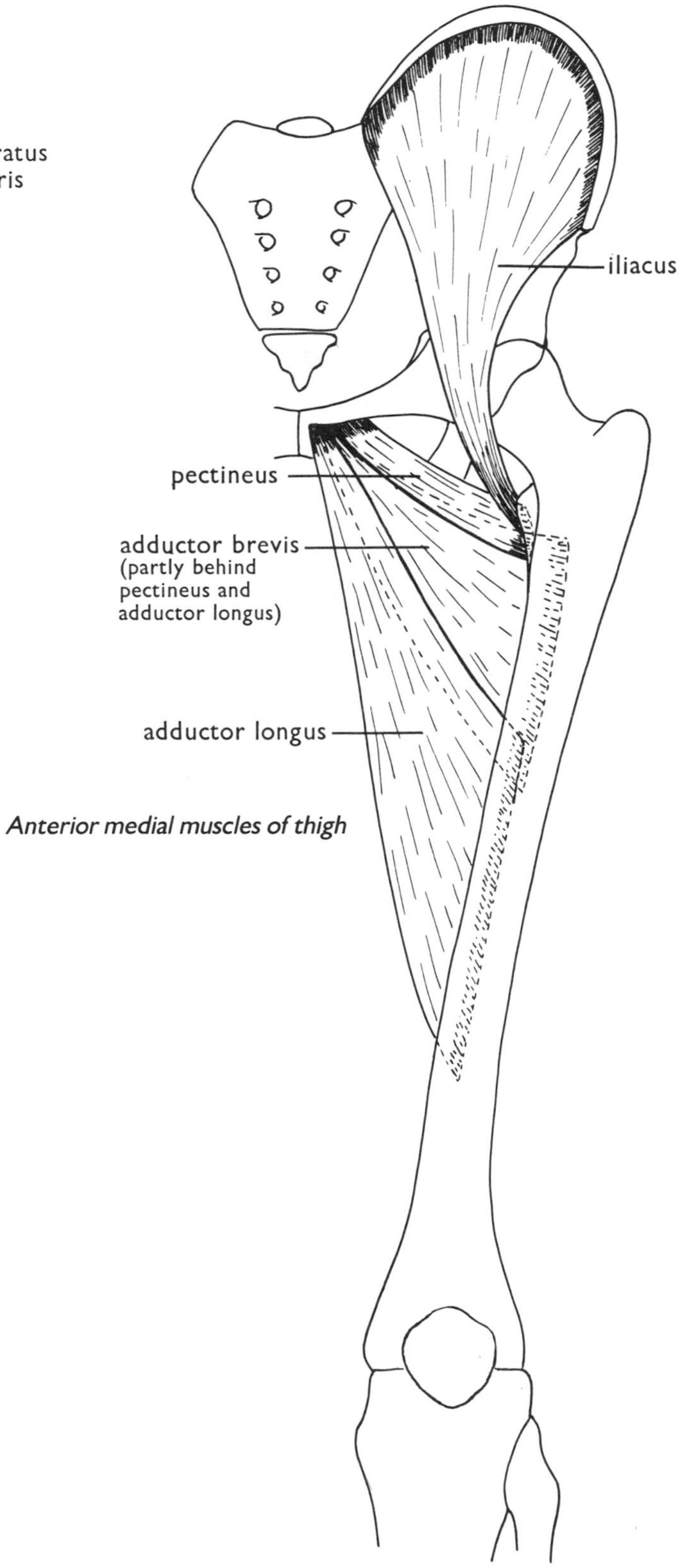

iliacus
pectineus
adductor brevis (partly behind pectineus and adductor longus)
adductor longus
Anterior medial muscles of thigh

Hip and thigh—*anterior view 2*

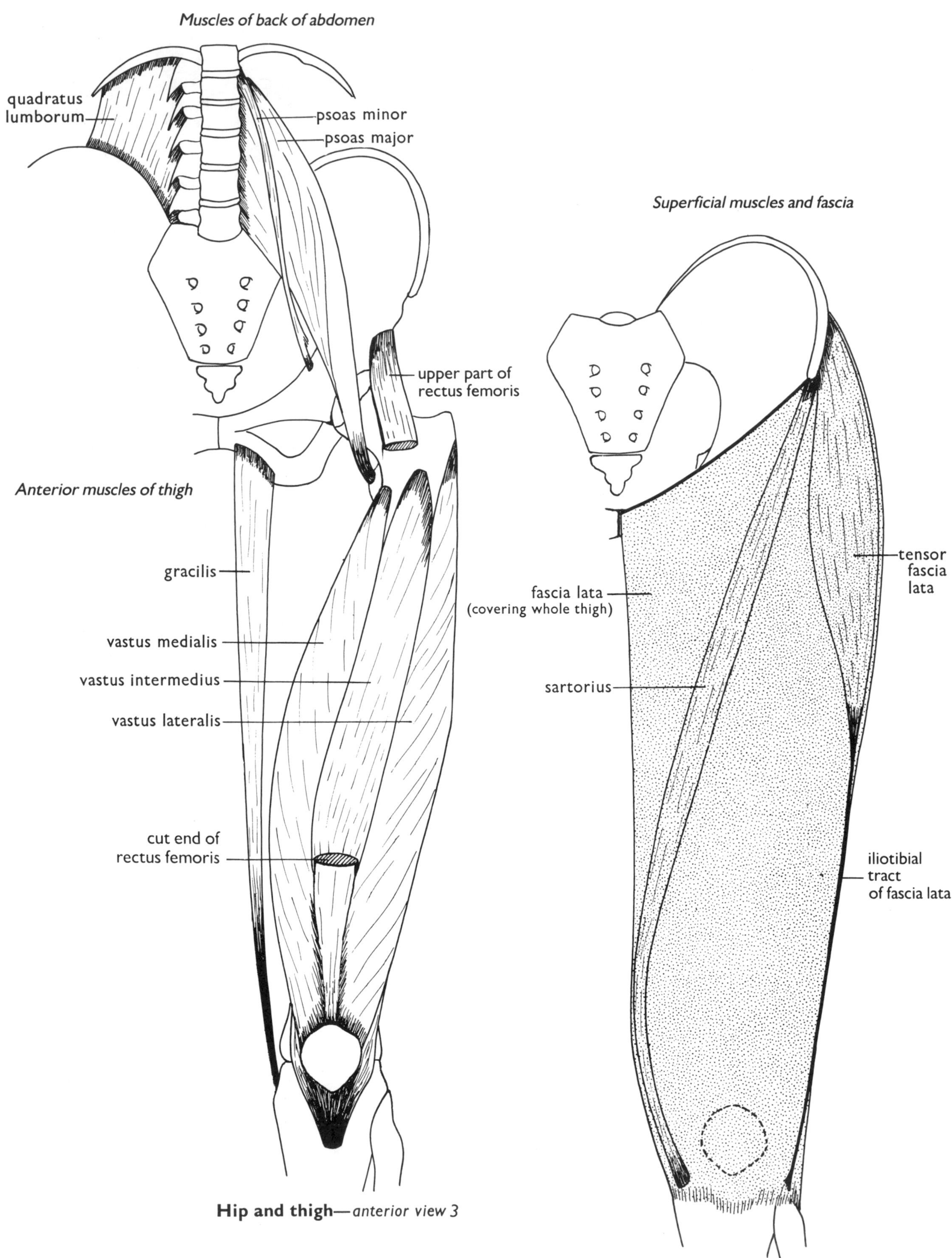

Hip and thigh—*anterior view 3*

Hip and thigh—*anterior view 4*

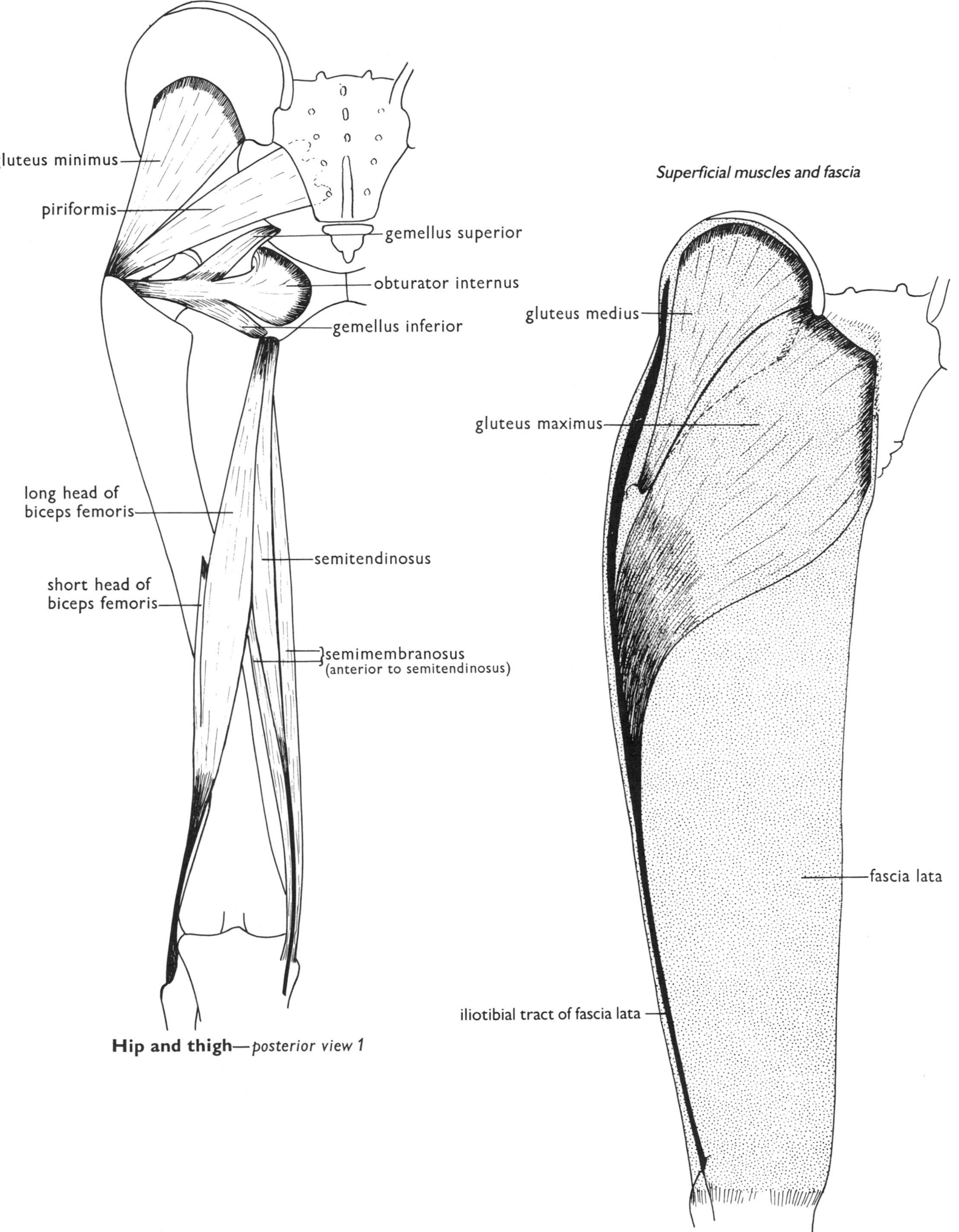

Hip and thigh—*posterior view 1*

Hip and thigh—*posterior view 2*

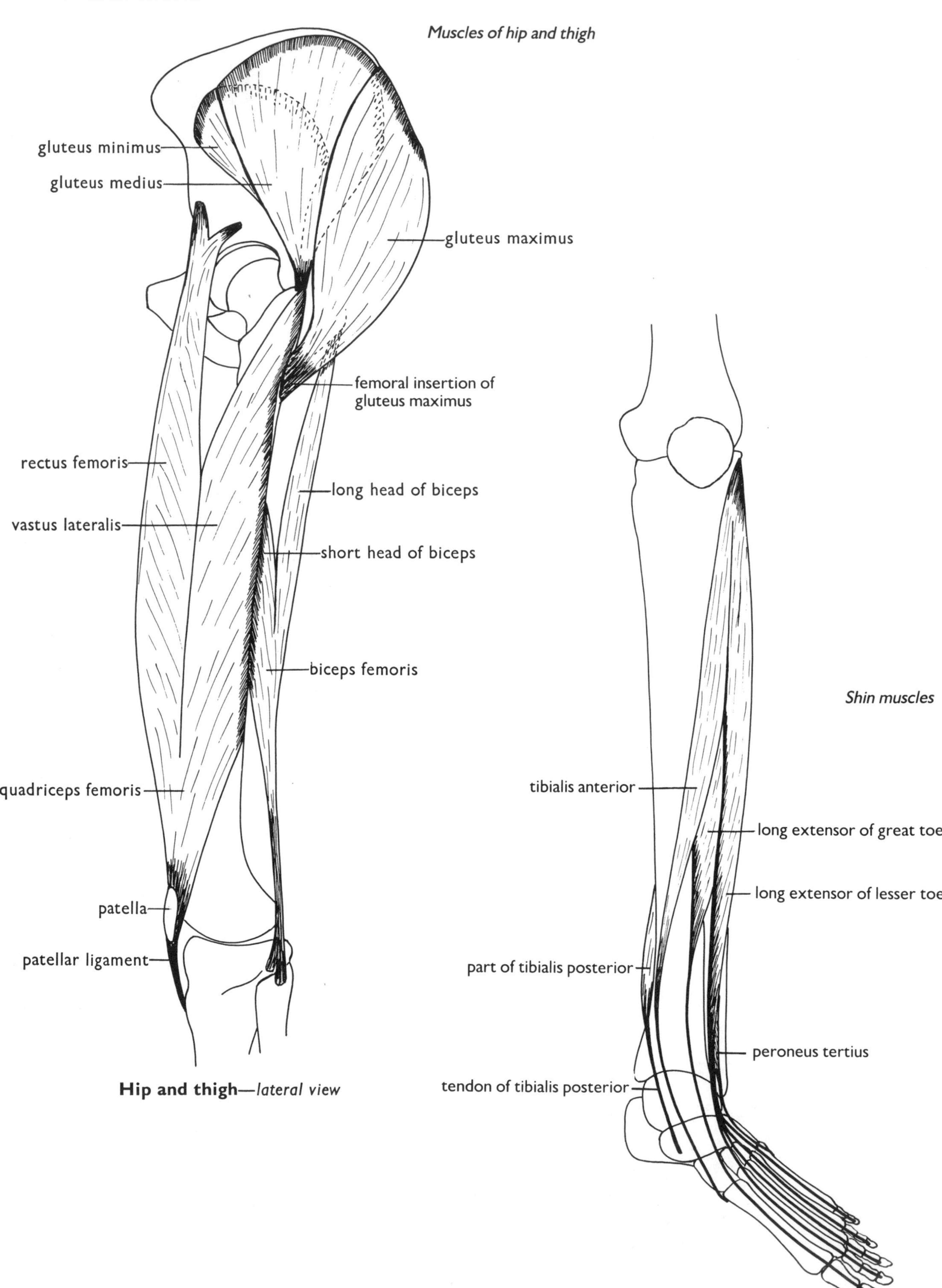

Hip and thigh—*lateral view*

Shin and foot—*antero-medial view*

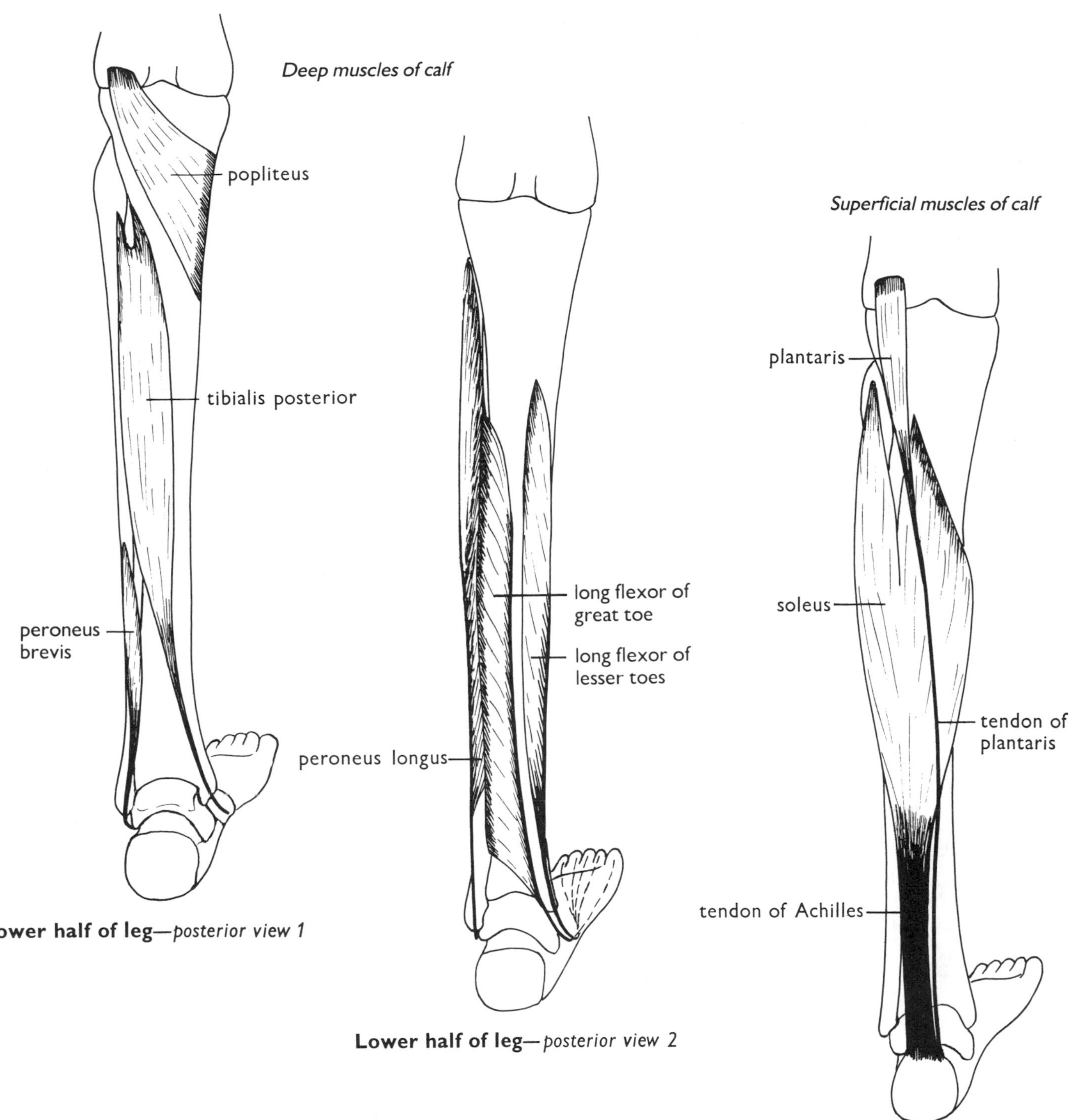

Lower half of leg—*posterior view 1*

Lower half of leg—*posterior view 2*

Lower half of leg—*posterior view 3*

The chief muscles which move the lower limb—continued from page 48

Muscle	Origin	Insertion	Action	Notes
Popliteus	Lateral condyle of femur	Shaft of tibia	Flexes knee and rotates tibia medially	
Gastrocnemius	Lateral and medial condyles of femur	Through tendon of Achilles on calcaneum	Flexes knee and plantar-flexes ankle	These muscles give the force when walking, running, etc. Plantarflexion is equivalent to extension of other joints. It causes pointing of the foot and raising of the body on the toes.
Plantaris	Above lateral condyle of femur	Long tendon joins tendon of Achilles	Accessory to gastrocnemius	
Soleus	Fibula and tibia	Through tendon of Achilles on calcaneum	Plantarflexes ankle	

continued on page 54

The chief muscles which move the lower limb—continued

Muscle	Origin	Insertion	Action	Notes
Tibialis posterior	Tibia and fibula	Navicular	Inverts foot	These muscles help balance when standing. Their tendons are held in place at the ankle by fibrous bands, the superior and inferior extensor retinacula and the flexor retinaculum.
Tibialis anterior	Lateral condyle and shaft of tibia, interosseus membrane	Medial cuneiform and 1st metatarsal	Inverts foot and rotates it laterally, dorsiflexes ankle	
Peroneus (three muscles)	Fibula	5th metatarsal and across sole of foot to 1st metatarsal	Evert foot and rotate it medially, dorsiflex ankle	
Long extensors of toes	Fibula and lateral condyle of tibia	Distal phalanges of toes	Extend toes and dorsiflex ankle	
Long flexors of toes	Back of tibia and fibula	Distal phalanges of toes	Flex toes and help to plantarflex ankle	

Note. There are also short extensors and flexors, adductors and abductors of the toes in the foot.

Superficial muscles of the calf

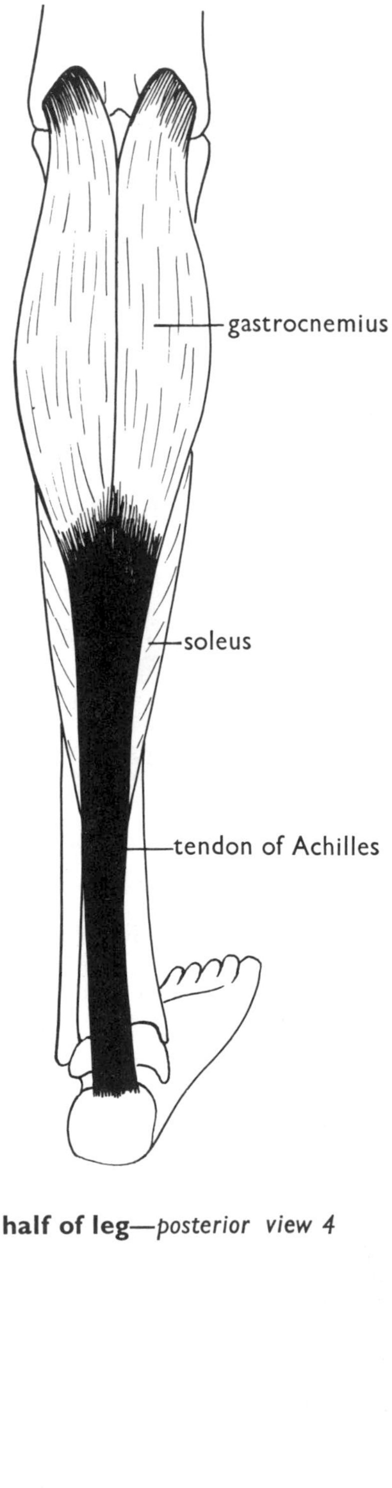

gastrocnemius
soleus
tendon of Achilles

Lower half of leg—*posterior view 4*

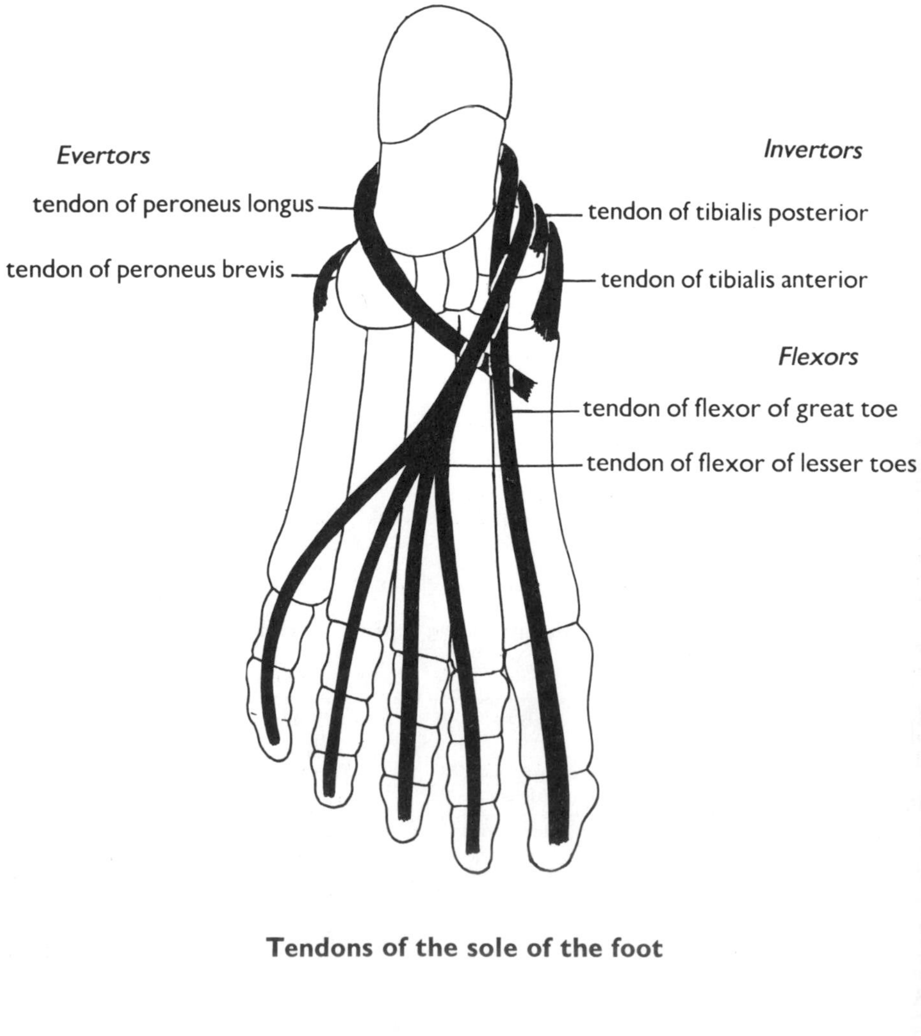

Evertors
tendon of peroneus longus
tendon of peroneus brevis
Invertors
tendon of tibialis posterior
tendon of tibialis anterior
Flexors
tendon of flexor of great toe
tendon of flexor of lesser toes

Tendons of the sole of the foot

The muscles of the abdominal wall

1. **Posterior group** The back of the abdomen is supported by vertebrae but the quadratus lumborum and psoas muscles contribute to its wall—see page 40.

2. **Lateral and anterior group**. The sides and front of the abdomen are formed of a continuous sheet of muscle and aponeurosis, the main parts of which have three distinct layers.

Muscle	Origin	Insertion	Action	Notes
Transversus	Inguinal ligament, iliac crest, lumbar fascia and cartilages of lower six ribs	Conjoint tendon and linea alba through abdominal aponeurosis	Supports the viscera, helps in vomiting, micturition, defaecation, parturition and forced expiration	The lumbar fascia lies around the muscles of the back in the lumbar region.
Internal oblique	Inguinal ligament, iliac crest and lumbar fascia	Conjoint tendon and linea alba through abdominal aponeurosis and lower 3 ribs	Internal oblique of one side with external oblique of the other produce rotation, together they flex the trunk ventrally—see page 40	
External oblique	Lower 8 ribs	Iliac crest and linea alba through abdominal aponeurosis		
Rectus abdominis	Pubis	5th, 6th and 7th costal cartilages	Supports viscera, helps vomiting, etc., and flexes trunk ventrally	The upper ⅔ of this muscle lie in the aponeurotic sheath but the lower ⅓ lies internal to the whole aponeurosis.
Pyramidalis	Pubic symphysis	Linea alba	Tenses linea alba	

3. **Inferior group**. The pelvic basin is lined with muscles many of which are concerned with movements of the thigh—see page 50, but in addition there is a muscular sheet which supports the pelvic viscera. The anus and urethra have sphincters of voluntary muscle which control defaecation and micturition respectively—see pages 90 and 117.

Muscle	Origin	Insertion	Action	Notes
Levator ani	Spine of ischium and tendinous arch of obturator fascia	Coccyx, perineal body and fibres of muscle of opposite side	Constricts rectum and vagina	These muscles form the pelvic floor.
Coccygeus	Spine of ischium	Sacrum and coccyx	Pulls coccyx forwards after defaecation	

4. **Superior group**. The upper wall of the abdomen is formed by the diaphragm—see pages 58 and 59.

The abdominal aponeurosis

The abdominal aponeurosis is a flattened sheet of tendinous fibres. Laterally it is in three layers attached to the external oblique, the internal oblique and the transversus muscles respectively, but medially the fibres from the two sides are interwoven to form a strong tendinous band called the **linea alba** which extends from the xiphoid process to the pubic symphysis.

The lower edge of the external layer of the aponeurosis is thickened to form the **inguinal ligament** which extends from the anterior superior iliac spine to the pubic tubercle. The middle layer of the aponeurosis is split above the **arcuate line**, i.e. for the upper three-quarters of its length, to form a sheath around the rectus abdominis muscle. The external part of this layer merges with the external layer of the aponeurosis and the internal part merges with the internal layer, the lower edge of which forms the **conjoint tendon** inserted on the pubic crest. The inguinal and conjoint tendons support the abdominal wall where there is greatest weakness, i.e. where it is perforated by the inguinal canals.

Each **inguinal canal** is about 40 mm long and lies parallel to and slightly above corresponding inguinal ligament. The inner end of the canal opens into the abdominal cavity at a **deep inguinal ring**, a perforation in the transversalis fascia half-way between the anterior superior iliac spine and the pubic symphysis. The outer end of each canal opens at a **superficial inguinal ring** which perforates the aponeurosis just above and lateral to the crest of the pubis. The conjoint tendon lies deep to this ring.

In the male the inguinal canals are wide because they transmit the spermatic cords—see page 125—but in the female they are narrow, because they transmit only the round ligaments of the uterus—see page 127. As a result of this there is greater weakness in the regions of the inguinal rings of the male than of the female and inguinal hernia is much more common in men than in women. A further position of weakness is the **umbilicus** where the linea alba is perforated by the umbilical blood vessels during foetal life. Though the perforation normally closes soon after birth umbilical hernia occasionally occurs, especially in babies.

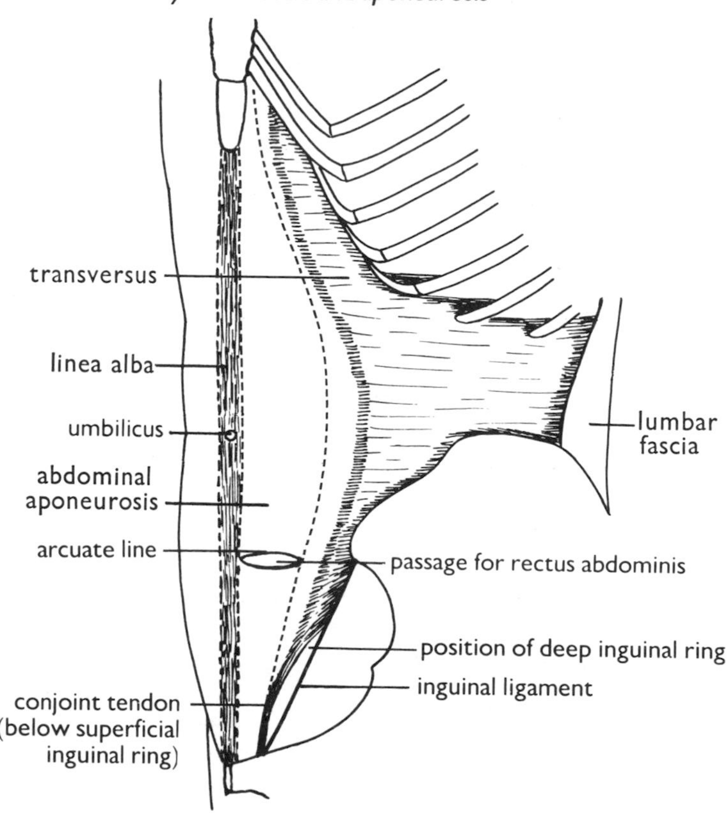

Abdominal wall—1

Middle layer of muscle and aponeurosis

tendinous intersections
rectus abdominis
umbilicus
linea alba
abdominal aponeurosis
conjoint tendon
internal oblique
lumbar fascia
inguinal line

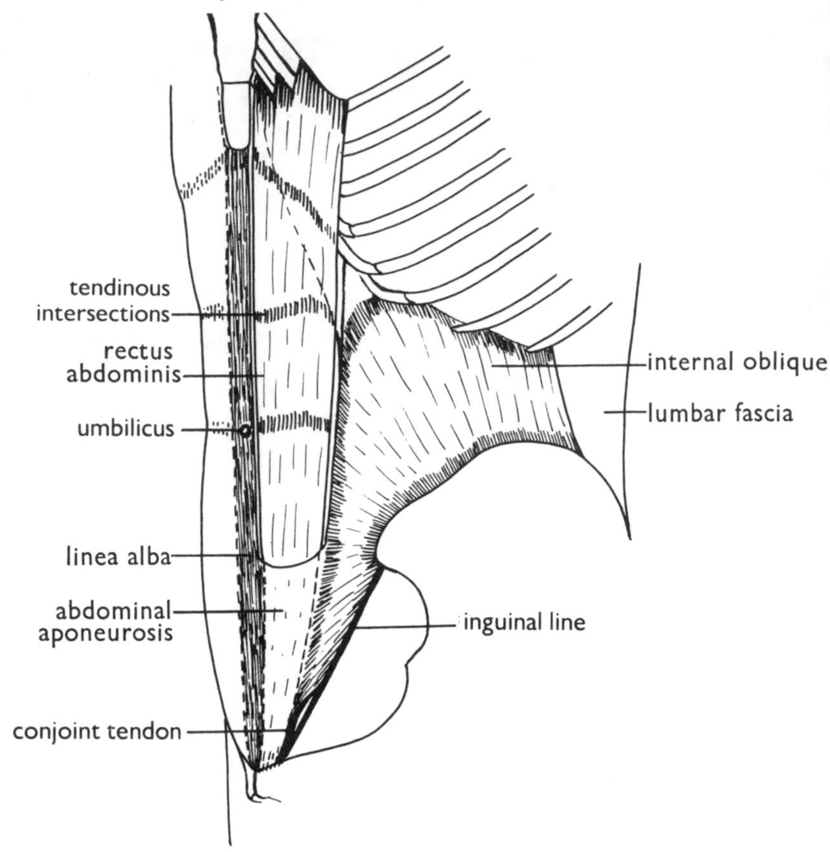

Abdominal wall—*2*

Outer layer of muscle and aponeurosis

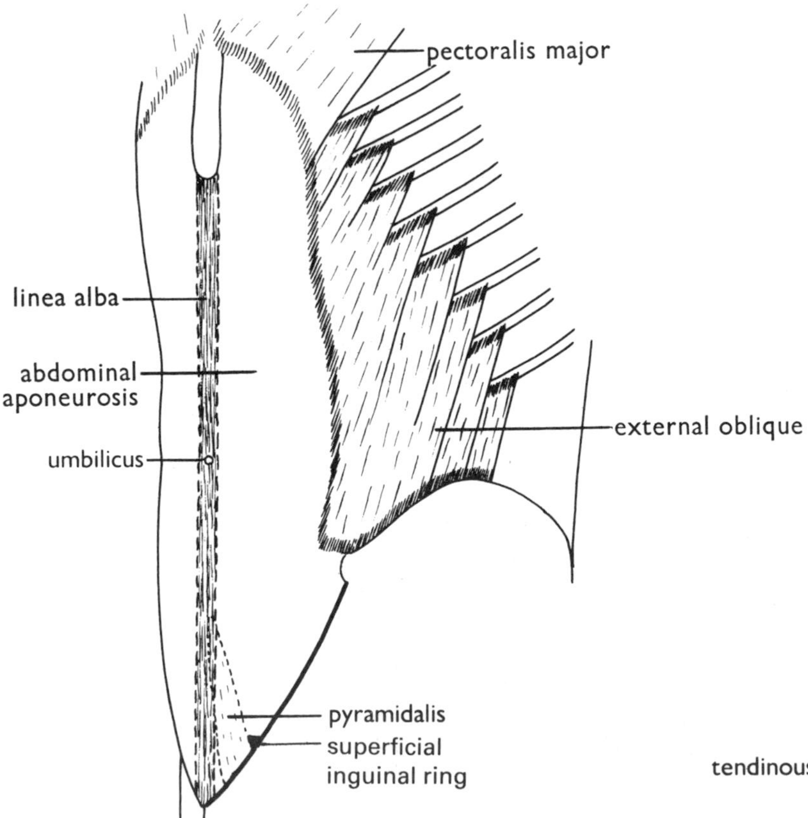

pectoralis major
linea alba
abdominal aponeurosis
umbilicus
external oblique
pyramidalis
superficial inguinal ring

Abdominal wall—*3*

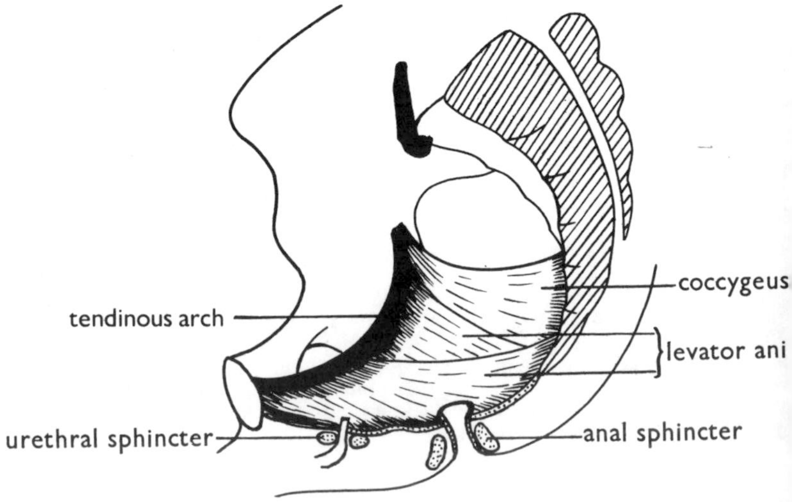

coccygeus
tendinous arch
levator ani
urethral sphincter
anal sphincter

Pelvic floor

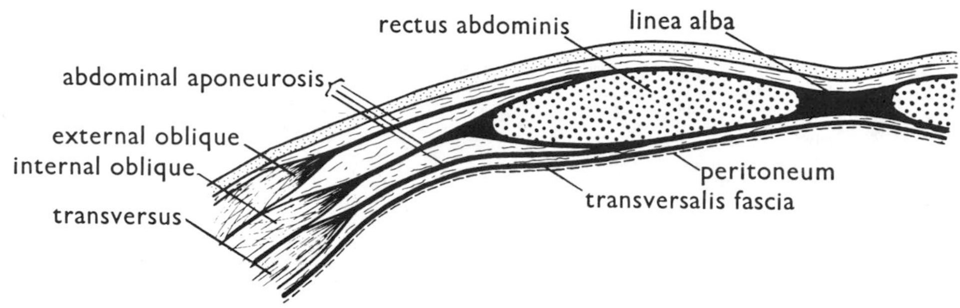

Section of front of abdominal wall above arcuate line

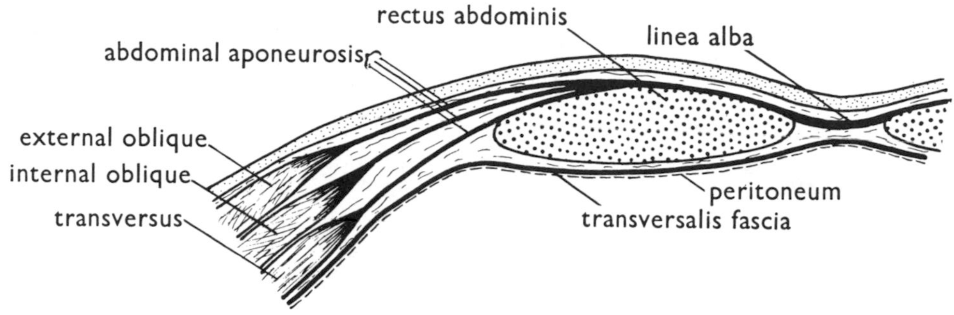

Section of front of abdominal wall below arcuate line

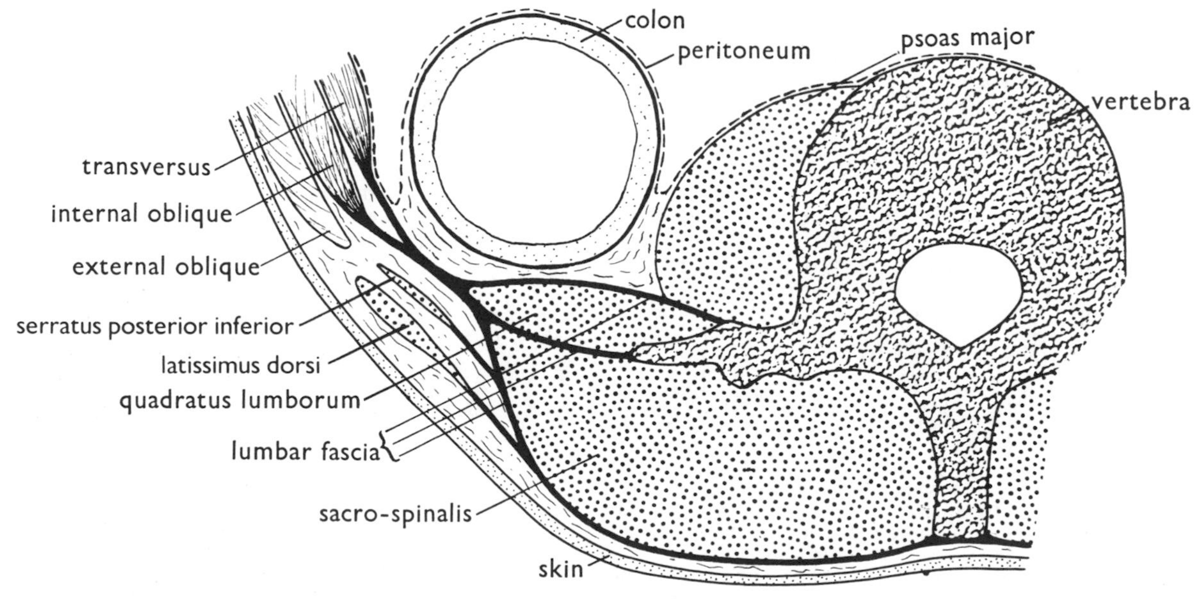

Section of back of abdominal wall

The diaphragm

The diaphragm is a dome-shaped partition between the **thoracic cavity** (chest cavity) and the **abdominal cavity**—see page 82. It is formed of radially arranged muscle fibres which take origin from the sternum, ribs and costal cartilages, the arcuate ligaments and the second and third lumbar vertebrae, and which are inserted on a **central tendon**. The sternal and costal parts of the diaphragm and the fibres from the arcuate ligaments are spread out into **flat sheets**, but the fibres originating from the vertebrae form two pillars called the **crura**.

The upper surface of the diaphragm is covered by parts of the membranous sacs, which enclose the lungs and the heart. The lower surface is covered by part of the lining of the abdominal cavity and has the liver suspended from it—see page 83.

The diaphragm is perforated to allow passage between the thoracic and abdominal cavities.

1. The **aortic opening** transmits the aorta—see page 104, and the thoracic duct—see page 115.
2. The **oesophageal opening** transmits the oesophagus—see page 84, and the vagus nerves—see page 64.
3. The **vena caval opening** transmits the inferior vena cava—see page 109.

Simultaneous contraction of all parts of the diaphragm pulls down the central tendon and reduces the concavity of the dome, thus increasing the capacity of the thorax. This produces **suction** in the thoracic cavity, which draws air into the lungs during **inspiration** (breathing in). As the diaphragm contracts it presses on the abdominal viscera, i.e. the organs in the abdominal cavity. Further downward movement is resisted by tone of the abdominal muscles. Continued contraction then causes the sternum to move forwards assisted by the intercostal muscles and the levatores costarum. The quadratus lumborum and serratus posterior inferior muscles fix the lower ribs and therefore hold the lower edge of the diaphragm in place. **Expiration** (breathing out) involves relaxation of the diaphragm with **elastic recoil** aided by pressure of the abdominal viscera against its under side. Expulsive actions such as coughing, vomiting, passing of faeces and urine, and the birth of a baby, involve contraction of the diaphragm and of the abdominal muscles.

Summary of muscles involved in breathing movements

Quiet inspiration (gentle breathing in):
diaphragm
external and internal intercostals
levatores costarum
quadratus lumborum, serratus posterior inferior } fix lower ribs
Deep inspiration—as above plus:
scaleni
sternomastoid
serratus posterior superior
sacro-spinalis—straightens back
Forced inspiration—as above plus:
serratus anterior, pectoralis minor } raise ribs while scapulae fixed
trapezius, levator scapulae, rhomboideus } fix scapulae
pectoralis major—if arms fixed in raised position
Quiet expiration (gentle breathing out):
elastic recoil of diaphragm
abdominal muscles—press viscera upwards
sternocostalis—depresses ribs
Forced expiration—as above plus:
abdominal muscles—stronger contraction, trunk flexed
latissimus dorsi, serratus posterior inferior } depress ribs

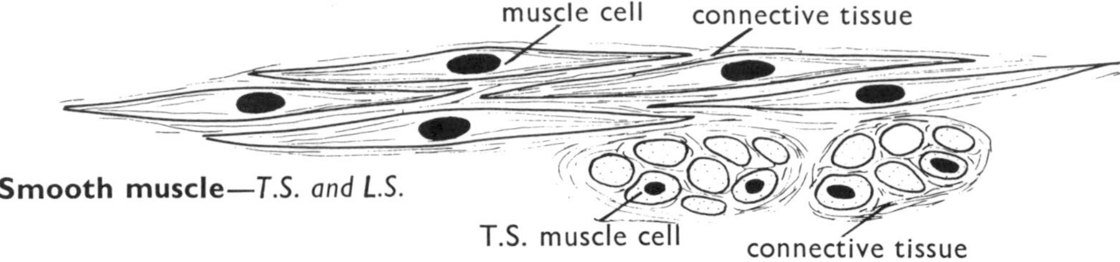

Smooth muscle—*T.S. and L.S.*

SMOOTH MUSCLE

Smooth muscle is associated with the internal organs and the blood vessels. It is composed of minute spindle-shaped cells, 2–5 μm in diameter and 30–200 μm in length. The muscle cells are bound together into sheets by areolar connective tissue which often contains many yellow elastic fibres. In most places where this muscle occurs, the sheets form separate circular and longitudinal coats whose actions are antagonistic.

Each smooth muscle cell has a single nucleus, poor sarcoplasmic reticulum and no transverse tubules—features related to the smaller size compared with a skeletal muscle cell. The actin and myosin fibrils are not arranged in sarcomeres so that there is no obvious striation, and this type of muscle is also known as **unstriated muscle**. The actin fibrils are attached to dark bodies on the surface and there are numerous small vesicles under the sarcolemma.

The mechanism of contraction is the same as for skeletal muscle—see page 39, with release of calcium ions triggering formation of cross-bridges which pull the actin fibrils along the myosin ones to produce shortening. Release of Ca^{2+} may be brought about in a number of ways:

1. electrically through simple neuromuscular junctions (cf. skeletal muscle)—see page 69;
2. chemically or hormonally—see page 69;
3. by stretching;
4. by change in oxygen levels;
5. by change in surrounding osmotic pressure.

Contraction of smooth muscle is evenly spread and slower than that of skeletal muscle. It does not show fatigue. It is normally controlled by **involuntary nerves** of the **autonomic system**—see pages 69 and 70—but tone is present even when the main nervous connections are severed. Differences in these connections result in two types of smooth muscle:

1. **Unitary smooth muscle**, also known as visceral muscle, in which the sheet as a whole is enervated, but not the individual fibres, e.g. the alimentary canal, uterus, ureters and small blood vessels. Initiation of contraction is frequently chemical or hormonal. Action spreads from one fibre to the next by stretching, thus producing waves of motion, e.g. peristalsis—see page 85.
2. **Multiunit smooth muscle** in which each fibre is individually enervated, e.g. the larger blood vessels and the muscles of the iris and ciliary body of the eye—see pages 78 and 79.

Note. The smooth muscle of the bladder can be stimulated by stretching as well as by nerves.

CARDIAC MUSCLE

Cardiac muscle is only found in the **heart**. It is composed of polygonal cells which anastomose with one another to form a continuous mass with very little connective tissue. Between the connecting processes there are irregular transverse thickenings of the sarcolemma called **intercalated discs** which add strength and aid conduction. Each cardiac muscle cell has a single nucleus and the fibrils are arranged to produce striations though these are finer than in skeletal muscles.

Cardiac muscle contracts rhythmically even without nervous stimulation, but the rhythm is normally controlled by nervous impulses received by the **sinuatrial node**—see page 113, which consists of specially sensitive cardiac muscle cells. The stimulus to contract is relayed through the rest of the tissue from cell to cell. The majority of cardiac cells have no nervous connections. Cardiac muscle does not fatigue readily, but can fatigue if the rate of heart beat is much increased for a long period so that there is insufficient rest between each contraction.

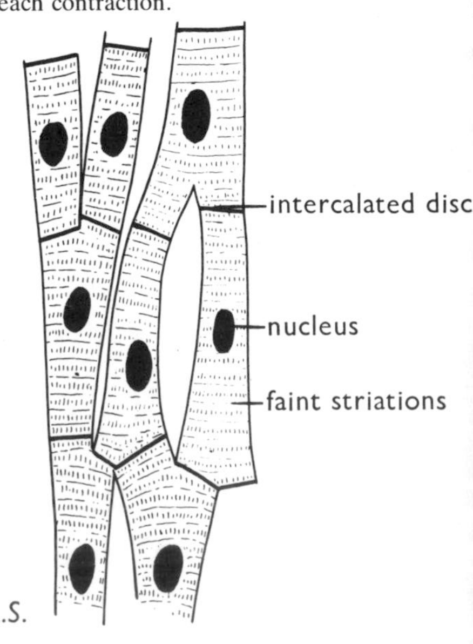

Cardiac muscle—*L.S.*

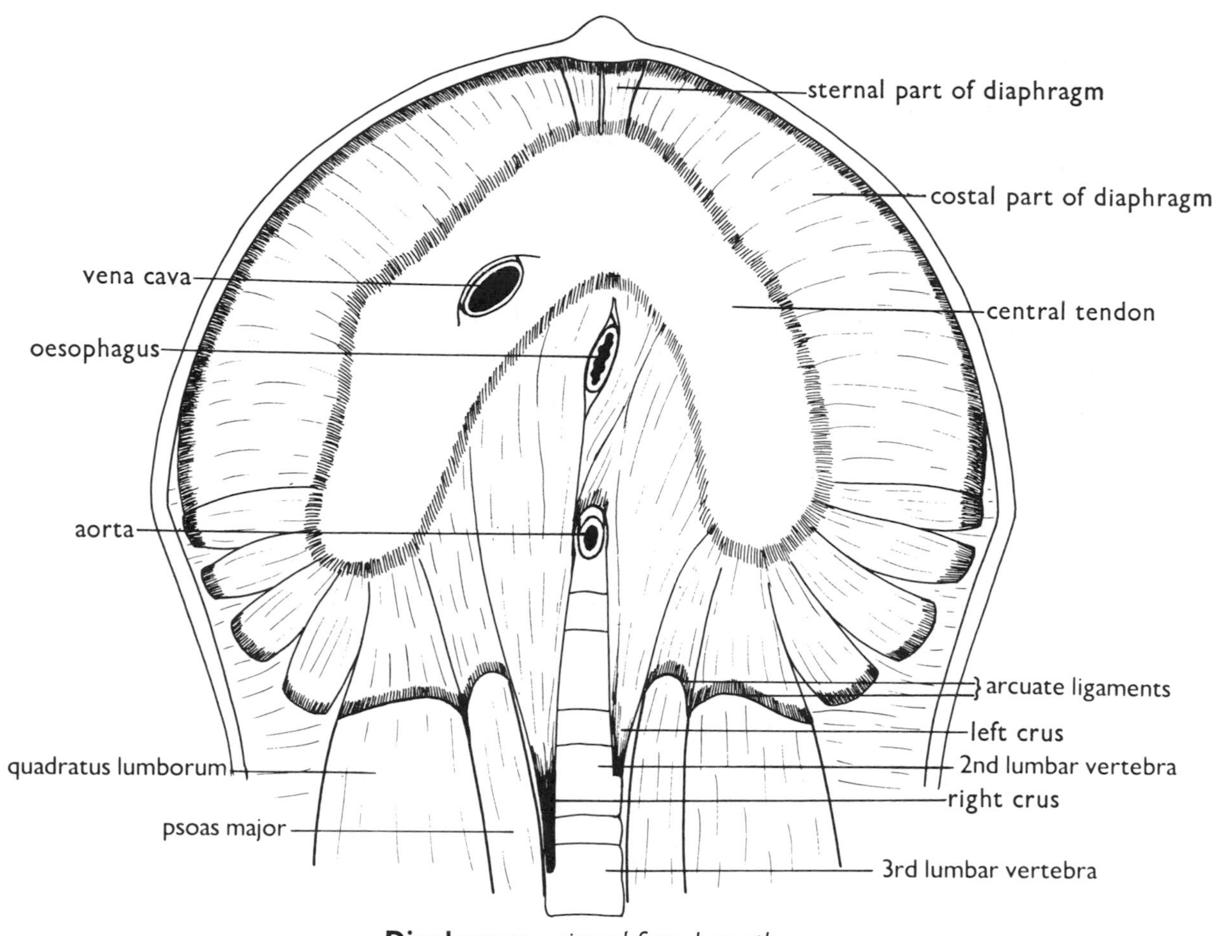

Diaphragm—*viewed from beneath*

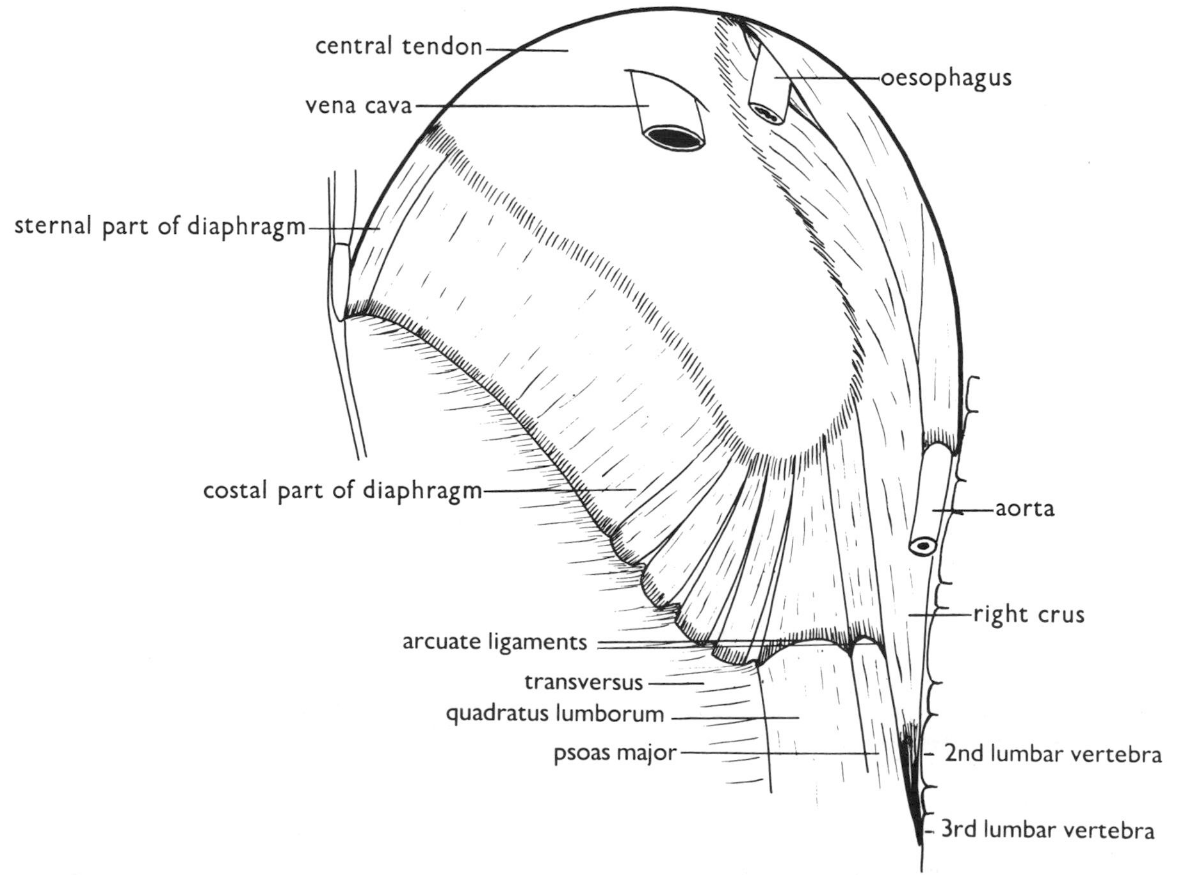

Diaphragm—*right half viewed from the side*

The Nervous System

The nervous system deals with rapid **conduction** of messages from one part of the body to another and the **co-ordination** of the body's activities. It is formed of specially reactive cells called **neurones**, supported by a meshwork of non-nervous **neuroglial cells**, small blood vessels and soft matrix. The neurones are so arranged that they make up a **central nervous system (CNS)** consisting of **brain** and **spinal cord** and a **peripheral system** of **nerves** and nerve masses called **ganglia**.

NEURONES

Functionally neurones are of three types:

1. **sensory** or **afferent neurones** which receive stimuli and pass impulses to the spinal cord and brain;
2. **association** or **internuncial neurones** which relay impulses and thus bring about distribution and amplification;
3. **motor** or **efferent neurones** which carry impulses from the CNS to muscles and glands.

Connections between the neurones are called **synapses**.

In shape the motor and internuncial neurones are **multipolar**, with many processes attached to a cell body or **perikaryon** (literally the area round the nucleus). The many processes which carry impulses towards the cell body are called **dendrites**, while the single larger process which carries impulses away from the cell body is the **axon**. The axon arises from a thickening called the **axon hillock**. Sensory neurones have only two main processes, one dendrite and one axon. These may meet the cell body separately in **bipolar neurones** or by a common stalk in **unipolar neurones**. All dendrites have numerous tributaries. Axons may have a few branches known as **collaterals** and end in **terminal arborisations** of fine processes called **telodendria** each of which bears a **synaptic knob**.

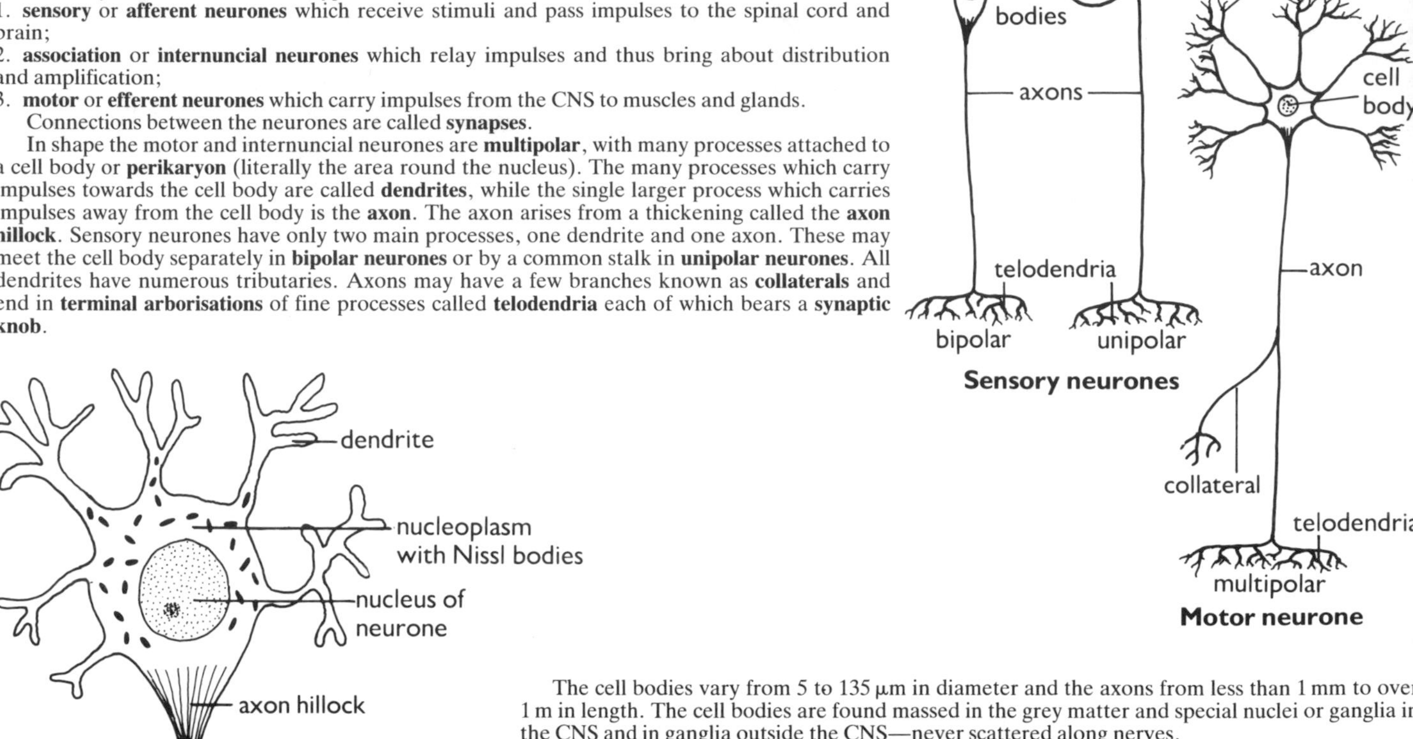

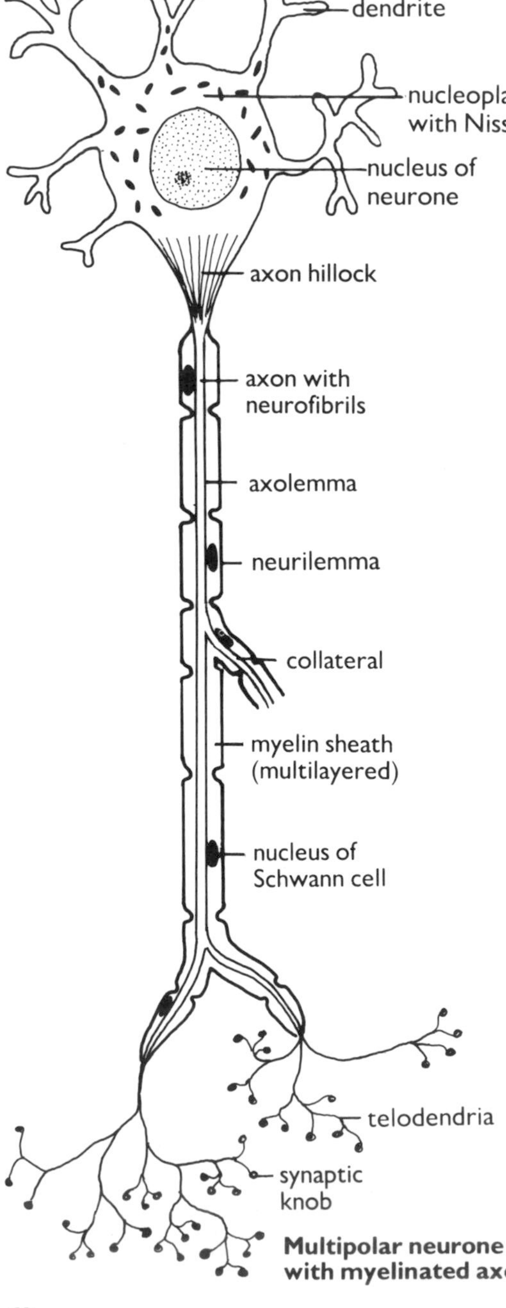

Multipolar neurone with myelinated axon

The cell bodies vary from 5 to 135 μm in diameter and the axons from less than 1 mm to over 1 m in length. The cell bodies are found massed in the grey matter and special nuclei or ganglia in the CNS and in ganglia outside the CNS—never scattered along nerves.

Each cell body has all the components of an active cell, i.e. a single nucleus, cytoplasm, mitochondria, Golgi apparatus and lysosomes, and an endoplasmic reticulum, the granular portion of which is in the form of ribosome-studded vesicles called **Nissl bodies**. The mitochondria and Nissl bodies extend into the dendrites of the multipolar neurones, but the Nissl bodies are not found in the axon or in the dendrites of bipolar and unipolar neurones which are axon-like in structure. Instead proteins synthesised in the cell body are transmitted by slow **axoplasmic flow** or by faster **axonic transport** in a system of microtubules called **neurofibrils**. Expended materials can be returned to the cell body for destruction or recycling.

Each nerve process has its own plasma membrane. In addition the principal processes may have **neurilemma** and/or **myelin sheath**. Both of these are derived from neuroglial cells. Outside the CNS the cells concerned are called **Schwann cells**. They lie wrapped closely round the nerve fibres. The neurilemma is a simple covering of Schwann cells with their associated nuclei and cytoplasm. Fibres with neurilemma only appear greyish and serve involuntary muscle and glands. Schwann cells associated with sensory nerve fibres and fibres serving voluntary (skeletal) muscles become coiled many times around the main nerve processes. The inner layers so formed become impregnated by **phospholipid**, lecithin, and are said to be **myelinated** (**medullated**). The myelin sheath is in segments with gaps called **nodes of Ranvier**. Myelinated fibres appear white. Inside the CNS similar myelination is produced by multiprocessed neuroglial cells called **oligodendrocytes**. There are longer gaps between the nodes of Ranvier and the neurones have no neurilemma. Tracts of such fibres appear white. The majority of small internuncial fibres have neither myelin sheath nor neurilemma and appear grey.

During the early months of life additional neurones are formed from undifferentiated cells called **neuroblasts**, but soon the power of mitosis is lost—see page 124—and damaged cell bodies cannot be replaced. In the peripheral system, however, processes can be regenerated, assisted by the Schwann cells, provided that the cell bodies are intact. Inside the CNS the oligodendrocytes cannot assist regeneration and scar tissue forms readily so that injury to the CNS is permanent, while in the peripheral system some function can be restored.

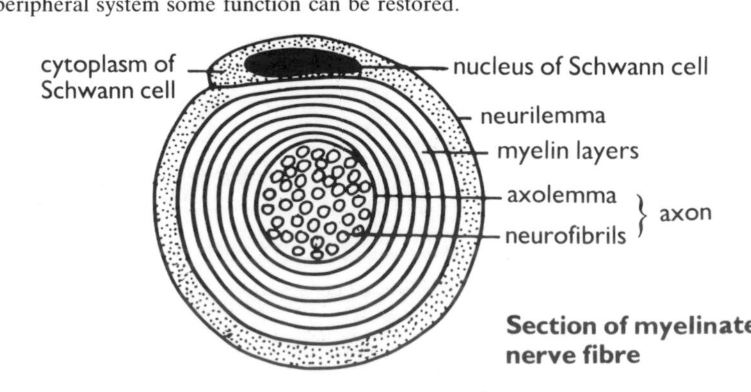

Section of myelinated nerve fibre

NERVE IMPULSES

At rest the semipermeable surface membranes of neurones are **polarised**, i.e. electrically charged, due to unequal distribution of Na^+ and K^+ ions on opposite sides of the membrane and the impermeability to the negative protein ions inside. The normal **resting potential** of the membrane is −70 mV (millivolts). On stimulation the permeability of the membrane to Na^+ is greatly increased and these ions diffuse inwards down the concentration gradient, more than neutralising the negative ions and creating a potential of +40 mV. This is the **action potential**. It sends an electric current to the adjacent piece of membrane, which becomes **depolarised** in its turn. The wave of depolarisation is the **nerve impulse**, which runs down the length of the nerve fibre in an **all-or-none** manner. Thus each impulse is an electric discharge.

Repolarisation follows rapidly. In its first stage the membrane becomes more permeable to K^+ and less to Na^+. The K^+ ions move outwards restoring the negative state inside the fibre. Further recovery is brought about by the activity of the **sodium/potassium pump**, which removes Na^+ from the fibre and brings in K^+. The mechanism involves a carrier and enzymes and uses energy from ATP. There is usually a small overshoot of the resting potential, but ultimately the concentration differential established has about 14 times as much Na^+ outside as inside and about 30 times as much K^+ inside as outside. As there is always a slight permeability to these ions, particularly K^+, the sodium/potassium pump must work continuously in the resting fibre, and nervous tissue cannot live long without supplies of oxygen for energy renewal.

In non-myelinated fibres conduction speed is about 0.5 m/s. In myelinated fibres depolarisation occurs only at the nodes of Ranvier and the electric current between these flows in the extracellular fluid and cytoplasm. This is known as **saltatory conduction** and results in much faster conduction—up to 130 m/s in the largest fibres. It also conserves energy because less repolarisation is needed. In spite of the differences in speed of transmission down nerve fibres, ultimately the time taken from original stimulus to response mainly depends on the number of synapses (joins) involved. At these transmission is chemical—not electrical—see page 69.

depolarisation
Na^+ → in
stimulated

recovery
out ← K^+
wave of depolarisation
impulse passed on

out ← Na^+
K^+ → in
sodium/potassium pump active
resting state resumed and maintained

Changes at surface membrane during passage of an impulse

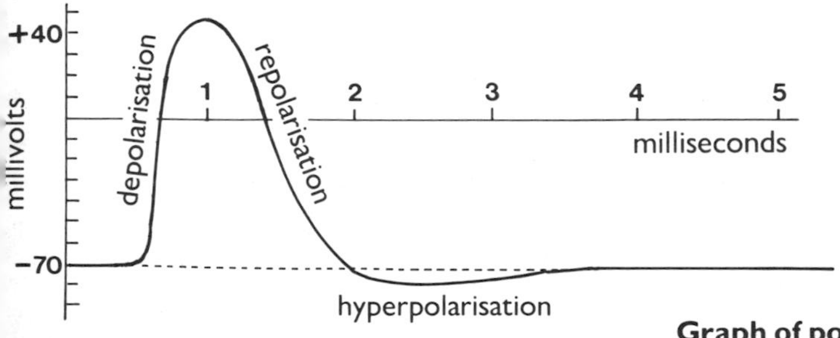

Graph of potential changes

Some terms associated with neurotransmission

Threshold—minimal stimulus needed to produce a response.
Summation—adding of two or more subliminal (less than threshold) stimuli:
(a) temporal summation—stimuli occur at the same site in close succession;
(b) spacial summation—stimuli occur at different sites simultaneously.
Inhibition—blocking of response by rise in threshold.
Accommodation—rise in threshold after a response—stimulus can be ignored.
Absolute refractory period—duration of depolarisation during which no further response is possible.
Relative refractory period—duration of recovery during which the threshold is raised.

Note. Fibres with small cross section have longer refractory periods than large ones and can therefore carry fewer impulses in a given time. In the body as a whole the normal range is 10 to 500 impulses/s. On the basis of size the fibres are classified as A, B or C. A and B fibres are myelinated; C fibres are not.

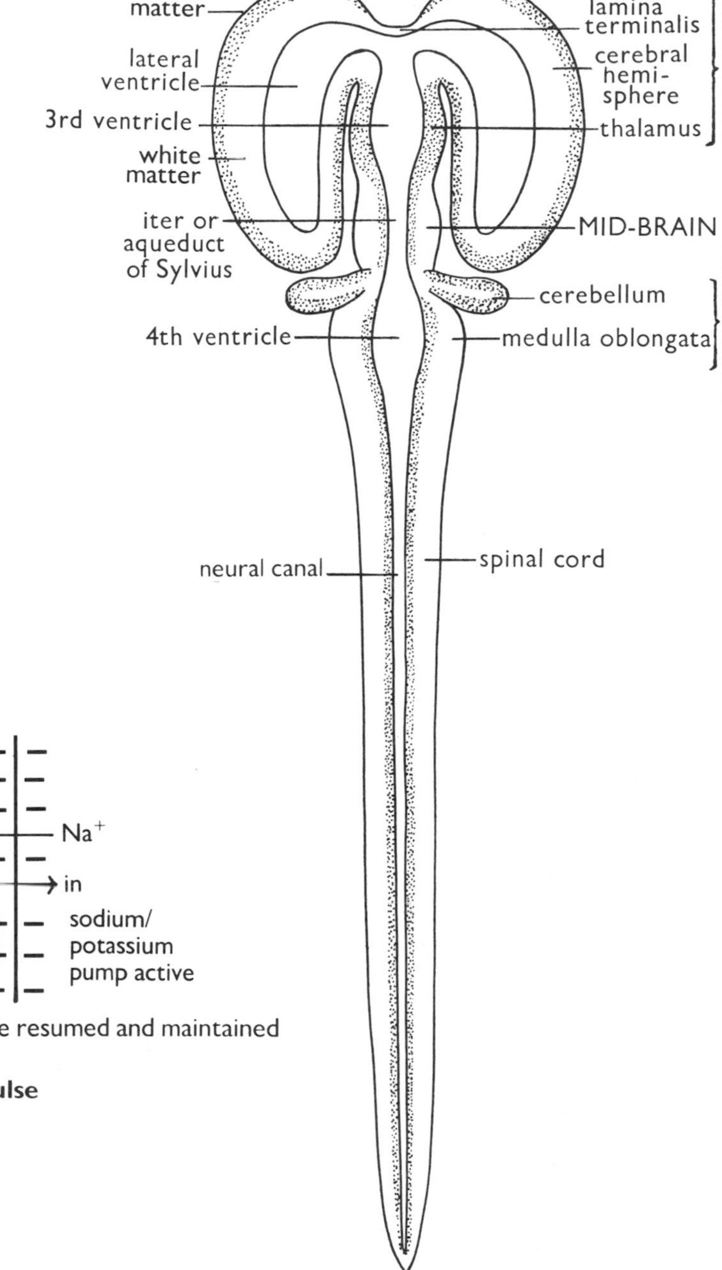

Diagram of the derivatives of the neural tube

THE CENTRAL NERVOUS SYSTEM

The central nervous system (brain and spinal cord) starts its development as a continuous neural tube. Three hollow swellings in the head region are known as the fore-brain, mid-brain and hind-brain rudiments. The relationship of the principal parts of the brain to the neural tube is shown in the diagram above. At the stage shown, the neural canal has become distended into two lateral ventricles and the medial third and fourth ventricles. The distribution of the grey matter (cell bodies and internuncial neurones) as opposed to white matter (tracts of myelinated nerve fibres) is also indicated.

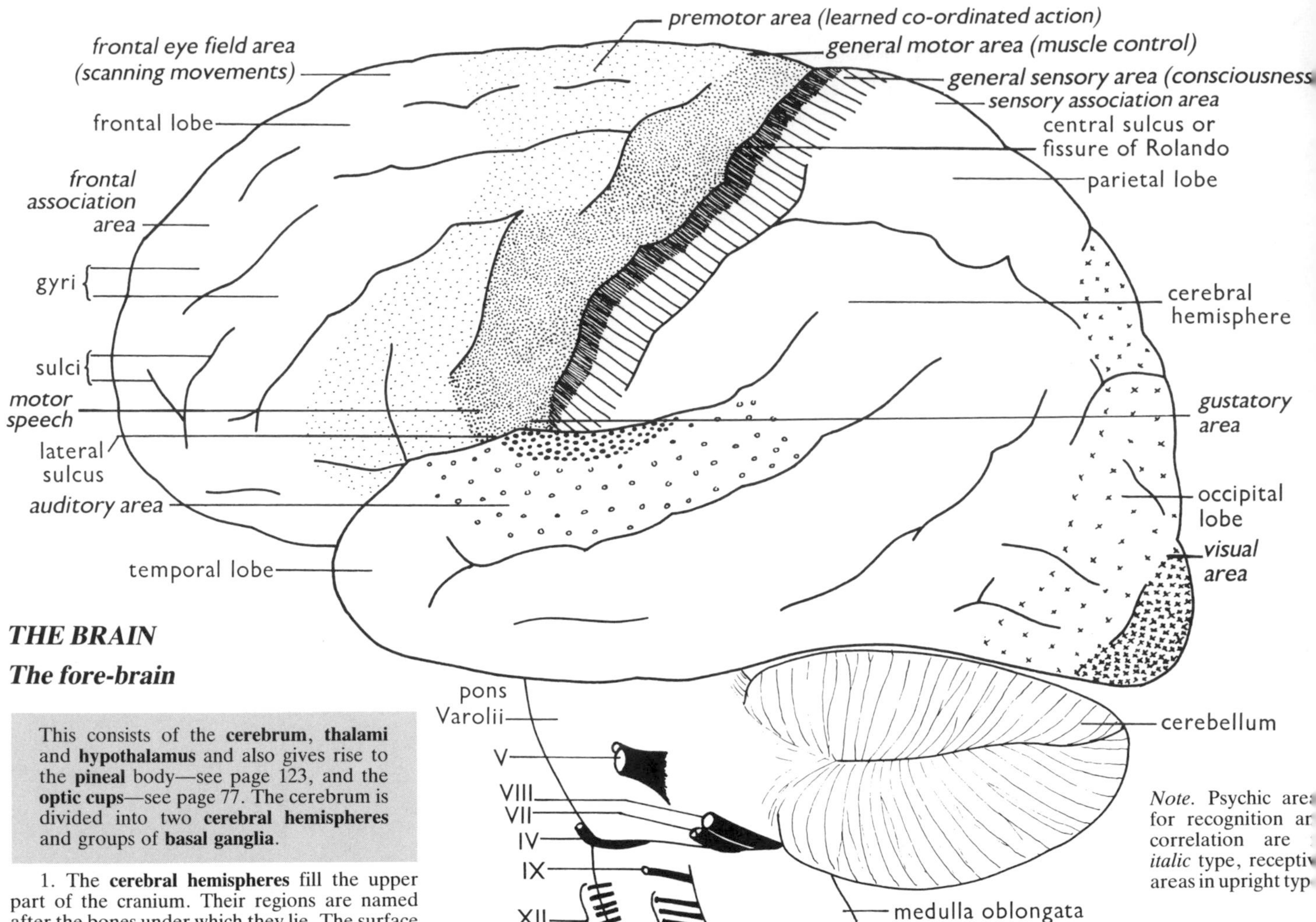

Note. Psychic are
for recognition ar
correlation are
italic type, recepti
areas in upright typ

Brain—*lateral vie*

THE BRAIN

The fore-brain

This consists of the **cerebrum**, **thalami** and **hypothalamus** and also gives rise to the **pineal** body—see page 123, and the **optic cups**—see page 77. The cerebrum is divided into two **cerebral hemispheres** and groups of **basal ganglia**.

1. The **cerebral hemispheres** fill the upper part of the cranium. Their regions are named after the bones under which they lie. The surface of each hemisphere is convoluted with folds called **gyri** and grooves called **sulci**, which allow a much greater area for the **cerebral cortex**. This consists of a 2–4 mm thick layer of grey matter with white cerebral matter below. In the latter there are (a) **association fibres** making connections within the hemisphere, (b) **commissural fibres** forming the corpus callosum and fornix between the hemispheres, (c) **projection fibres** connecting with other parts of the brain and spinal cord.

Areas of the cortex concerned with special sensory and motor functions have been mapped out. Those associated with the head are near the central sulcus and disproportionately large. Speech centres are normally present on the left side only, while the right side is more important for spacial perception and creative functions, e.g. art and music. Other areas are concerned with memory, emotions, reasoning, will, judgement, personality traits and intelligence.

2. The **basal ganglia** are masses of grey matter on the floor of each lateral ventricle. The largest masses are the corpora striata, formed of **caudate** and **lentiform nuclei**. The latter are subdivided into **putamen** concerned with subconscious movements of skeletal muscles, and the **globus pallidus** concerned with regulation of muscle tone and suppression of certain types of muscle actions. Corpora striata in general may be concerned with instinctive behaviour.

3. The **thalami** form the side walls of the third ventricle. They are joined together by the **interthalamic connection** and contain nuclei concerned with interpretation of some sensory impulses, e.g. pain, temperature, pressure, and with relay of others, e.g. hearing, vision, taste. The **reticular nuclei** receive impulses from the cerebrum and relay them to other thalamic nuclei, modifying their activity. The motor connections from the thalamus to the cerebrum influence emotions and memory.

4. The **hypothalamus** lies on the floor of the third ventricle. It contains the main centres which control the **homeostatic functions**, which maintain a constant environment for the body cells. It acts through nervous and chemical connections with the pituitary body—see page 121, and by nerves to the mid-brain and limbic system. This system includes the **mammillary body** and adjacent parts of the cerebrum (**limbic lobe**, **hippocampus** and **amygdaloid nucleus**) and the **anterior nucleus** of the thalamus. It is associated with emotions, particularly those which promote survival behaviour.

The mid-brain

This consists of the **cerebral peduncles**, the **superior** and **inferior colliculi** (also known as the corpora quadrigemina) and a number of small nuclei including the **red nuclei.**

1. The **cerebral peduncles** are bundles of ner fibres joining the fore-brain to the hind-bra and spinal cord.
2. The **superior colliculi** control movement the eyeballs and head in response to visu stimuli.
3. The **inferior colliculi** act similarly for audito stimuli.
4. The **red** and **other nuclei** contain cell bodi of some spinal tracts and cranial nerves.

The hind-brain

This consists of the **cerebellum** and **pons Varolii** and the **medulla oblongata**.

1. The **cerebellum** has a central constricted **vermis** and two **cerebellar hemispheres**, the surfaces which consist of much folded cortex. The tracts within form the **arbor vitae** and there are some de nuclei. **Superior cerebellar peduncles** connect each hemisphere with the red nucleus and thalamus the opposite side. **Inferior cerebellar peduncles** connect with the medulla obongata and spinal cor The **middle cerebellar peduncles** encircle the cerebral peduncles to form the pons Varolii. This co tains the **nuclei pontis** in which tracts from the cerebral cortex connect with tracts from the opposi cerebellar hemisphere. It also has connections with the thalamus and spinal cord and contains nuc of some cranial nerves including those concerned with balance.

The cerebellum is concerned mainly with subconscious control of skeletal muscles, comparing i tent with performance to produce accurate co-ordinated movement.

2. The **medulla oblongata** or **spinal bulb** is like an enlarged region of the spinal cord, with which is continuous. Spinal sensory information is relayed to the opposite side on its way to the thalamus a cerebral cortex. The majority of descending motor tracts, i.e. those to the spinal cord, cross to t opposite side (**decussate**) forming the **pyramidal tracts** and there are nuclei of several cranial nerv These nuclei include the **vital centres** controlling rate and depth of breathing, rate of heart beat a diameter of blood vessels and thus blood pressure. They also include **non-vital centres** involved swallowing, vomiting, coughing and sneezing.

Note. The medulla oblongata, pons Varolii and mid-brain together form the **brain stem**.

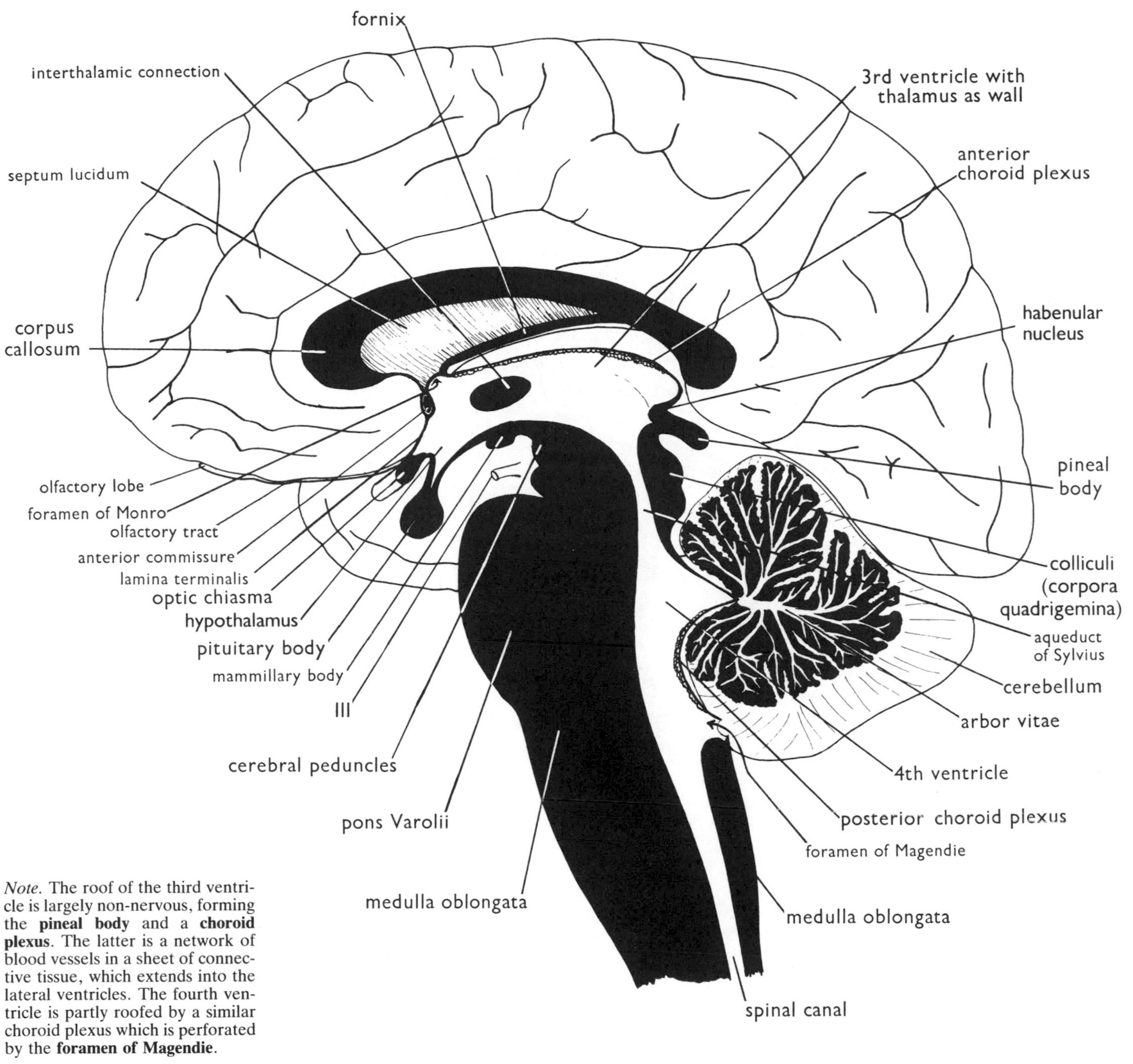

Note. The roof of the third ventricle is largely non-nervous, forming the **pineal body** and a **choroid plexus**. The latter is a network of blood vessels in a sheet of connective tissue, which extends into the lateral ventricles. The fourth ventricle is partly roofed by a similar choroid plexus which is perforated by the **foramen of Magendie**.

Brain—*V.S.*

THE CRANIAL NERVES

There are twelve pairs of cranial nerves attached to the brain stem at different levels. Some of these nerves contain sensory fibres only and some motor fibres only, while the remainder are mixed nerves but have separate sensory and motor roots. The first two cranial nerves differ from the rest in being non-segmental in origin—i.e. not related to the embryological segmentation of the head.

All the sensory nerves have ganglia outside the brain, while all the motor nerves originate from nuclei inside the brain.

No.	Name	Distribution
I	Olfactory	Sensory nerves of smell from the nose
II	Optic	Sensory nerves of sight from the eyes. Many fibres cross in the optic chiasma
III	Oculomotor	Motor nerves to the superior, inferior and medial rectus and inferior oblique muscles of the eye
IV	Trochlear	Motor nerves to the superior oblique muscles of the eyes
V	Trigeminal (three branches)	Ophthalmic—sensory nerves from above and around the orbits and parts of the nasal cavities Maxillary—sensory nerves from around and below the orbits and from upper jaw and teeth Mandibular—sensory nerves from lower part of face, lower jaw and teeth and from temples and pinnae; also motor nerves to the muscles of mastication

continued on page 64

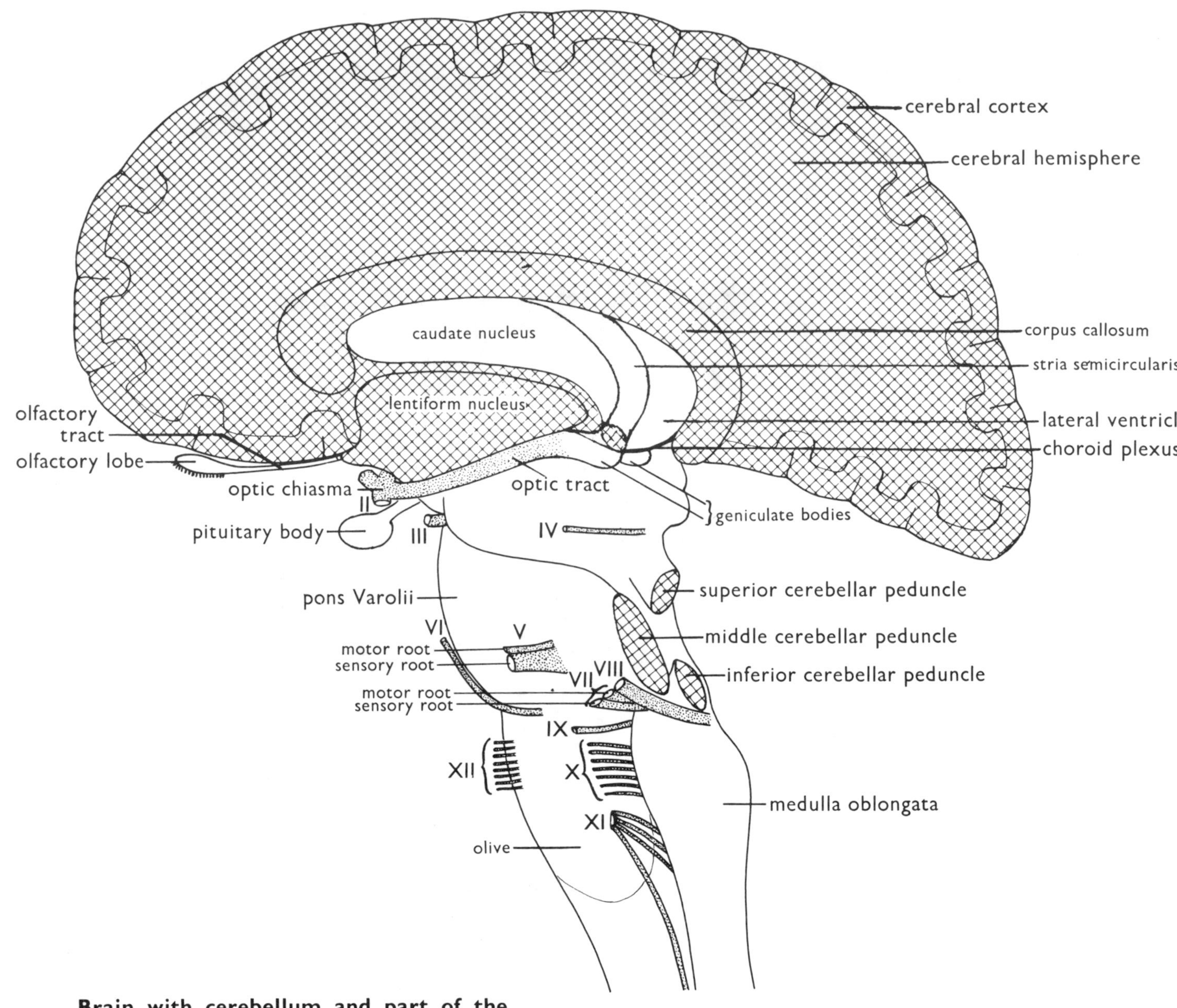

Brain with cerebellum and part of the left cerebral hemisphere removed

The cranial nerves—continued

No.	Name	Distribution
VI	Abducent	Motor nerves to the lateral rectus muscles of the eyes
VII	Facial	Sensory nerves of taste from the anterior part of the tongue Motor nerves to the muscles of expression, scalp, pinnae and neck
VIII	Auditory (two branches)	Cochlear—nerves of hearing from the cochlea Vestibular—nerves of balance from the semicircular canals and vestibule
IX	Glosso-pharyngeal	Sensory nerves of taste from the posterior part of the tongue and from the pharynx Motor nerves to the muscles of the pharynx and to the parotid gland
X	Vagus	Sensory nerves from the larynx, trachea, lungs, oesophagus, stomach, intestines, gall bladder and large arteries and veins Motor nerves consisting chiefly of parasympathetic (involuntary) fibres to the pharynx, larynx, trachea, oesophagus, stomach, small intestine, pancreas, liver, spleen, ascending colon, kidneys, heart and visceral blood vessels
XI	Accessory (two parts)	Cranial—motor nerves joining the vagus to supply the pharynx and larynx Spinal—motor nerves arising from the spinal cord and entering the skull, then leaving it again to the muscles of the neck
XII	Hypoglossal	Motor nerves to the muscles of the tongue and hyoid region

Positions of the cranial nerve ganglia

I in olfactory lobes.

II in the retinae of the eyes—see page 79.

III and IV in the mid-brain.

V, VI, VII and parts of VIII concerned with balance in the pons Varolii.

Parts of VIII concerned with hearing and balance, IX, X, XI and XII in the medulla oblongata.

XI also has nuclei in the spinal cord.

Note. XI and XII carry some sensory fibres from proprioceptors in the muscles which they serve.

III, IV and VI carry involuntary motor fibres to the iris and ciliary body of the eye—see page 79.

VII and IX carry involuntary motor fibres to the salivary glands—see page 90.

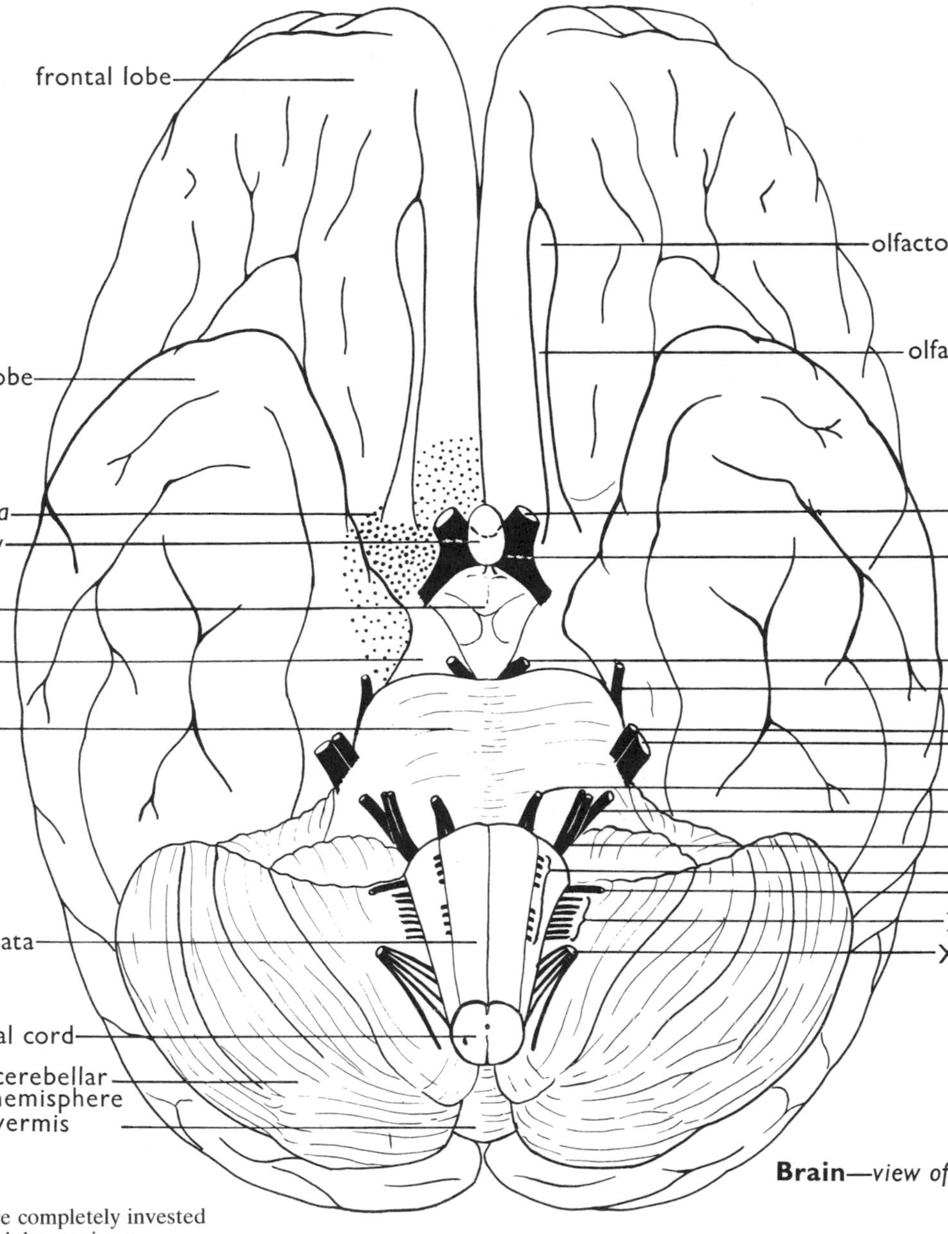

frontal lobe
olfactory lobe
olfactory tract
temporal lobe
olfactory area
pituitary body
optic nerve II
optic chiasma
mammillary body
cerebral peduncle
III
IV
pons Varolii
V
VI
VIII
VII
XII
IX
X
XI
medulla oblongata
cut spinal cord
cerebellum
cerebellar hemisphere
vermis

Brain—*view of base*

THE MENINGES

The brain and spinal cord are completely invested in three layers of tissue called the meninges.

1. The **dura mater** is the tough fibrous outermost layer. It lines the bones of the cranium and the canal formed by the vertebrae. It forms also the **falx cerebri** between the cerebral hemispheres and the **tentorium cerebelli** separating the cerebral hemispheres from the cerebellum. The main venous sinuses inside the cranium lie in the dura mater—see page 108.

2. The **arachnoid mater** is a delicate membrane separated from the dura mater by the subdural space in which there is a thin film of serous fluid. It encloses the **subarachnoid space**, which is traversed by fine trabeculae of connective tissue and which contains cerebro-spinal fluid and the larger blood vessels of the brain.

3. The **pia mater** is the very delicate inner membrane which closely invests the brain and spinal cord and supports a network of fine blood vessels, including those of the choroid plexuses which roof the ventricles. It follows every convolution of the surface of the nervous tissue and sheathes the roots of the cranial and spinal nerves.

dura mater (lining skull bones)
venous sinus
dura mater (forming falx cerebri)
subdural space
arachnoid mater (lining dura mater)
subarachnoid space or theca (with cerebro-spinal fluid)
pia mater (covering brain)

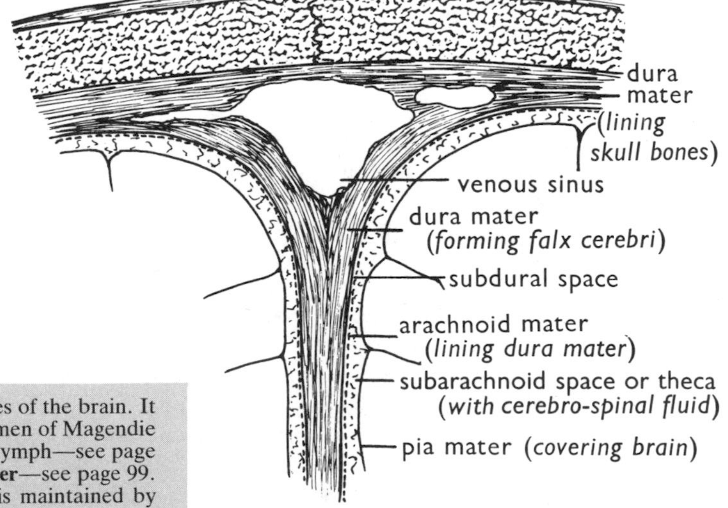

Meninges

CEREBRO-SPINAL FLUID

Cerebro-spinal fluid is secreted by the choroid plexuses into the ventricles of the brain. It fills the cavities of the brain and spinal cord and passes through the foramen of Magendie to fill the **subarachnoid spaces**. It is of the same general composition as lymph—see page 114, but with less protein and fewer cells owing to the **blood–brain barrier**—see page 99. Little fluid enters the spinal region, where constancy of composition is maintained by diffusion and by alteration in posture. There is a distinct, though slow, flow into the cranial region, where fluid re-enters the blood through the arachnoid villi, which are small papillae projecting into the superior sagittal venous sinus—see page 108.

The functions of the cerebro-spinal fluid are:
1. to support the delicate nervous tissue; 2. to protect it against shock; 3. to maintain a uniform pressure around it; 4. to supply it with food and oxygen by direct bathing of the cells in parts where no blood vessels penetrate.

THE SPINAL CORD

The spinal cord lies in a canal formed by the neural arches of the vertebrae—see diagram on page 68. In the three-month-old foetus, it stretches from the foramen magnum of the skull to the lower end of the canal, but it grows more slowly than the vertebral column, and therefore in the adult it reaches only the upper margin of the second lumbar vertebra. It is connected to the lower end of the canal by the non-nervous **filum terminale**. The average length of the cord is about 430 mm and thickness about 20 mm, but it has two wider parts, the **cervical** and **lumbar enlargements**, where the nerves of the limbs originate. Because of the shortness of the spinal cord, nerves from the lumbar enlargement pass for some distance down the vertebral canal before emerging segmentally. The bundle of nerves with the filum terminale forms the **cauda equina**.

There is a fine central canal throughout the cord, continuous with the ventricles of the brain and like them filled with cerebro-spinal fluid. The grey matter lies around this canal and extends as two **posterior** and two **anterior horns** in which there are cell bodies of association and motor neurones respectively. The white matter lies external to the grey matter and forms **ascending** and **descending tracts** of fibres through which messages are passed up and down the cord to different levels and to and from the brain.

Functions of the spinal cord

1. To relay impulses coming in and going out at the same level.
2. To relay impulses up and down the cord to other levels.
3. To relay impulses to and from the brain.

THE SPINAL NERVES

There are 31 pairs of spinal nerves, each of which is attached to the spinal cord by a **posterior** or **sensory root**, on which there is a ganglion containing the cell bodies of the sensory nerves, and an **anterior** or **motor root** with cell bodies in the anterior horns—see diagram on next page. The two roots join where the nerves leave the vertebral canal. The mixed nerves so formed soon divide into posterior and anterior branches or **rami**. The posterior primary rami are small and serve the skin and muscles of the back. The anterior primary rami are larger and have a varied distribution to the skin and muscles of the sides and front of the trunk and limbs. Many of the anterior rami form complicated **plexuses** by means of which fibres from different levels of the cord can serve the same region of the body.

Also connected to the anterior primary rami are the **rami communicantes**. These are bundles of fibres connected with chains of ganglia forming the **sympathetic cords**—see diagrams on pages 70 and 72. The white rami communicantes, found in the thoracic and lumbar regions only, contain myelinated motor fibres from the CNS to the sympathetic ganglia. The grey rami communicantes found on all spinal nerves contain non-myelinated post-ganglionic fibres, which are distributed with the voluntary nerves, to serve the skin and blood vessels of the various regions. From the sympathetic ganglia there are additional post-ganglionic nerves to groups of interconnected ganglia called plexuses: the **cardiac plexus** near the heart, the **coeliac** or **solar plexus** near the stomach, the **mesenteric plexuses**, near the intestines, and the **hypogastric plexus** in the pelvic region. The sacral nerves carry parasympathetic fibres to the pelvic organs not served by the cranial parasympathetic fibres of the vagus nerves.

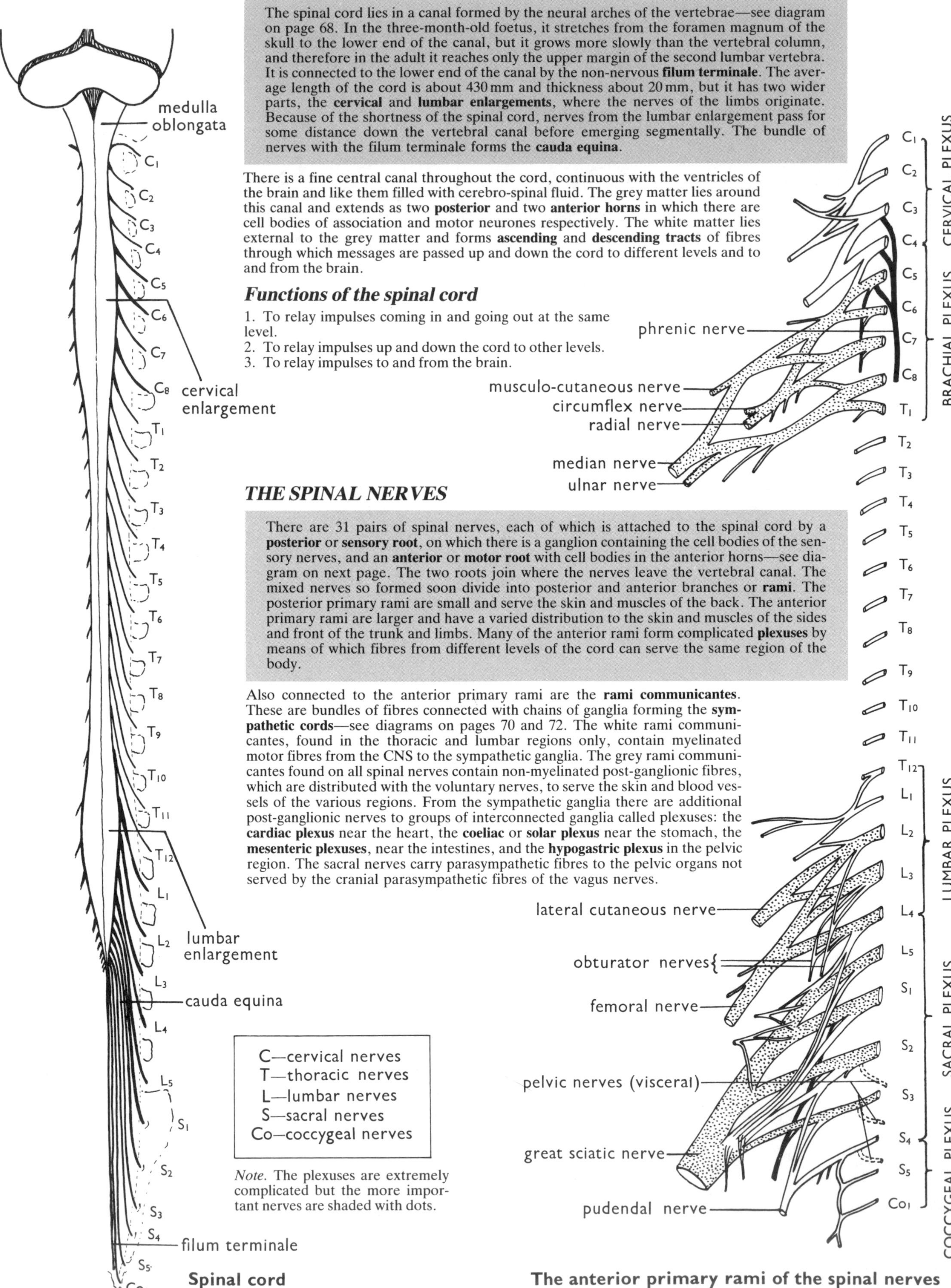

Note. The plexuses are extremely complicated but the more important nerves are shaded with dots.

Spinal cord

The anterior primary rami of the spinal nerves

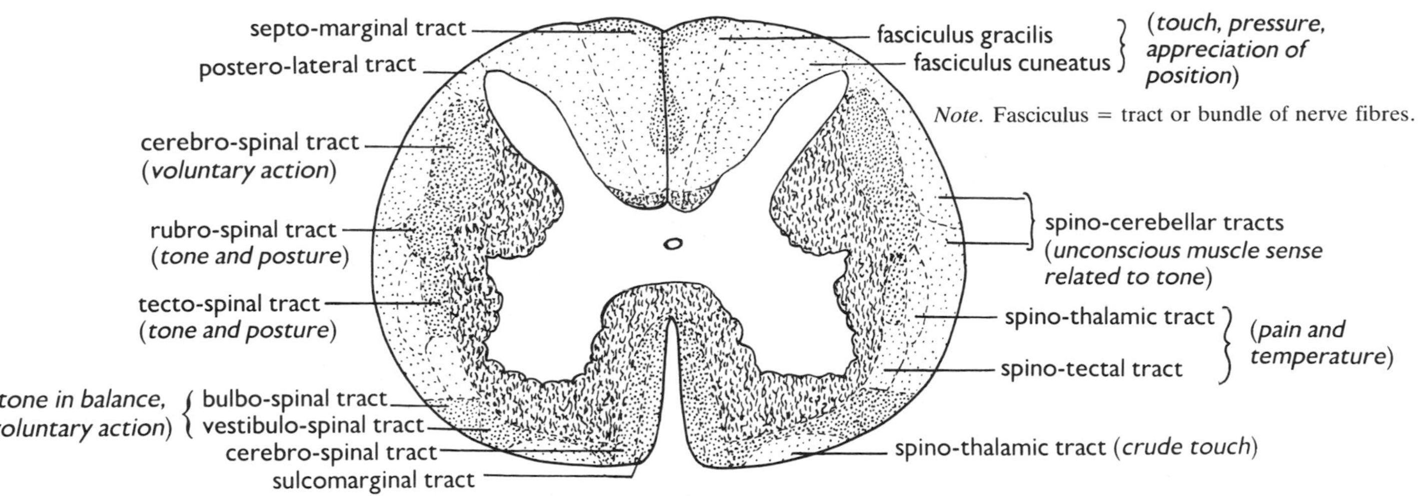

The arrangement of the nerve tracts in the spinal cord

Note. Type of information carried in the main tracts is indicated.

Summary of the distribution of cranial and spinal nerves to some of the chief voluntary muscles

Note. In the case of spinal nerves, the posterior primary rami are marked (P); the anterior primary rami are unmarked.
Cr=cranial; C=cervical; T=thoracic; L=lumbar; S=sacral

Muscles	Nerves
Adductor brevis and longus	Obturator L2 and L3
Adductor magnus	Sciatic L4
Anal sphincter	Pudendal L4
Anconeus	Radial C7 and C8
Biceps brachii	Musculocutaneous C5 and C6
Biceps femoris	Sciatic (medial and lateral popliteal branches) L5–S3
Brachialis	Musculocutaneous C5 and C6
Brachioradialis	Radial C5 and C6
Coccygeus	S4 and S5
Constrictors of pharynx	Glossopharyngeal (pharyngeal plexus) Cr9
Coracobrachialis	Musculocutaneous C7
Deltoid	Circumflex C5 and C6
Diaphragm	Phrenic C3–C5 and T6 or T7 or T7–T12
Extensors of fingers	Interosseus C7
Extensors of toes	Sciatic (lateral popliteal branch) L5 and S1
Extensors of wrist	Radial C6 and C7 and median C7
External oblique	T7–T12
Facial expression muscles	Facial Cr7
Flexors of fingers	Median C7 and C8
Flexors of toes	Sciatic (medial popliteal branch) L5–S2
Flexors of wrist	Median C6 and C7 and ulnar C8
Gastrocnemius	Sciatic (medial popliteal branch) L5 and S1
Gemellus inferior	L4–S1
Gemellus superior	L5–S2
Gluteus maximus	Inferior gluteal L5–S2
Gluteus medius and minimus	Superior gluteal L4 and L5
Gracilis	Obturator L3 and L4
Iliacus	Femoral L3
Inferior oblique	Oculomotor Cr3
Infraspinatus	Suprascapular C5
Intercostals	Intercostals T1–T11
Internal oblique	T6–L1
Latissimus dorsi	C7 and C8
Levator ani	L4 and pudendal S2–S4
Levatores costarum	T1–T12 (P)
Levator palpibrae	Oculomotor Cr3
Levator scapulae	C3–C5
Masseter	Trigeminal (mandibular branch) Cr5
Nasal muscles	Facial (upper buccal branch) Cr7
Obturator externus	Obturator L3 and L4
Obturator internus	L5–S2
Occipitofrontalis	Facial Cr7
Orbicularis oculi	Facial (temporal and zygomatic branches) Cr7

Muscles	Nerves
Pectineus	Femoral L2 and L3
Pectoralis major	Pectoral C6–C8
Pectoralis minor	Pectoral C7 and C8
Peroneus	Sciatic (lateral popliteal branch) L5 and S1
Piriformis	S1 and S2
Plantaris	Sciatic (medial popliteal branch) L5 and S1
Platysma	Facial (cervical branch) Cr7
Popliteus	Sciatic (medial popliteal branch) L5 and S1
Pronator quadratus	Median C8
Pronator teres	Median C6
Psoas major	L2 and L3 or L4
Psoas minor	L1
Pterygoid	Trigeminal (mandibular branch) Cr5
Pyramidalis	Subcostal T12
Quadratus femoris	L4–S1
Quadratus lumborum	T12–L3 or L4
Quadriceps femoris	Femoral L3 and L4
Rectus abdominis	T6 or T7–T12
Rectus externus, internus and superior	Oculomotor Cr3
Rectus inferior	Abducens Cr6
Rhomboideus major and minor	C5
Sacro-spinalis	Lower C, all T and upper L (P)
Sartorius	Femoral L2 and L3
Scalenus	C4–C8
Semimembranosus	Sciatic (medial popliteal branch) L4 and L5
Semitendinosus	Sciatic (medial popliteal branch) L5–S2
Serratus anterior	C5–C7
Serratus posterior inferior	T9–T12
Serratus posterior superior	T2–T5
Soleus	Sciatic (medial popliteal branch) L5–S2
Sternomastoid	Accessory Cr11 and C2
Subclavius	C5 and C6
Subscapularis	Subscapular C6
Superior oblique	Trochlear Cr4
Supinator	Interosseus C5 and C6
Supraspinatus	C5
Temporalis	Trigeminal (mandibular branch) Cr5
Tensor fasciae latae	Superior gluteal L4–S1
Teres major	Subscapular C6
Teres minor	Circumflex C5
Tibialis anterior	Sciatic (lateral popliteal branch) L4 and L5
Tibialis posterior	Sciatic (medial popliteal branch) L4 and L5
Tongue muscles	Hypoglossal Cr12
Transversus	T6–L1
Trapezius	Accessory Cr11 and C3–C4
Triceps	Radial C7
Urethral sphincter	Pudendal S2–S4

Diagram to show the relationship of the spinal cord and its membranes and of a typical spinal nerve to a vertebra

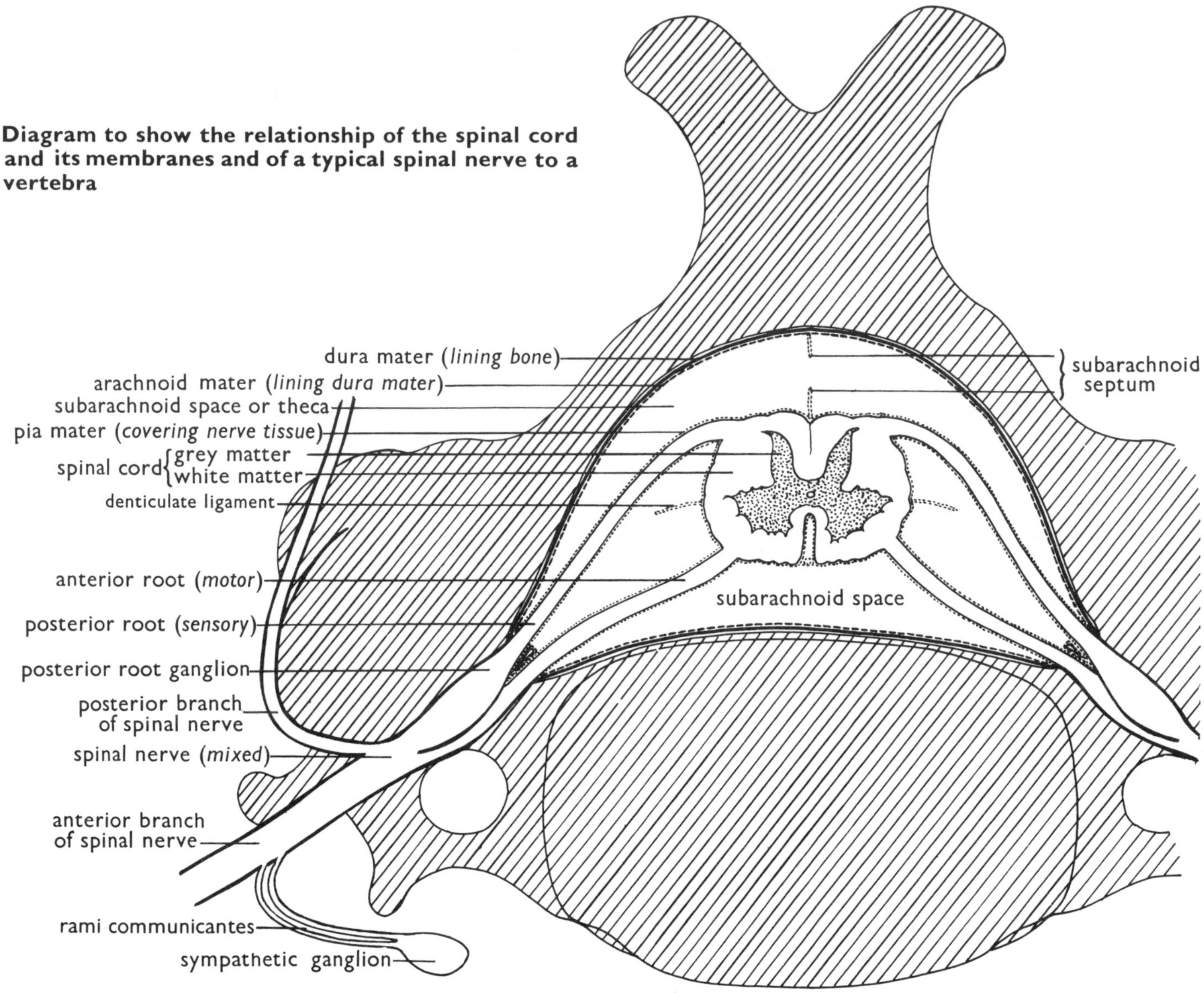

dura mater (*lining bone*)
arachnoid mater (*lining dura mater*)
subarachnoid space or theca
pia mater (*covering nerve tissue*)
spinal cord {grey matter, white matter}
denticulate ligament
anterior root (*motor*)
posterior root (*sensory*)
posterior root ganglion
posterior branch of spinal nerve
spinal nerve (*mixed*)
anterior branch of spinal nerve
rami communicantes
sympathetic ganglion
subarachnoid septum
subarachnoid space

NEURONAL CONNECTIONS

Neurones are connected to one another at **synapses**, which are small gaps, 200 Å across. A **pre-synaptic neurone** may make a synapse with a variety of points on the **post-synaptic neurone**: (a) on a dendrite—**axodendritic**, (b) on the cell body—**axosomatic**, (c) on the axon—**axoaxonic**. The transmission is always in the same direction. Even if there is backward spread of an impulse up a fibre, it is always blocked when it reaches the synapses. In a chain of neurones each post-synaptic neurone becomes the pre-synaptic neurone for the next link in the chain, until the **effector neurone** is reached. This makes its **neuromuscular** or **neuroglandular synapse** with the **effector organ**.

Because each neurone has a large number of terminal synaptic knobs—see page 60, it is possible to have a great variety of circuits, the main types of which are:

1. **Divergent circuits**. The impulse from a single pre-synaptic neurone makes synapses with a number of other neurones. The post-synaptic neurone most directly in the path of the pre-synaptic one may receive most stimulation and be **depolarised** (discharge an impulse) while others nearby are **facilitated** (made ready to discharge). The post-synaptic neurones may go down several pathways or continue on a common pathway.

2. **Convergent circuits**. Impulses are received from several pre-synaptic neurones. When these are from different sources, the same ultimate response can be made to different original stimuli. When they are from the same source there may be either **reinforcement** or **inhibition**.

3. **Reverberating circuits**. Branches from the post-synaptic neurones **feed back** to the pre-synaptic neurone to reinforce its activity. Once fired the signal may continue from a few seconds to many hours, producing co-ordinated activities, steady rate of breathing, waking versus sleeping and short-term memory.

divergent
convergent
reverberating
parallel after-discharge

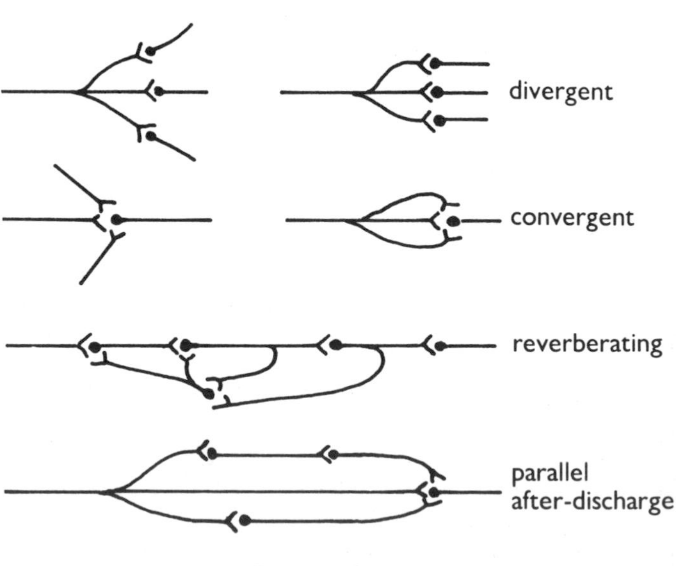

Types of circuit

NEUROTRANSMISSION

When the pre-synaptic impulse reaches a pre-synaptic knob, this becomes temporarily more permeable to Ca^{2+} ions. These stimulate a few of the vesicles in the knob to burst, liberating their chemical contents, the **neurotransmitter** substance. This diffuses across the 200 Å gap and is picked up by binding sites on the post-synaptic neurone. If sufficient neurotransmitter is received to exceed the **threshold**, an action potential is produced and **electrical discharge** takes place, spreading rapidly down the post-synaptic neurone. **Inhibitory transmission**, producing increased or **hyperpolarisation**, is also possible. As each post-synaptic neurone normally has synapses with many pre-synaptic ones, the ultimate response is the **summation** of all the excitatory and inhibitory stimuli received. Continuous transmission is prevented by rapid destruction or removal of the neurotransmitter.

The commonest neurotransmitter is **acetylcholine**. It is produced by some neurones inside the CNS and many outside. It is also released at neuromuscular junctions with skeletal muscles, some cardiac and smooth muscles and some glands. It is rapidly destroyed by **choline esterase** so that concentrations do not build up in the synapses. **Noradrenaline (norepinephrine)** acts similarly in some parts of the brain, at junctions with some cardiac and smooth muscle and with some glands. Once used it is pumped back and destroyed or recycled in the pre-synaptic knobs.

Gamma-aminobutyric acid has been identified as inhibitory in the CNS and **glycine** has similar effects in the spinal cord. Altogether over 40 neurotransmitter substances have been recognised, including **serotonin** found in the hypothalamus and basal ganglia and believed to effect behaviour, and **dopamine** possibly associated with memory storage in the brain.

More general chemical transmitters include **encephalins** and **endorphins**, which are believed to act as pain killers and to be linked with learning, sexual activity and temperature control.

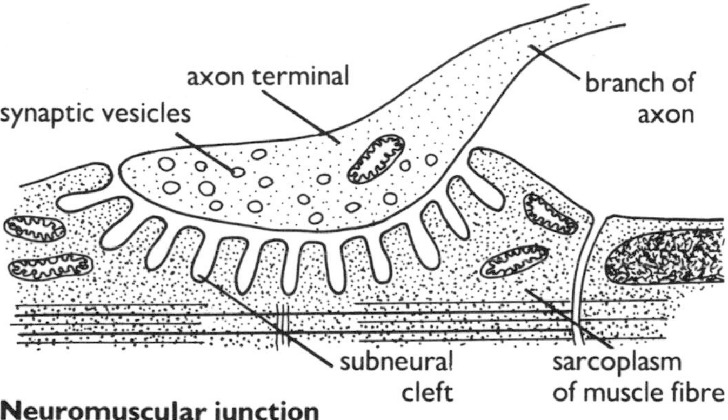

Neuromuscular junction

Neuromuscular transmission

Muscle cells and neurones are the only cells in the body which have surface potentials. In a neuromuscular junction the synaptic knob of the neurone is closely applied to the sarcolemma, with a synaptic gap similar to that between neurones but with deep subneural clefts on the muscle surface. The neurotransmitter released becomes bonded with the sarcolemma and sets up a wave of potential similar to that in a nerve fibre. As already described—see page 38, this penetrates a skeletal muscle fibre via the transverse tubules and sarcoplasmic reticulum, where it initiates the chain of events leading to muscle shortening. In the case of the much smaller smooth muscle fibres, the impulse remains superficial and the action potential can spread from fibre to fibre without direct nerve contact.

Neuroglandular transmission

The nerve impulse releases a neurotransmitter substance at a neuroglandular junction. The neurotransmitter has the direct effect of stimulating or inhibiting release of secretion by the gland as a whole.

VOLUNTARY AND INVOLUNTARY ACTION

Voluntary action involves skeletal muscles. The effector neurones have their cell bodies inside the CNS and the nerve fibres are myelinated. Most of the action produced is consciously recognisable, even if not consciously initiated, and can be voluntarily controlled—see page 38.

Involuntary action may involve cardiac muscle, smooth muscle or glands. The effector neurones have their cell bodies outside the CNS and the fibres are non-myelinated. The ganglia in which these cell bodies lie are served by myelinated fibres from the CNS so that overall central nervous control occurs. Involuntary action is the result of the activation of the involuntary or autonomic system.

Note. M=mitochondrion

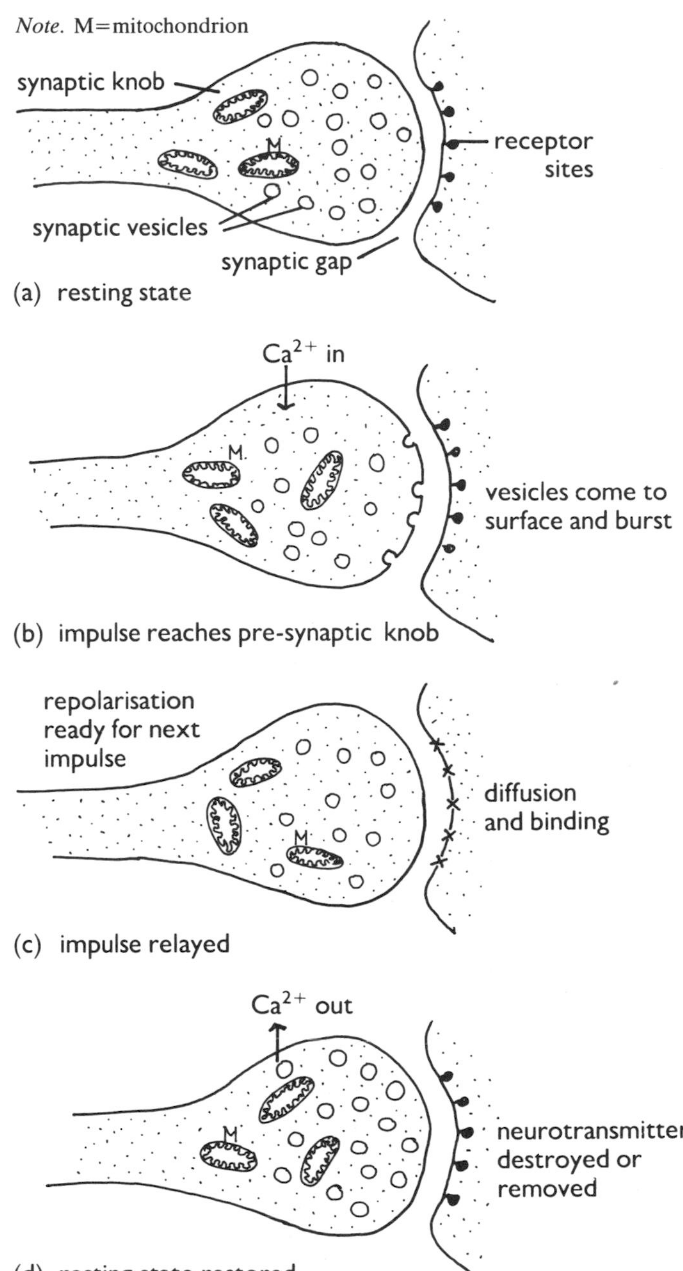

Stages in neurotransmission

THE AUTONOMIC SYSTEM

The autonomic system is in two parts:

1. The **parasympathetic** or **cranio-sacral system**. In this system the preganglionic fibres are long and the postganglionic fibres are short. The ganglia of the latter lie in the organs served, allowing them to be stimulated individually. The preganglionic parasympathetic fibres accompany the 3rd, 7th, and 9th cranial nerves, form the greater part of the 10th cranial nerves (vagus nerves) and accompany the 2nd, 3rd and 4th sacral nerves. The short postganglionic fibres produce acetylcholine as neurotransmitter.

2. The **sympathetic** or **thoracico-lumbar system**. In this system the preganglionic fibres are short. They come from the thoracic and first three lumbar nerves in white rami communicantes to chains of **sympathetic ganglia**, from which there are plexuses of fibres to further ganglia near the viscera. Very long postganglionic fibres pass through grey rami communicantes to serve the skin and the blood vessels of the skeletal muscles.

The system tends to act as a whole because each preganglionic fibre relays to many postganglionic ones, most of which produce noradrenaline (norepinephrine). The action of this is supplemented by noradrenaline and adrenaline from the adrenal medulla—see page 122. Sympathetic fibres to the sweat glands and to blood vessels of the skin and skeletal muscles produce acetylcholine. The effects of noradrenaline build up more slowly, but last longer than those of acetylcholine. Thus sympathetic stimulation produces rapid initial sweating and redistribution of blood between the skin and the skeletal muscles followed by more prolonged changes—see page 70.

THE EFFECTS OF THE AUTONOMIC SYSTEM

Parts affected	Para-sympathetic	Sympathetic	General result
Pupils	Contracted	Dilated	Controls the amount of light entering the eyes
Ciliary muscles	Contracted	Relaxed	Controls accommodation of the eye
Blood vessels	Arterioles of glands and viscera dilated	Arterioles of alimentary canal and skin constricted; those of skeletal muscles dilated or constricted by different fibres; tone raised in walls of larger vessels	Adjusts the blood pressure and the distribution of the blood
Spleen	Dilated	Constricted	Adjusts the quality and quantity of blood in circulation
Heart beat	Slowed and weakened	Hastened and strengthened	Adjusts the rate of the heart beat according to the blood pressure and varying muscle activity
Bronchioles	Constricted	Dilated	Adjusts ease of breathing to requirements
Sweat glands		Activity increased	Produces extra sweat in anticipation of heat production during activity
Adrenal medulla		Activity increased	Reinforces direct effects mentioned above
Peristalsis of the alimentary canal	Increased	Decreased	Controls the speed of passage of food and the rate of digestion
Sphincters	Relaxed	Contracted	*Note.* Digestion is slow when body activity is great and vice versa
Digestive glands	Activity increased	Activity decreased	

Stress

Stress is any stimulus which tends to change the internal environment, whether from outside, e.g. heat, cold, lack of oxygen, unpleasant noise or sight, or from inside, e.g. high blood pressure, pain, unpleasant thoughts. Much normally occurring stress can be compensated. Extreme stress, e.g. by wounds, surgical operations, poisons and overexposure, can produce conditions outside the limits of tolerance and result in catastrophic changes known as **shock**—see page 102.

The healthy body shows regular **biological rhythms**, e.g. circadian (daily) variations in body temperature, sleeping and waking and monthly period in women. Interference with these rhythms by emotional stress can lead to the so-called **psychosomatic disorders**, with genuine physical symptoms, due to upset of the homeostatic balance.

The **parasympathetic system** serves all the autonomically controlled organs except the adrenal medulla, spleen, sweat glands and blood vessels of skin and skeletal muscles. It helps to create the conditions needed for **rest**, **sleep** and **digestion**.

The **sympathetic system** supplies all the organs that are supplied by the parasympathetic providing antagonistic control. It also supplies the adrenal medulla, sweat glands and blood vessels of skin and skeletal muscle, where control is simple on/off. It helps to prepare the body for '**fight or flight**' and acts in conjunction with **adrenaline** to create conditions needed for **physical activity**.

Though there can be some response to sensory input by afferent nerves from the viscera and the larger blood vessels, overall control is from the hypothalamus, which balances activity to need. This regulates rate of heart beat, movement of food in the alimentary canal, contraction of the bladder and hormonal activity and produces a condition of homeostasis.

Homeostasis

Homeostasis is the state of stability of the internal environment. The extracellular fluid is maintained within the limits which living cells can tolerate.

Temperature, pressure and concentration of gases, nutrients, salts and water are allowed to vary only very slightly before remedial action occurs. Much of the control is completely unconscious—see temperature, page 81, blood pressure, page 110, and osmoregulation, page 118. In the case of hunger and thirst conscious remedial action may be needed. As the hunger centre senses lowered blood sugar, impulses are relayed from the hypothalamus to the cerebrum producing the desire to eat. When food stretches the stomach or blood sugar rises hunger ceases. The thirst centre is turned on by increase in concentration of dissolved substances and turned off by drinking.

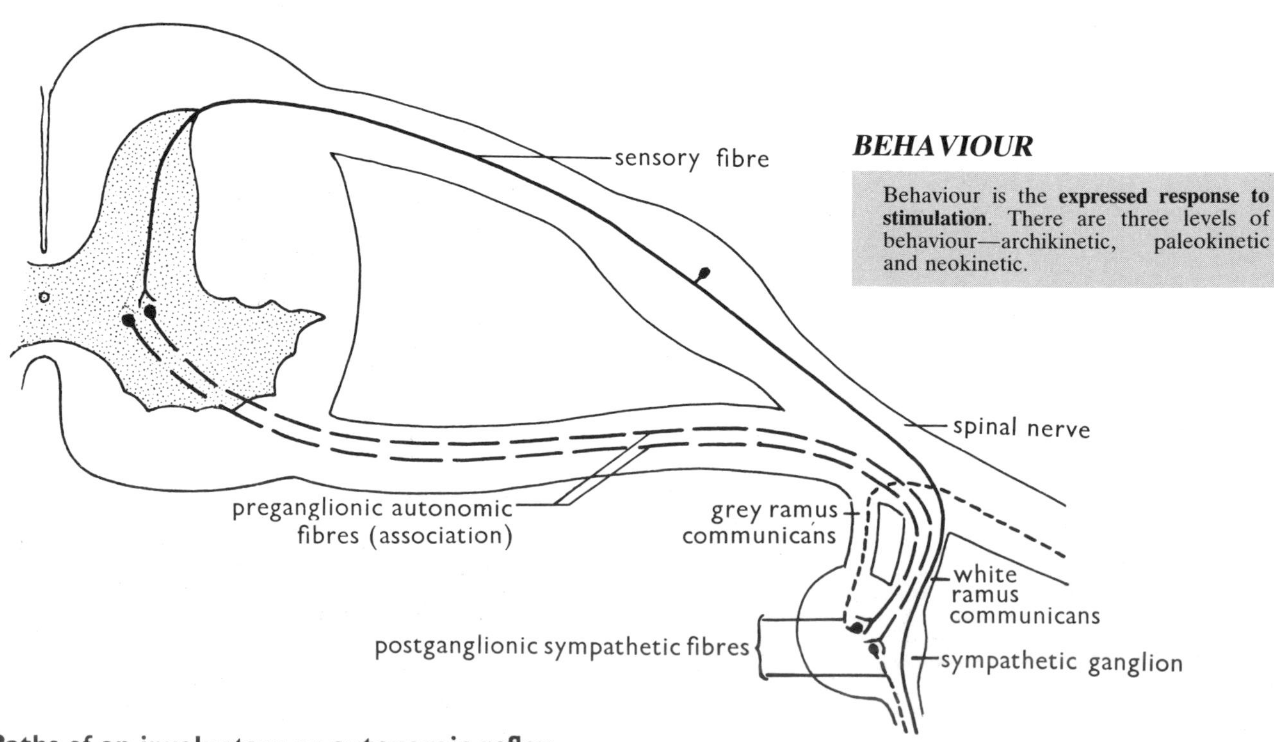

Paths of an involuntary or autonomic reflex

BEHAVIOUR

Behaviour is the **expressed response to stimulation**. There are three levels of behaviour—archikinetic, paleokinetic and neokinetic.

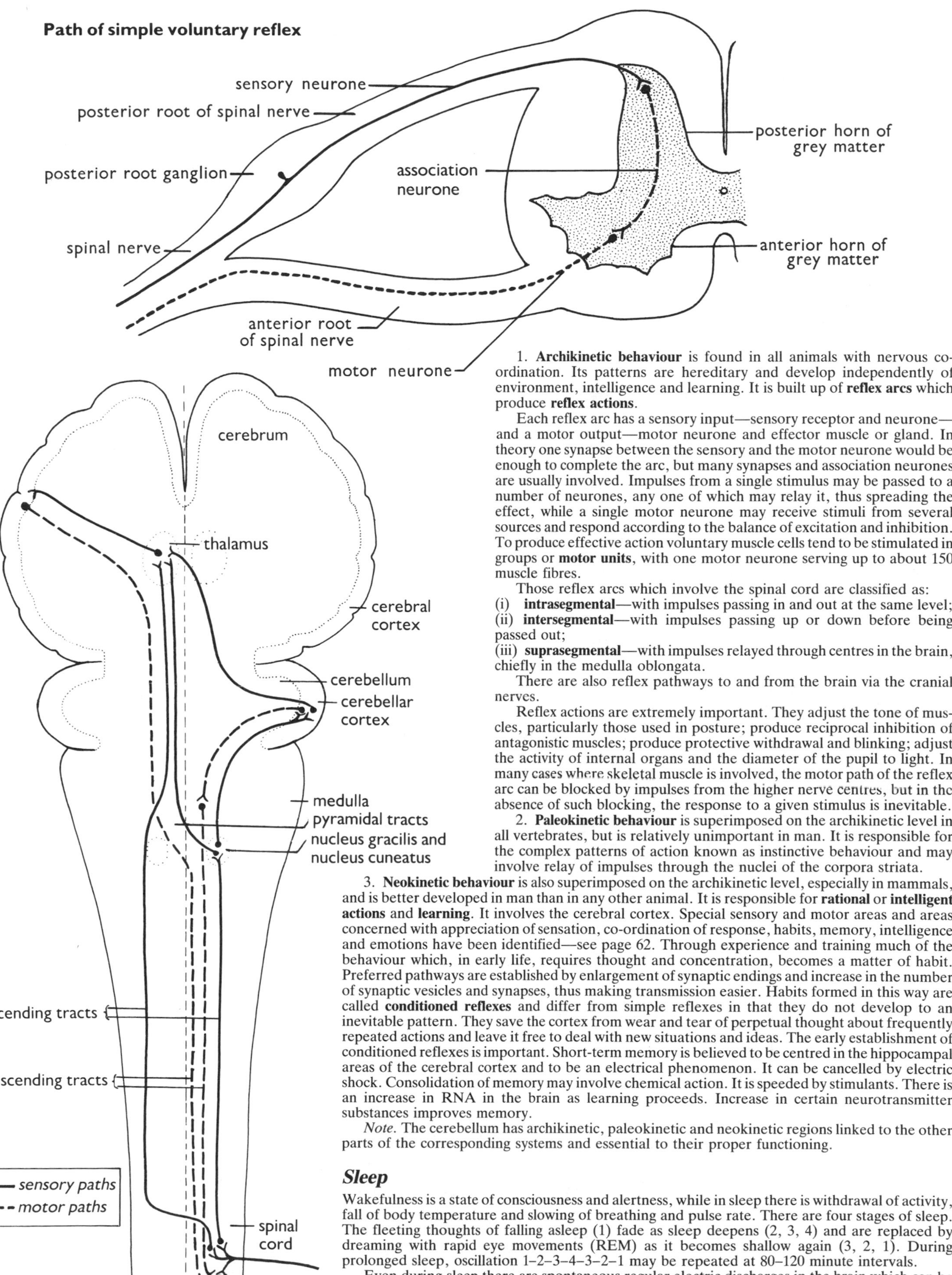

1. **Archikinetic behaviour** is found in all animals with nervous co-ordination. Its patterns are hereditary and develop independently of environment, intelligence and learning. It is built up of **reflex arcs** which produce **reflex actions**.

Each reflex arc has a sensory input—sensory receptor and neurone—and a motor output—motor neurone and effector muscle or gland. In theory one synapse between the sensory and the motor neurone would be enough to complete the arc, but many synapses and association neurones are usually involved. Impulses from a single stimulus may be passed to a number of neurones, any one of which may relay it, thus spreading the effect, while a single motor neurone may receive stimuli from several sources and respond according to the balance of excitation and inhibition. To produce effective action voluntary muscle cells tend to be stimulated in groups or **motor units**, with one motor neurone serving up to about 150 muscle fibres.

Those reflex arcs which involve the spinal cord are classified as:
(i) **intrasegmental**—with impulses passing in and out at the same level;
(ii) **intersegmental**—with impulses passing up or down before being passed out;
(iii) **suprasegmental**—with impulses relayed through centres in the brain, chiefly in the medulla oblongata.

There are also reflex pathways to and from the brain via the cranial nerves.

Reflex actions are extremely important. They adjust the tone of muscles, particularly those used in posture; produce reciprocal inhibition of antagonistic muscles; produce protective withdrawal and blinking; adjust the activity of internal organs and the diameter of the pupil to light. In many cases where skeletal muscle is involved, the motor path of the reflex arc can be blocked by impulses from the higher nerve centres, but in the absence of such blocking, the response to a given stimulus is inevitable.

2. **Paleokinetic behaviour** is superimposed on the archikinetic level in all vertebrates, but is relatively unimportant in man. It is responsible for the complex patterns of action known as instinctive behaviour and may involve relay of impulses through the nuclei of the corpora striata.

3. **Neokinetic behaviour** is also superimposed on the archikinetic level, especially in mammals, and is better developed in man than in any other animal. It is responsible for **rational** or **intelligent actions** and **learning**. It involves the cerebral cortex. Special sensory and motor areas and areas concerned with appreciation of sensation, co-ordination of response, habits, memory, intelligence and emotions have been identified—see page 62. Through experience and training much of the behaviour which, in early life, requires thought and concentration, becomes a matter of habit. Preferred pathways are established by enlargement of synaptic endings and increase in the number of synaptic vesicles and synapses, thus making transmission easier. Habits formed in this way are called **conditioned reflexes** and differ from simple reflexes in that they do not develop to an inevitable pattern. They save the cortex from wear and tear of perpetual thought about frequently repeated actions and leave it free to deal with new situations and ideas. The early establishment of conditioned reflexes is important. Short-term memory is believed to be centred in the hippocampal areas of the cerebral cortex and to be an electrical phenomenon. It can be cancelled by electric shock. Consolidation of memory may involve chemical action. It is speeded by stimulants. There is an increase in RNA in the brain as learning proceeds. Increase in certain neurotransmitter substances improves memory.

Note. The cerebellum has archikinetic, paleokinetic and neokinetic regions linked to the other parts of the corresponding systems and essential to their proper functioning.

Sleep

Wakefulness is a state of consciousness and alertness, while in sleep there is withdrawal of activity, fall of body temperature and slowing of breathing and pulse rate. There are four stages of sleep. The fleeting thoughts of falling asleep (1) fade as sleep deepens (2, 3, 4) and are replaced by dreaming with rapid eye movements (REM) as it becomes shallow again (3, 2, 1). During prolonged sleep, oscillation 1–2–3–4–3–2–1 may be repeated at 80–120 minute intervals.

Even during sleep there are spontaneous regular electric discharges in the brain which can be recorded. The pattern of discharges is altered when sense organs are stimulated. Coma is an unconscious state in which powerful stimuli cannot arouse response. It involves the brain stem. If the vital centres are damaged there is little chance of recovery. A life-support machine can keep the general body functions going. Brain death, when the electric waves in the brain cease, is the criterion for turning off the life support.

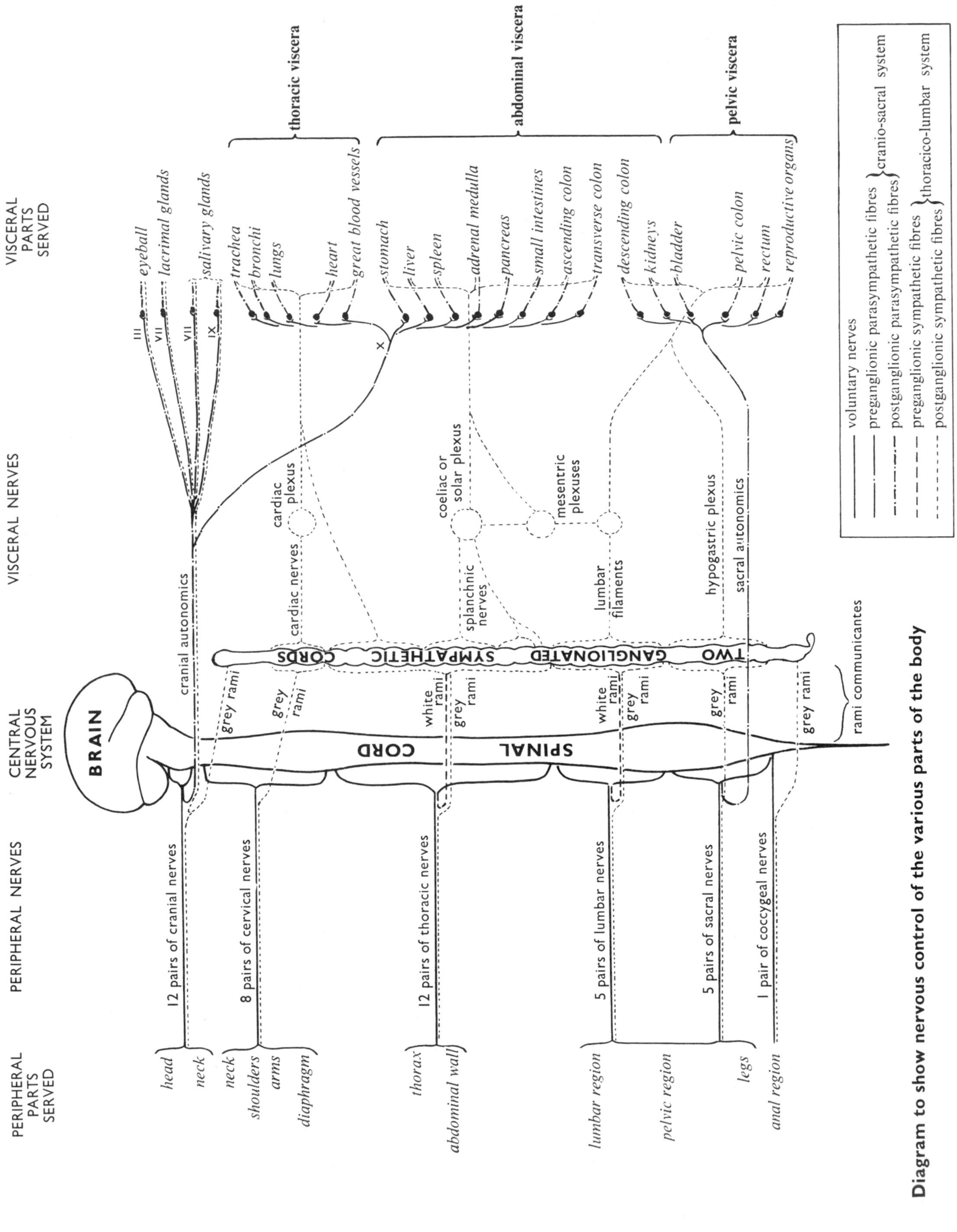

Diagram to show nervous control of the various parts of the body

Sense Organs

Sensory nerve endings fall into three categories:
1. **enteroceptors**, which sense stimuli in the viscera;
2. **proprioceptors**, which sense position and movement;
3. **exteroceptors**, which sense stimuli from outside the body.

Normally we are not conscious of the stimulation of enteroceptors and proprioceptors, except sometimes vaguely, but excessive stimulation may cause discomfort and pain. Enteroceptors set up autonomic reflexes to control visceral activity, e.g. movements of the alimentary canal. Most of the receptors are naked nerve endings, but some are encapsulated forming lamellated, **Pacinian corpuscles**. The latter are found in the walls of some of the major blood vessels—see page 110, where they act as **baroreceptors** helping to regulate blood pressure. Proprioceptors help to set up **kinaesthetic reflexes**, which co-ordinate action of antagonistic groups of skeletal muscles, maintaining posture and tone. Receptors in tendons and muscles are **neurotendinous** and **neuromuscular sensillae**. There are Pacinian corpuscles in the joint capsules.

Stimulation of exteroceptors may produce reflex action—see page 71, but whether they do or not, we are consciously aware of the stimuli and of their location. Interpretation of sensation depends on connections in the brain.

The brain is conscious of distinct types of sensation, grouped as:
1. **general surface senses**—pressure, touch, temperature and pain;
2. **chemical senses**—taste and smell;
3. **balance**—motion and gravitational sense;
4. **hearing**;
5. **sight**.

GENERAL SURFACE SENSES (somaesthesia)

Pressure, touch, temperature and pain receptors are widely but unevenly spread in the skin—see page 80 for diagrams.

Pressure

Pressure is sensed by the same type of lamellated corpuscles as occur internally. They lie deep in the dermis and are rare where the skin is hairy. They also occur in mammary glands and external genitalia.

Touch

Touch is sensed by naked nerve endings close to or penetrating the epidermis—see pages 80 and 81—and also by encapsulated fibres. Each touch capsule has a soft cellular core in which the dendrite ends after losing its myelinated sheath. Though detailed appearance varies—see diagram on page 80, four main types are recognised:

1. **Merkel's discs** are groups of minute plates usually surrounded by the basal layers of the epidermis and adapting slowly so that the effect of stimulation is persistent.

2. **Meisner corpuscles** may be oval, circular or lobulated with stacks of flattened cells. They lie in groups in the dermal papillae and form baskets round the hair follicles, picking up sensation from displacement of the hair. Each corpuscle may be served by several nerves and each nerve may serve several corpuscles, thus spreading the sensation, but, where densest, e.g. on the lips and fingertips, points only 1.2 mm apart can be discriminated. They adapt readily so that prolonged stimulation, e.g. by clothes, can be ignored.

3. **Krause corpuscles** may be cylindrical or spherical and are without capsules. The non-myelinated end branches of myelinated fibres are looped and coiled together. These corpuscles are found particularly in the conjunctiva (skin of the eyes), the lips, the tongue and the skin of the genitalia.

4. **Ruffini corpuscles** have large myelinated fibres ending in a non-myelinated network round a tubular core of cells. They record dermal distortion.

Temperature

Temperature receptors consist of **free nerve endings** encased only in **Schwann cell membrane**. They are sensitive to rate of change of temperature. The body has warmth and cold sensing regions. The warmth sensors respond between 20 and 45 °C with peak activity around 37.5–40 °C. The cold sensors normally respond between 10 and 41 °C with peak activity 15–20 °C, but they have a second peak 46–50 °C, when warmth is interpreted in the brain as cold.

Pain

Pain may occur when any of the above receptors are **over-stimulated**. The sharp pain of a vertical cut in the skin may be intense, but is short-lived. Damage such as burning affects more sense endings, builds up slowly and lasts longer. Its effects in terms of shock—see page 102—may be greater. Pain that seems to come from the skin may be due to intense stimulation of enteroceptors being referred, as close juxtaposition of neurones in the spinal nerves permits over-spill of impulses. Headache is a similar **referred pain** from a variety of causes.

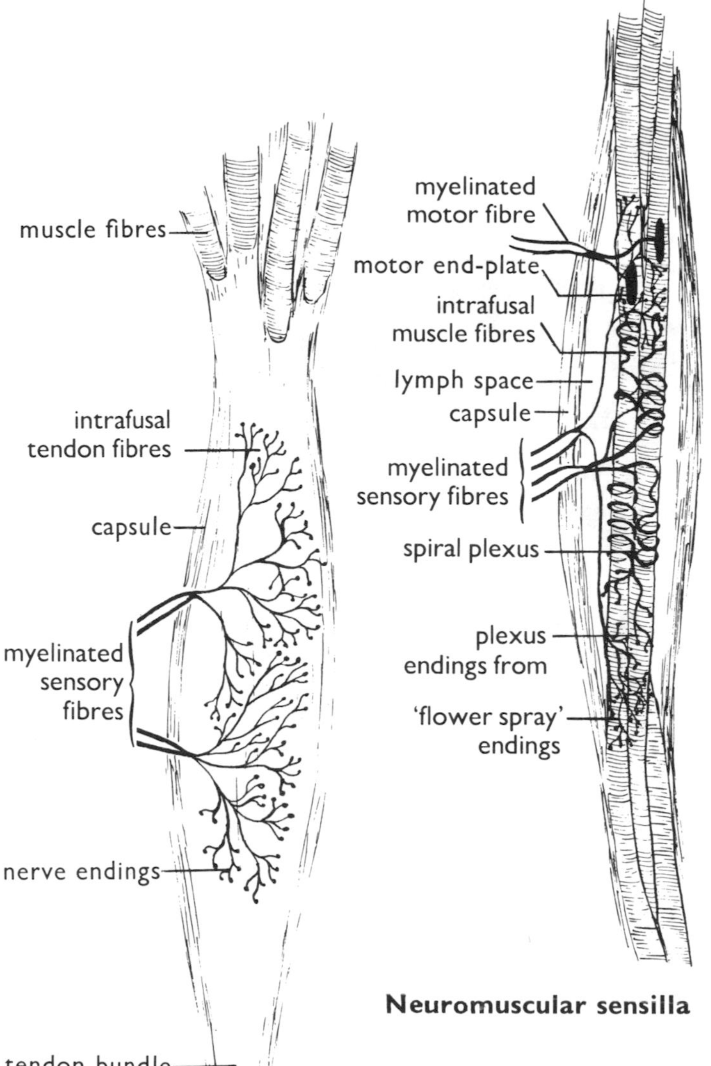

Neuromuscular sensilla

Neurotendinous sensilla

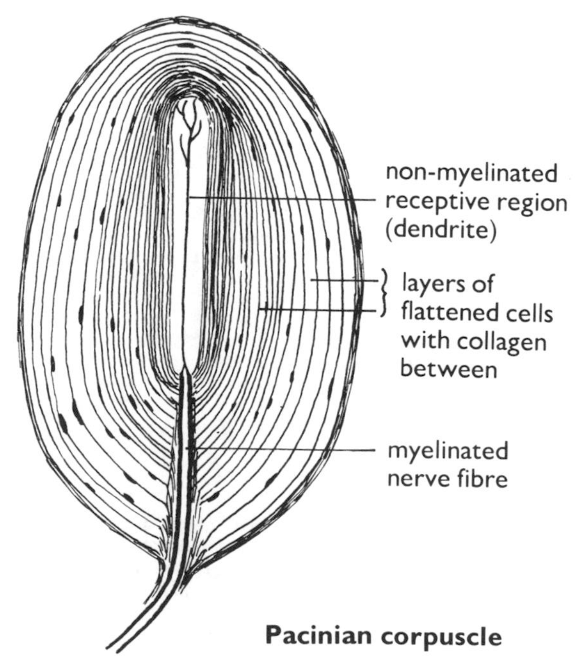

Pacinian corpuscle

Note. A rapidly moving touch can **tickle**. Rhythmic pressure is sensed as **vibration**. **Itch** is the result of minute chemical responses.

CHEMICAL SENSES

There are two types of **chemo-sensitivity**—taste and smell.

Taste requires the substance tasted to be in solution and is restricted to the tongue and palate. The taste buds are morphologically all alike but are physiologically differentiated into those sensitive to sweetness, saltness, sourness and bitterness. Each type of taste bud is found on a definite region of the tongue.

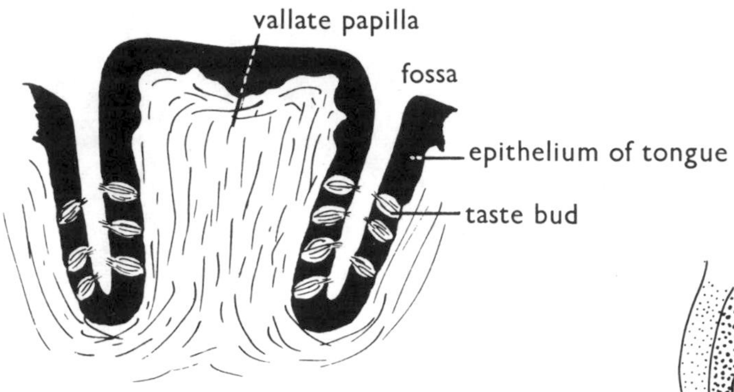

vallate papilla
fossa
epithelium of tongue
taste bud

Section of vallate papilla of tongue with taste buds

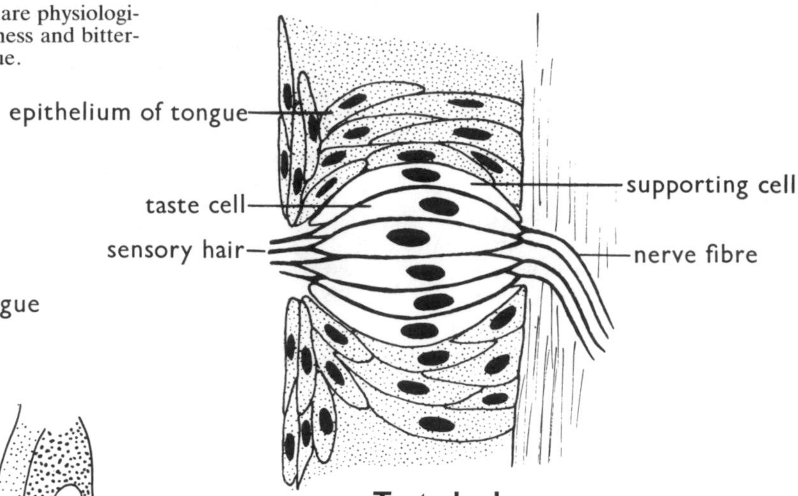

epithelium of tongue
taste cell
sensory hair
supporting cell
nerve fibre

Taste bud

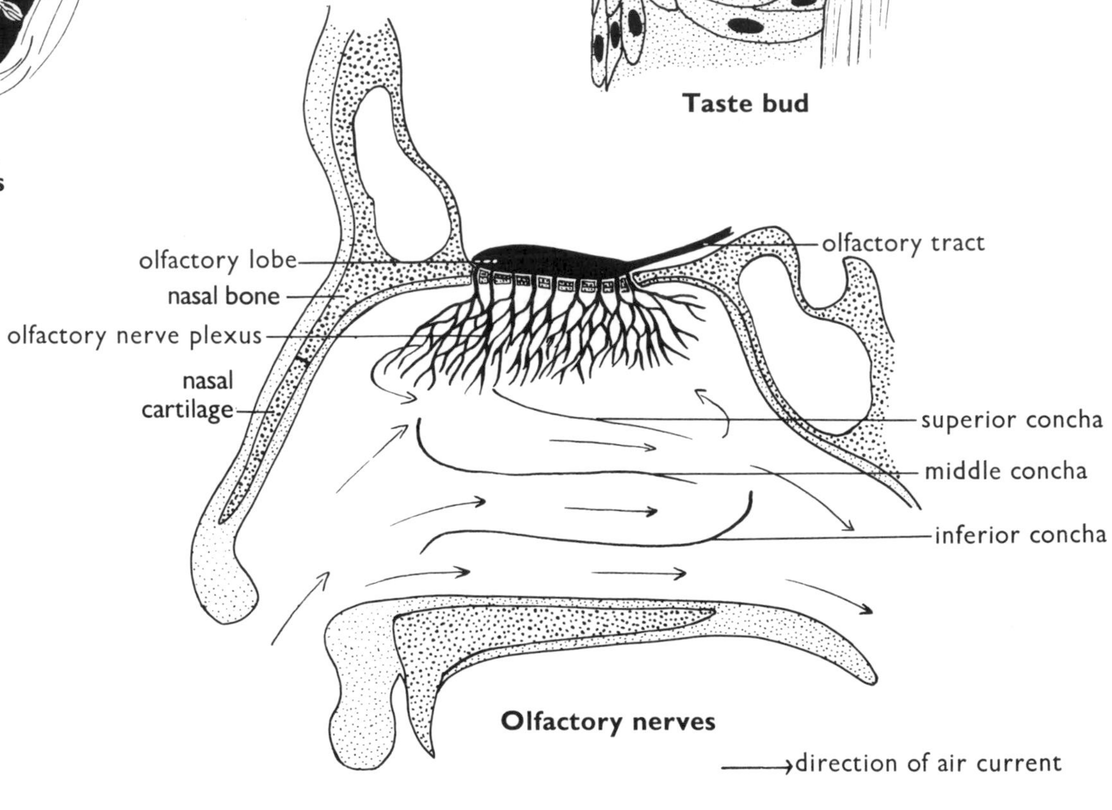

olfactory lobe
nasal bone
olfactory nerve plexus
nasal cartilage
olfactory tract
superior concha
middle concha
inferior concha

Olfactory nerves

→direction of air current

Smell does not require the substance to be in solution and is restricted to the epithelium of the upper regions of the nasal cavities. Many flavours are smelled rather than tasted. Impulses from olfactory stimulation, relayed through the habenular nuclei—see page 63, can produce visceral reflex responses, e.g. salivation, vomiting.

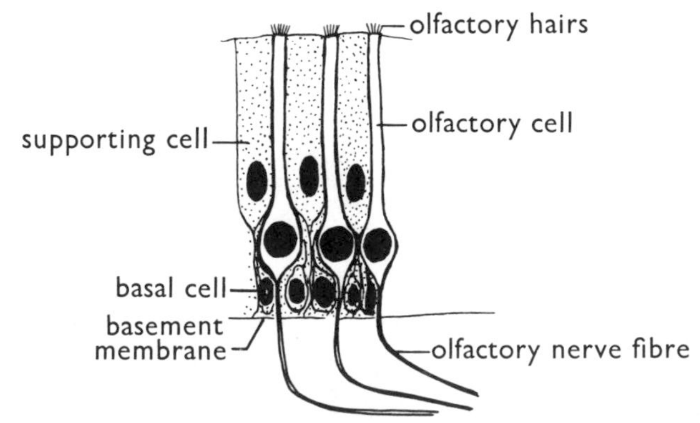

Olfactory mucous membrane

BALANCE AND HEARING

Both these senses are located in the **ears**. Each ear is divided into three regions, the outer ear, the middle ear and the inner ear.

The outer ear

The outer ear consists of a tube called the **external auditory meatus** whose inner end is closed by the **tympanic membrane** and whose outer end is surrounded by the **auricle** or **pinna**. The pinna is supported by cartilage which has extra elastic fibres giving it flexibility. It helps to collect sound waves and direct them along the meatus to the tympanic membrane. This membrane contains many fibres of different lengths so that it vibrates equally to sound waves of different frequencies. The walls of the external auditory meatus have **ceruminous (wax) glands**, the secretion from which helps to protect the tympanic membrane and keep it pliable. The wax is prevented from clogging by hairs, which also deter dust and insects from entering the outer ear.

The middle ear

The middle ear consists of the tympanic cavity which communicates with the nasopharynx through the **pharyngo-tympanic** or **Eustachian tube**. It contains air to equalise the pressure of the two sides of the tympanic membrane. Across the tympanic cavity is a chain of small bones called the **auditory ossicles** (see page 16). These bones relay the sound waves across the cavity. They rock in a lever-like manner—see page 39—and thus decrease the amplitude but increase the power of the vibrations.

Ear

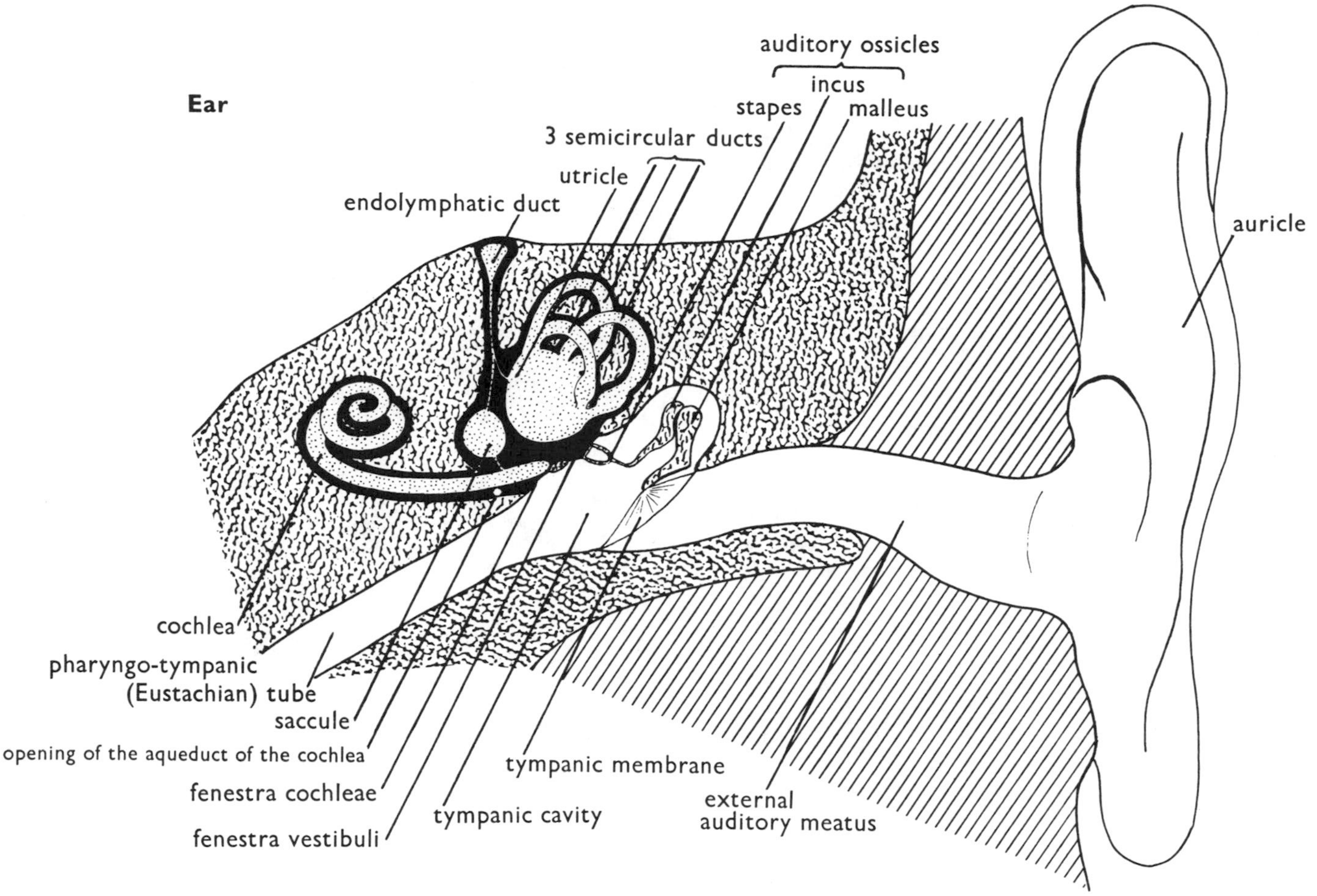

auditory ossicles
incus
stapes
malleus
3 semicircular ducts
utricle
endolymphatic duct
auricle
cochlea
pharyngo-tympanic (Eustachian) tube
saccule
opening of the aqueduct of the cochlea
fenestra cochleae
fenestra vestibuli
tympanic membrane
tympanic cavity
external auditory meatus

The inner ear

The inner ear consists of a **membranous labyrinth** lying in a bony labyrinth of similar shape. The membranous labyrinth is filled with fluid called endolymph, while the spaces between it and the bony labyrinth are filled with perilymph. The latter is continuous with the cerebro-spinal fluid through the **aqueduct of the cochlea**.

Two sacs, the **utricle** and the **saccule**, lie together in the vestibule and are both connected to the **endolymphatic duct**. Attached to the utricle there are three **semicircular ducts** each lying in a **semicircular canal**. Each duct and its canal has a swelling or **ampulla** at one end inside which is a **christa** or group of cells with hair-like processes embedded in a gelatinous **cupola** (hood). These are stimulated by movement of the endolymph in the ducts brought about by movements of the head. The semicircular canals are in mutually perpendicular planes so that movement in every plane can be perceived. The utricle and the saccule also contain patches of sensory hair cells called **maculae,** which are stimulated by **gravitational action** on the small crystals of calcium carbonate (**otoliths**) which adhere to them. The nerve fibres from these cells form the **vestibular nerves** and the information from impulses in these nerves is interpreted as a sense of **balance**.

Continuous with the saccule there is a spirally coiled duct called the **cochlea**, which is the organ of hearing—see diagram on page 76. The sensitive hair cells are arranged on the **basilar membrane** in the organ of Corti. Their processes are embedded in the **tectorial membrane** so that when the basilar membrane vibrates they are stimulated. The nerve fibres from these cells form the **cochlear nerve**.

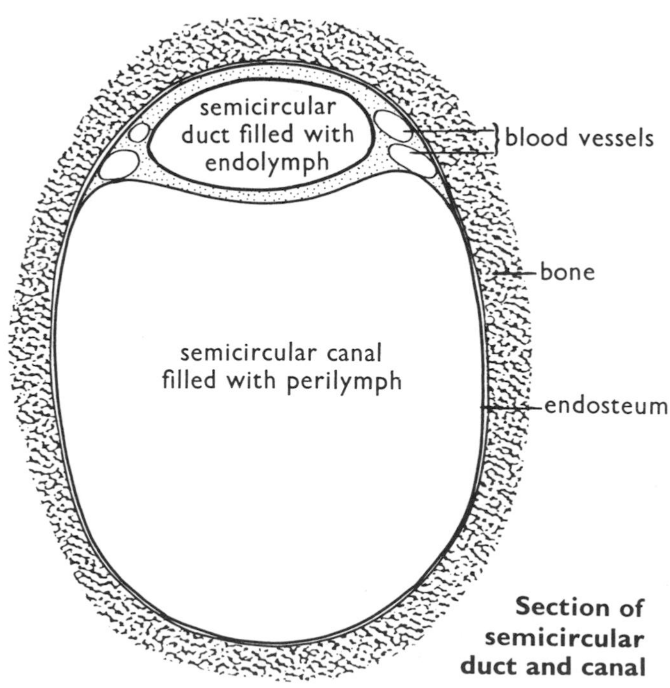

Section of semicircular duct and canal

The mechanism of hearing

The accepted theory of the mechanism of hearing is as follows.

The vibrations relayed by the auditory ossicles to the membrane over the fenestra vestibuli cause vibrations in the perilymph which are passed up the scala vestibuli of the cochlea and relayed successively through the vestibular membrane, the scala media, the basilar membrane and the scala tympani. The membrane over the fenestra cochleae (rotunda) at the end of the scala tympani allows equalisation of pressure.

The basilar membrane contains up to 30 000 fibres which vary in length, thickness and tension. Each fibre resonates to vibrations of a particular wavelength. Only in the vicinity of such resonance is the movement of the basilar membrane, and thus of the hair cells against the tectorial membrane, sufficient to produce stimulation. This enables pitch to be discriminated. High-pitch sounds, i.e. those with high-frequency vibrations, are picked up near the base of the cochlea, while low-pitch sounds, with low-frequency vibrations, are sensed near the apex.

The slight difference in time of arrival of sound waves at the two ears is appreciated by the brain as a sense of direction of the source of the sound.

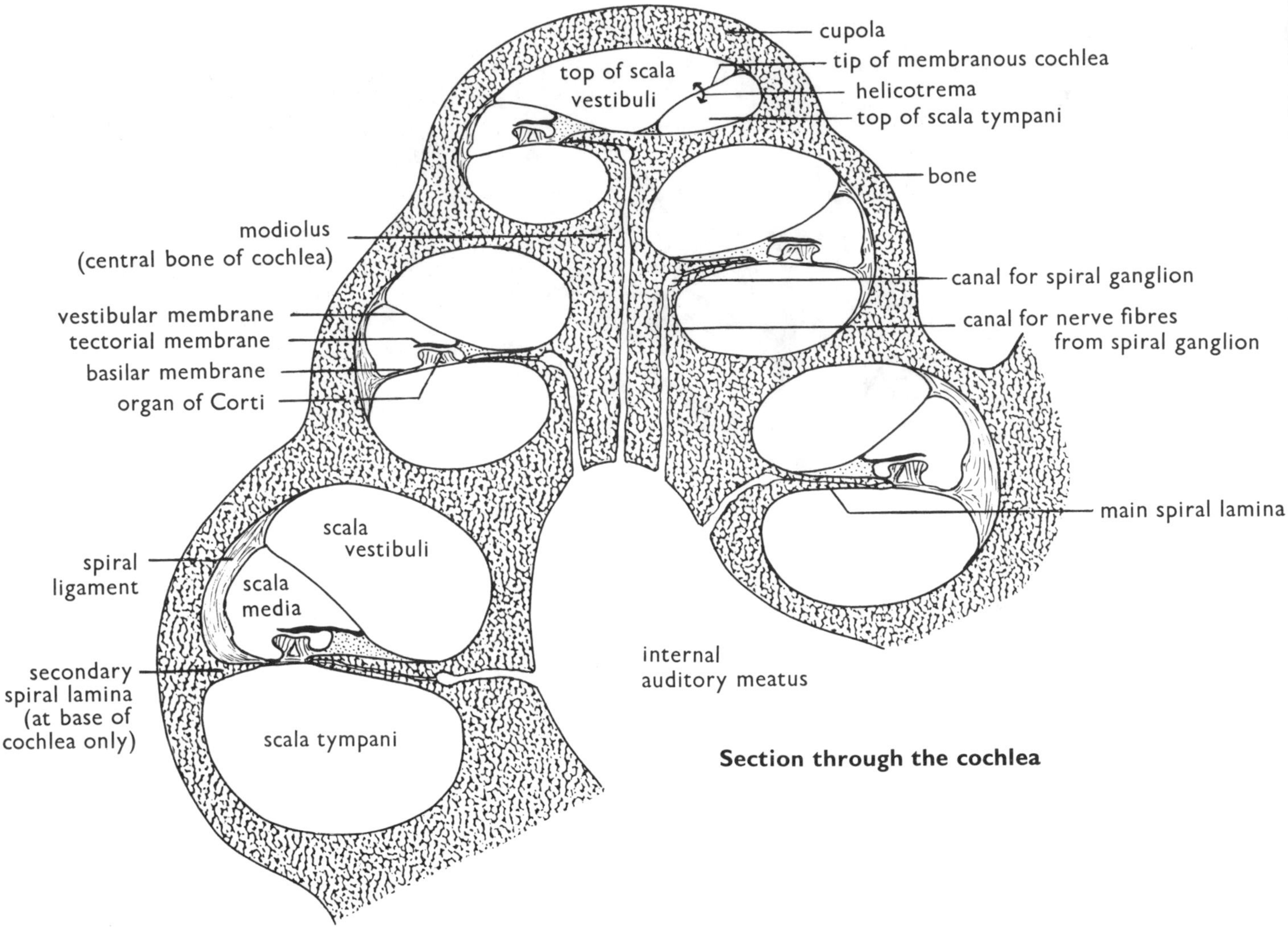

cupola
tip of membranous cochlea
top of scala vestibuli
helicotrema
top of scala tympani
bone
modiolus (central bone of cochlea)
canal for spiral ganglion
canal for nerve fibres from spiral ganglion
vestibular membrane
tectorial membrane
basilar membrane
organ of Corti
main spiral lamina
scala vestibuli
spiral ligament
scala media
secondary spiral lamina (at base of cochlea only)
internal auditory meatus
scala tympani

Section through the cochlea

inner hair cells
outer hair cells
tectorial membrane
supporting cells (cells of Hensen)
limbus
internal sulcus
cells of Claudius
nerves
supporting cells
foramen for nerves
inner rod
canal
outer rod
supporting cells (cells of Deiters)
basilar membrane

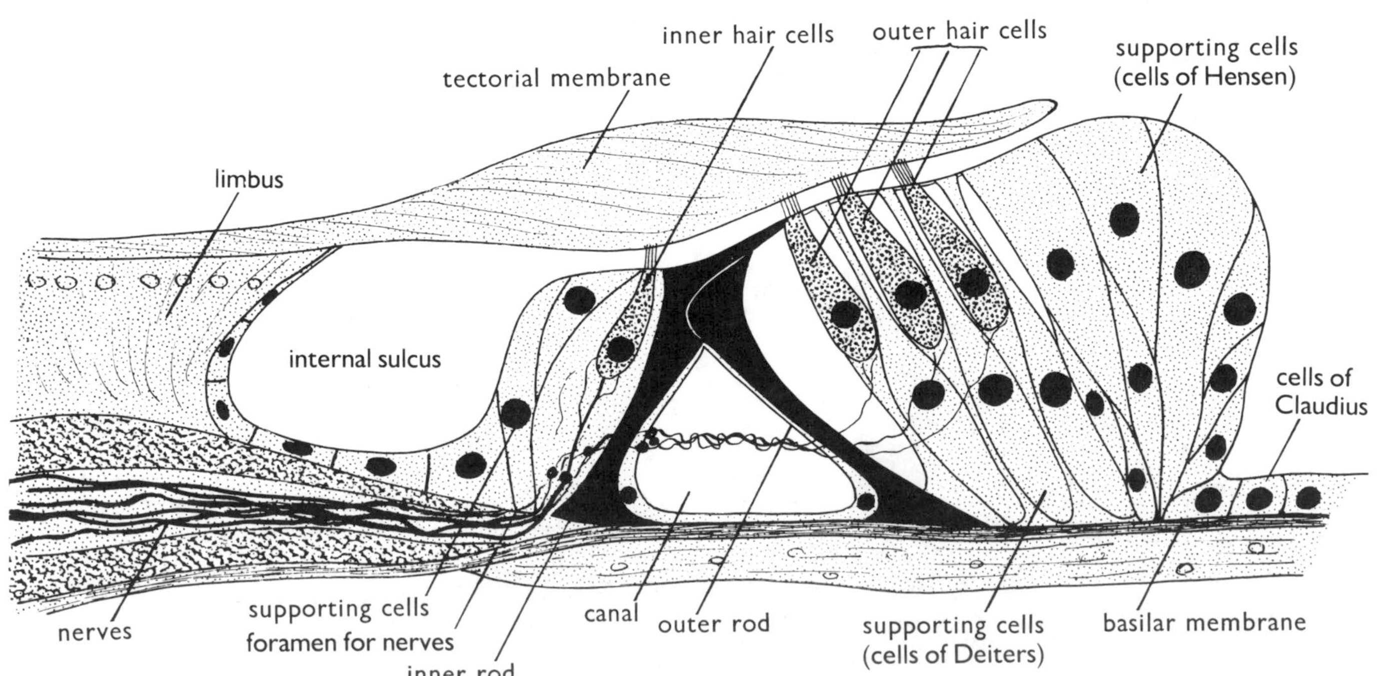

Section of organ of Corti

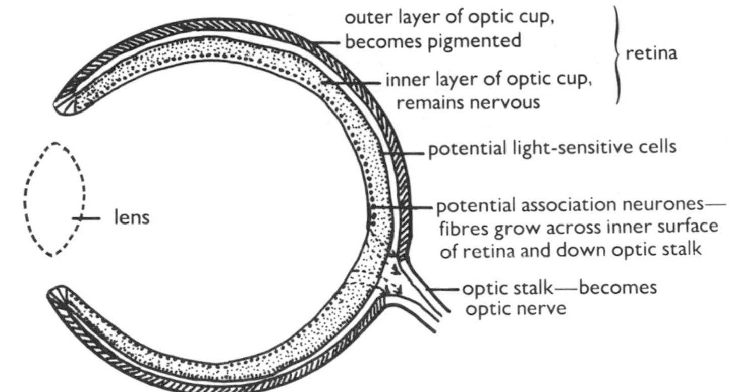

Development of optic cup

SIGHT

Light sensitivity is a feature of special cells in the eyes. These are derived from the same embryonic tissue as the brain—see page 61. Very early in development two cup-like structures grow outwards taking neurones with them. The linings of these **optic cups** eventually form the **retinae** consisting of light-sensitive cells and association neurones from which the tracts of fibres to the rest of the brain form the **optic nerves**. Many of these fibres cross over in the **optic chiasma**—see pages 65 and 79.

All other parts of the eyes are formed around the retinae and serve to (a) direct the gaze where required, (b) keep a clear window through which a controlled amount of light can pass, (c) produce a sharp image of the object viewed.

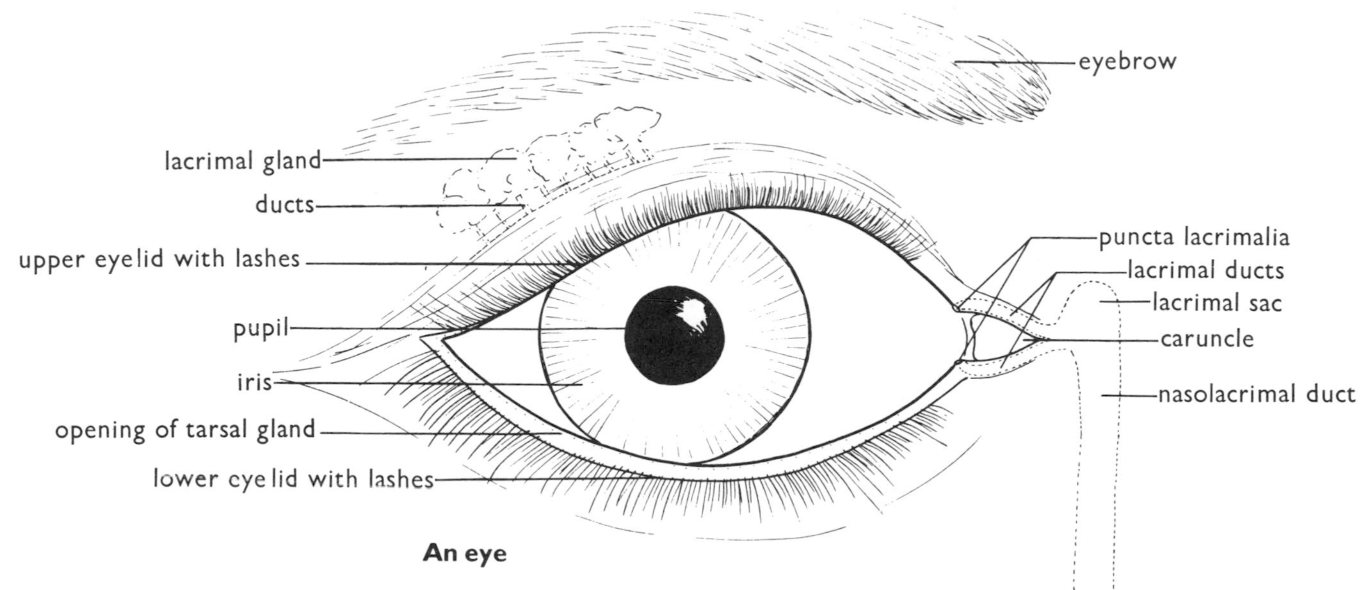

An eye

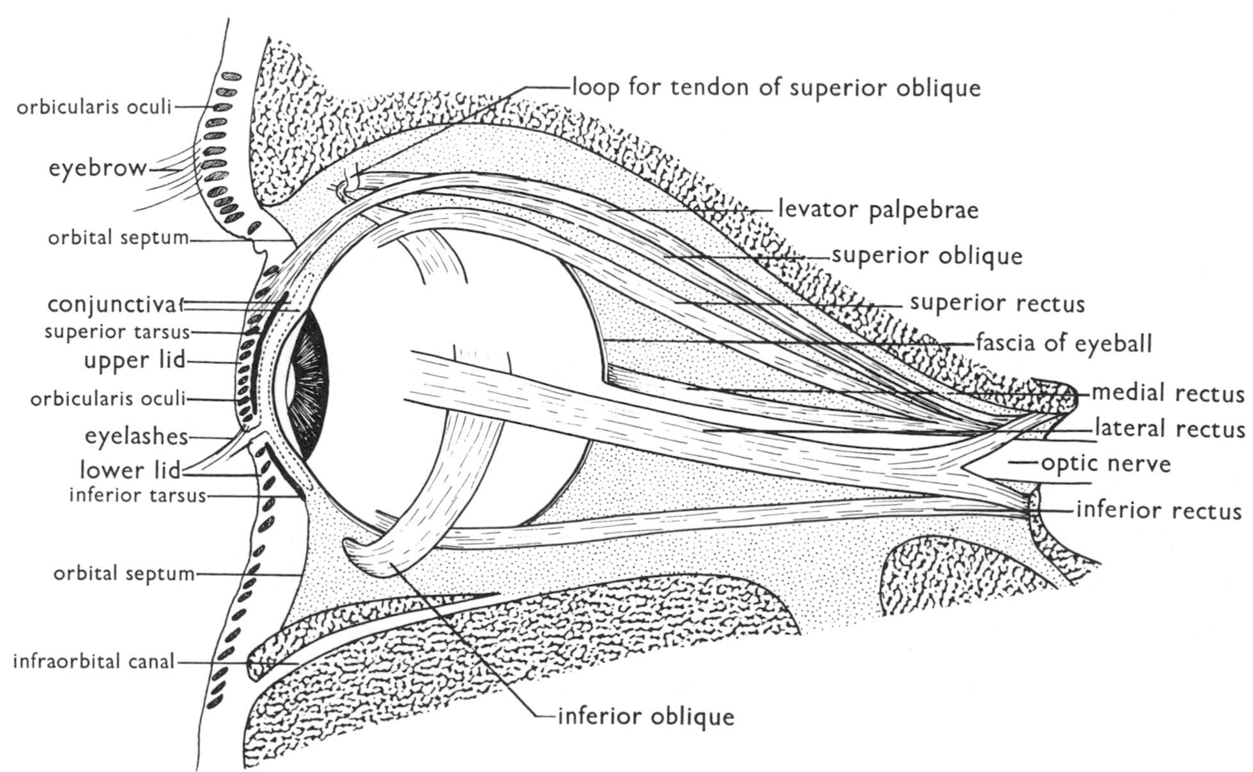

Eyeball and associated structures

The eyes

The eyes lie in bony **orbits**. Each eye is held in place and moved by four **rectus** and two **oblique muscles**. The back of the orbit is filled with fat. The front of the eye is covered with **conjunctiva**, thin transparent skin, which also lines the **eyelids**. The lids can be moved by muscles and are fringed with hairs. **Tarsal glands** secrete fluid to keep the lids from sticking together. A **lacrimal gland** lies in the outer corner of each orbit and opens by several ducts under the upper lid. The lacrimal secretion, containing lysozyme, bathes and disinfects the conjunctiva. It normally evaporates or is drained away through the **puncta lacrimalia**, **lacrimal ducts** and the **nasolacrimal ducts** into the nasal cavities behind the inferior nasal conchae—see page 12. Excess overflows as **tears**.

Each eyeball is approximately spherical but the front has a smaller radius of curvature than the rest and therefore bulges slightly. The wall of the eyeball is composed of three layers of tissue.

1. The **sclerotic** or **fibrous layer** forms a tough outer coat. It is transparent in front, forming the **cornea**, and opaque behind, forming the **sclera** into which the rectus and oblique muscles are inserted. It is continuous with the sheath round the optic nerve.

2. The **choroid** or **vascular layer** contains many blood vessels and some pigment. At the back of the eye it lines the sclera, but in front it is separated from the cornea and forms the iris and the ciliary body with ciliary processes. The **iris** is made opaque by pigment and contains radial and circular muscles which control the diameter of the pupil and thus control the amount of light entering the eye. The **ciliary body** contains circular and meridional muscles which affect the tension on the suspensory ligaments, and thus bring about accommodation of the eye to focus objects at different distances.

3. The **retina** consists of a pigmented layer and a nervous layer. The former lines the choroid right up to the edge of the iris and prevents reflection of light inside the eyeball. The latter lines the back and sides of the eyeball only and ends at the **ora serrata**. It contains the light-sensitive **rods** and **cones**. The tips of the rods contain visual purple or **rhodopsin**. Under the influence of light rhodopsin changes to **lumirhodopsin** and this by further molecular rearrangement to **metarhodopsin**, which then splits into colourless **retinine** (related to vitamin A) and a protein, **opsin**. This reaction causes depolarisation which is transmitted to the inner part of the rod and thence via relay neurones and ganglion cells to the brain. In darkness rhodopsin is regenerated from retinine and opsin. Only dim light is needed to produce the effect and the image is seen in black and white. Cones act similarly. They are of three types, each containing a special opsin and activated maximally by light of wavelength 445 nm (blue), 535 nm (green) and 570 nm (red) respectively. Brighter light is needed for colour vision and the whole visual spectrum is appreciated by proportional stimulation. Cones are concentrated at the **fovea** on the **optic axis** where vision is most acute. There are neither rods nor cones at the **blind spot** where the optic nerve leaves the eye.

Image formation

The biconvex **lens** is formed of layers of transparent cells enclosed in an elastic capsule. The space in front of the lens is filled with watery fluid, the **aqueous humour**. The space behind the lens is occupied by the semi-solid **vitreous body**, enclosed in a fine **hyaloid membrane**. This membrane continues down the hyaloid canal and forms part of the suspensory ligament which holds the lens in place. The fluids maintain the shape of the eyeball and with the lens refract light rays entering the pupil to form a focused image on the retina. At rest distant objects are in focus, but when the **ciliary muscles** contract, the tension on the suspensory ligament is reduced and the lens swells. This **accommodation** brings nearer objects into focus. The centre of the lens focuses more sharply than the periphery, therefore vision is more accurate when the pupil is small, i.e. in bright light.

Short sight or **myopia** is caused by either too curved a lens or too deep an eyeball. Long sight or **hypermetropia** is caused by either too flat a lens or too shallow an eyeball. Long sight of old age, **presbyopia**, is due to loss of elasticity of the lens, which hardens and holds a flattened shape. **Astigmatism** is caused by aberrations in the curvature of the cornea.

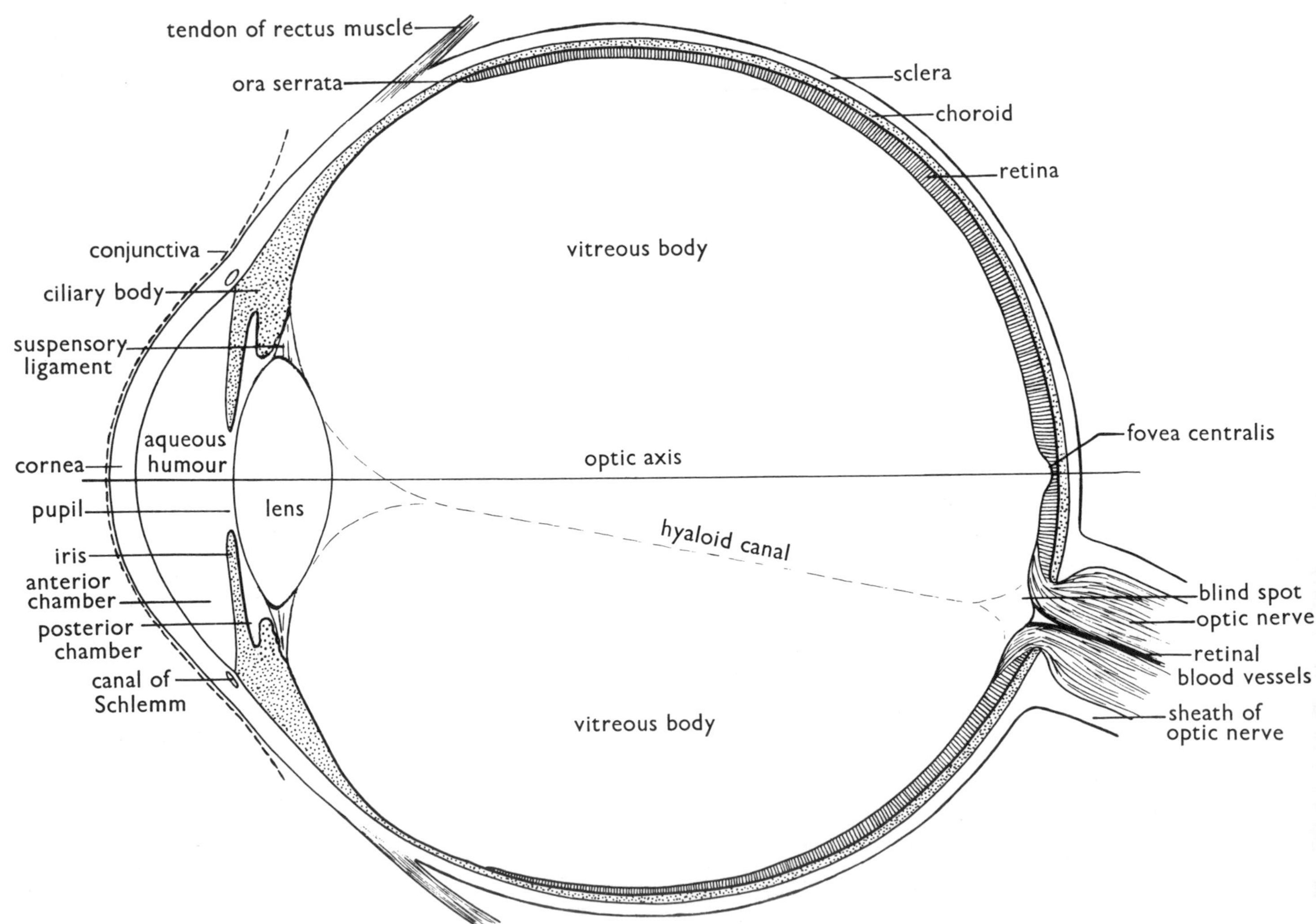

Eyeball—*V.S.*

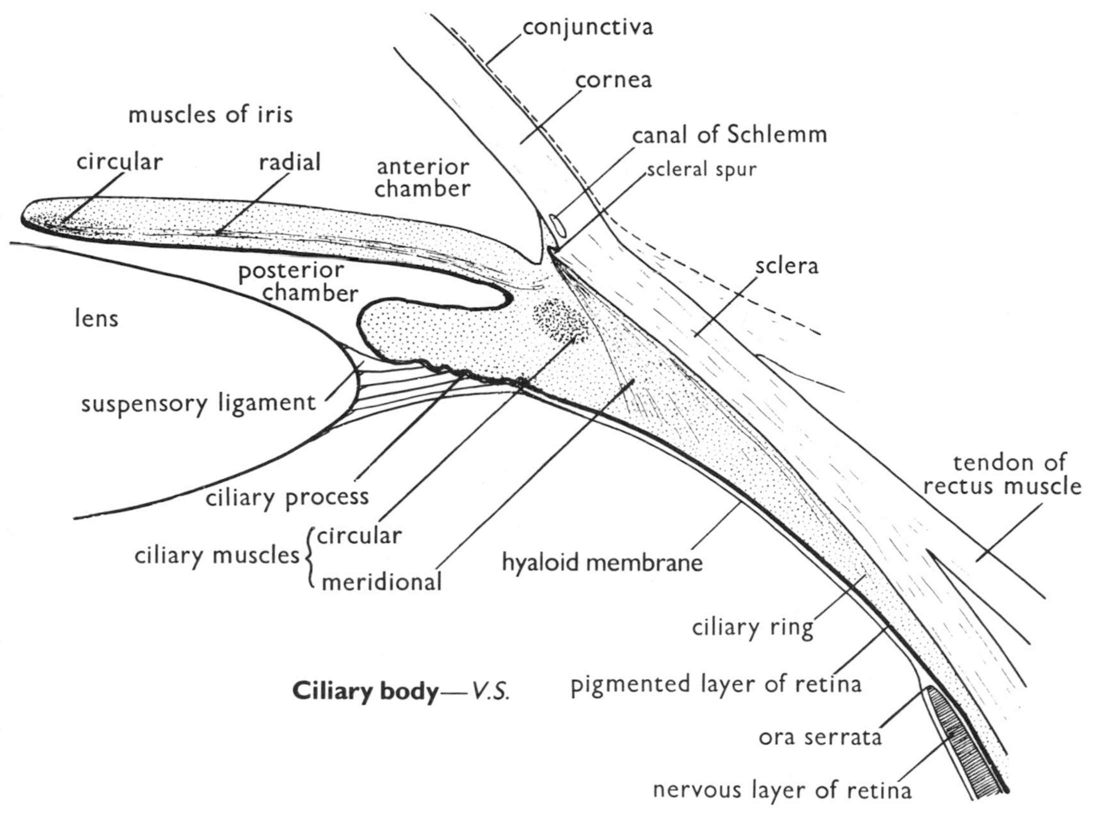

Ciliary body—*V.S.*

Binocular vision

The field of vision of each eye (i.e. the area seen by it), overlaps that of the other to a large extent, but the distance between the eyes causes objects to be viewed from slightly different angles. The brain interprets the difference between the images on the two retinae as a stereoscopic effect.

The lens system of the eye inverts the image on the retina, but as nervous transmission is not in image form, the object is seen the right way up as a learned experience. The nervous connections from the retina are such that the stimuli from the right half of the visual field are relayed to the visual cortex of the left cerebral hemisphere and vice versa.

All visual impulses are relayed in the **lateral geniculate bodies** and there are connections from these to the superior colliculi—see page 62, and thence to the nuclei of origin of the III, IV and VI cranial nerves—see page 64. These nerves contain involuntary motor fibres which control the ciliary muscles and muscles of the iris and also voluntary motor fibres which control the muscles of the orbit. Thus the light rays falling on the retina set up reflexes which control accommodation, the size of the pupil and maintenance of correct overlap of the visual fields. The retinae also have links, via the mid-brain, with the pineal body—see page 123. This is a vestige of the reflex path whereby light affects colouration in some vertebrates.

area seen by left eye only
area seen by both eyes
area seen by right eye only
optic nerve
optic chiasma
nerves with impulses from right half of visual field
nerves with impulses from left half of visual field
oculomotor ganglia
lateral geniculate body
superior colliculus
optic radiation
area striata (visual cortex)

Visual path

pigmented layer
rods
layer of rods and cones
cone
outer limiting membrane
outer nuclear layer (nuclei of rods and cones)
synapses
outer plexiform layer
relay cell
inner nuclear layer
inner plexiform layer
synapses
ganglionic layer
ganglion cell
layer of nerve fibres
inner limiting membrane

Structure of the retina

The Skin

The skin covers the entire outside of the body. It is formed of two types of tissue—surface **epidermis** and underlying **dermis**.

The **epidermis** is stratified (layered) epithelium (surface tissue). Identifiable layers are: (a) **stratum germinativum** made up of **basal cells** capable of cell division, and densely packed **prickle cells**, so called because they shrivel easily in stained preparations; (b) **stratum granulosum** in which cells become flattened, filled with granules of **keratohyalin**, and die; (c) **stratum lucidum** where the keratohyalin in the dead cells becomes translucent **eleidin**; (d) **stratum corneum** where, under the influence of secretion from lysosomes, the eleidin has become **keratin**, a horny material, scales of which are continuously rubbed off. Renewal of the cornified layer comes from the germinative layer. Friction stimulates this so that the epidermis becomes thicker where it has to withstand wear, e.g. on the soles of the feet and, when much used, the palms of the hands. The cornified layer of the epidermis is very slightly permeable to CO_2, but largely waterproof.

dead horny cells
granular cells
prickle cells
basal cells
stratum corneum
stratum lucidum
stratum granulosum
zona cornea
stratum germinativum—zona germinativa

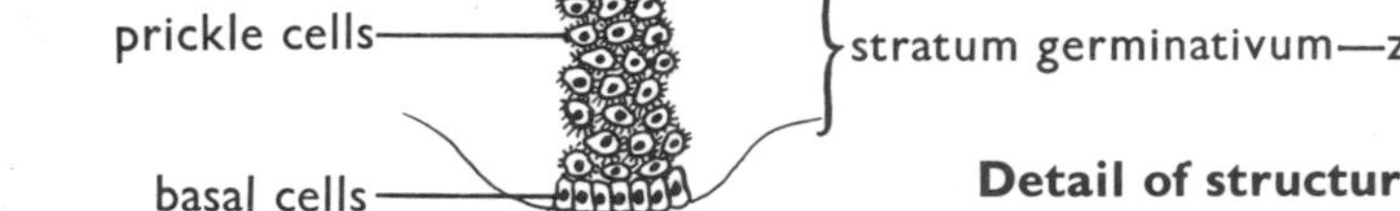

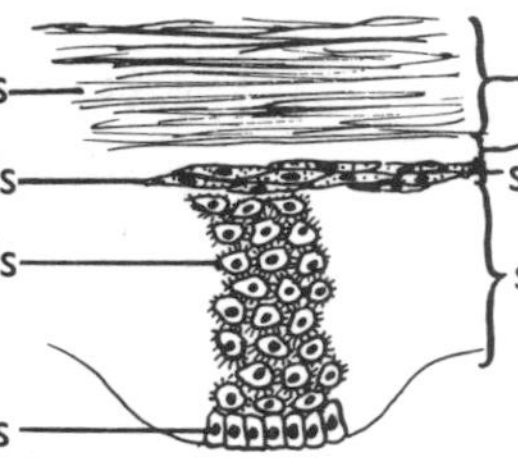

Detail of structure of epidermis

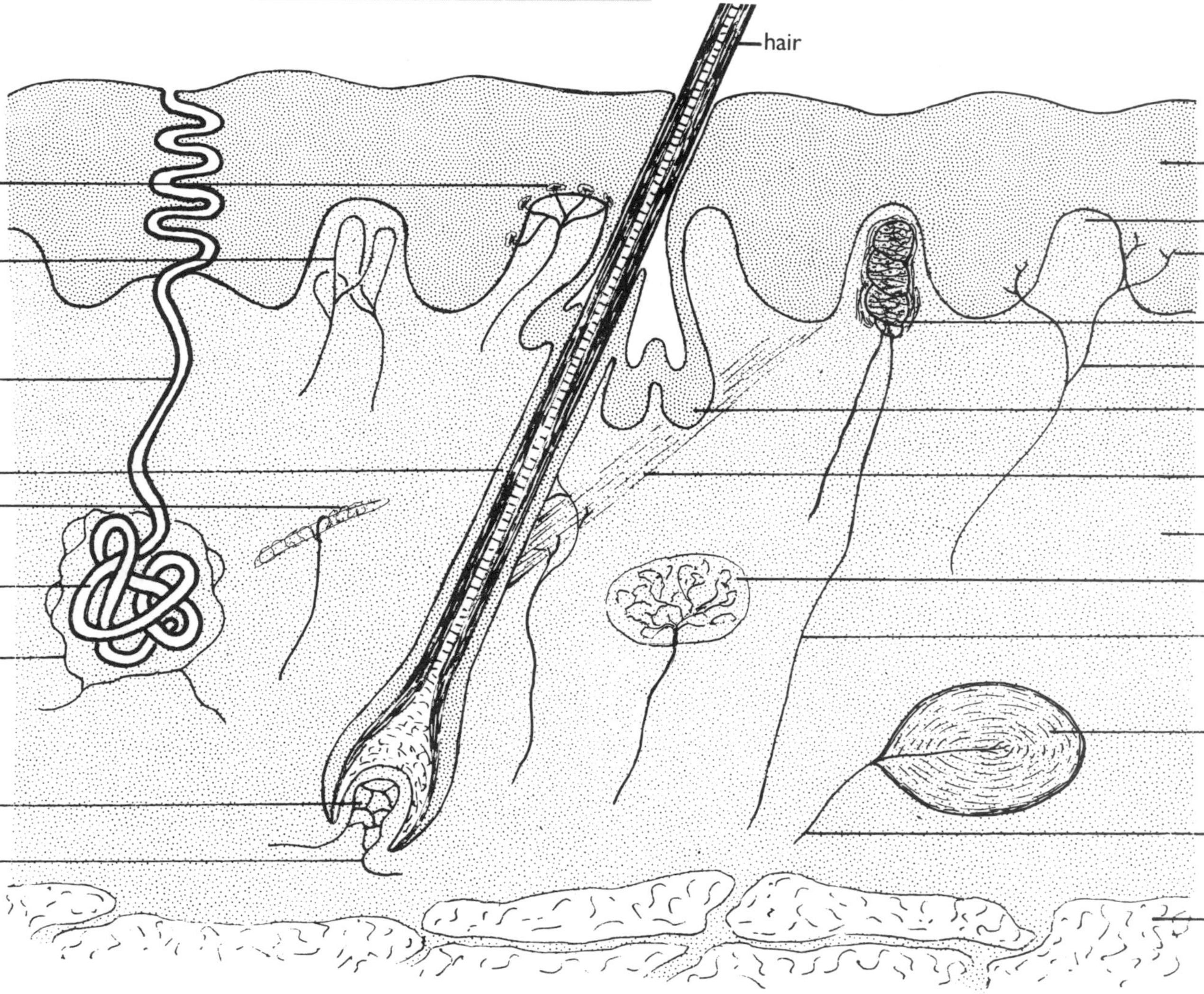
hair
Merkel's discs
capillaries near surface of dermis
duct of sweat gland
hair follicle
Ruffini corpuscle
coiled sweat gland
capillaries
hair papilla
capillaries
epidermis
dermal papilla
nerve endings
Meissner corpuscle
nerve fibre
sebaceous gland
erector pili muscle
dermis
Krause corpuscle
nerve fibre
Pacinian corpuscle
nerve fibre
subcutaneous layer with fat and loose connective tissue

Skin—*V.S.* (*highly diagrammatic*)

The epidermis protects the body against friction, water loss, and entry of germs. Its pigmentation shields deeper-seated tissues from ultraviolet light. The pigment, **melanin**, is formed in **melanocytes** between or just below the basal cells and passed through processes to the prickle cells, which take it up by phagocytosis. The amount of pigment varies with race, exposure to sunlight and production of a melanocyte-stimulating hormone (MSH)—see page 122. Some races also have **carotene** in the cornified layer, giving a yellowish colour. Albinos have no pigment.

The **dermis** connects the epidermis to the underlying structures. It consists of dense, tough but elastic, connective tissue with nerves and blood vessels. Some of the nerve endings lie in the epidermis, others are associated with hairs. They may be naked or have special corpuscles—see page 73. The blood vessels form an extensive fine capillary network in which the flow of blood is variable—see below. Beneath the dermis there is a **subcutaneous layer** of looser connective tissue with fat deposits in **adipose cells**. This gives insulation against heat loss and serves to a lesser extent as storage and padding.

The boundary between the epidermis and the dermis is undulating with protruding **dermal papillae** which help to hold the layers together. The epidermis also dips to line the hair follicles, sebaceous glands and sweat glands.

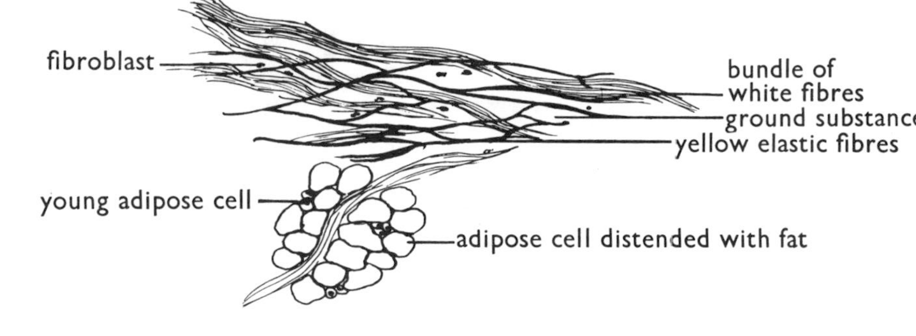

Connective tissue from dermis and subcutaneous layer

NAILS AND HAIRS

Nails and hairs are formed of the same horny material, keratin, as the surface of the epidermis. Nails develop from a specially active germinative layer called the **nail bed** and are pushed out from under a **cuticular fold**. Hairs grow from **follicles** at the base of each of which is an active germinative layer associated with a **hair papilla**, in which lies a knot of blood vessels. As the diagram shows the hair cells are arranged in layers. The cortical cells may contain pigment which gives the hair its colour.

The hairs in most mammals give insulation by entangling pockets of air, but in man they are too sparse and short over most of the body to be much use. The small **erector pili** muscle attached to each hair is a vestigeal structure retained from the time when there was a complete hairy coat and the hairs could be made to stand on end with cold, fear or anger, thus deepening the layer of air and making the ancestral (pre-human) animal look larger. Nerve fibres associated with the hairs are sensitive to hair movements.

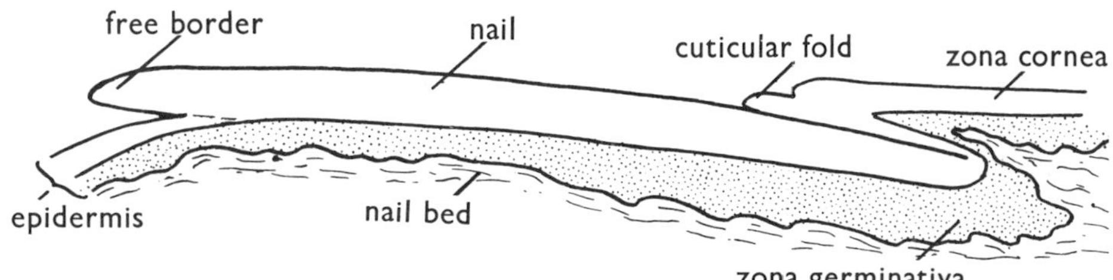

Nail—*L.S.*

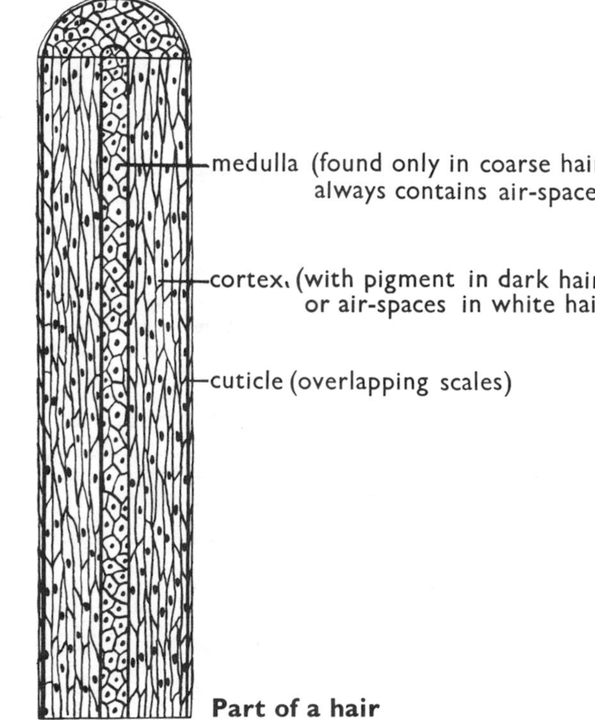

Part of a hair

SEBACEOUS AND SWEAT GLANDS

Sebaceous glands are often associated with hair follicles. They produce an oily secretion called **sebum**, which prevents the hairs and epidermis from becoming brittle and adds waterproofing. The **mammary glands**, which form milk after childbirth—see page 131, and the **ceruminous glands**—see page 74, are specialised sebaceous glands.

The sweat glands produce a slightly acid, watery solution of salts and nitrogenous waste and are therefore excretory, but the chief function of **perspiration** (**sweating**) is to help to regulate body temperature.

BODY TEMPERATURE

The average core body temperature taken under the tongue is 37 °C. Temperature in the limbs may be significantly lower, and the deep-seated active organs, e.g. the liver, may have somewhat higher temperatures, though the differences are kept minimal by continual circulation of the blood—see page 103.

If the body temperature deviates beyond the limits acceptable in the diurnal rhythms, centres in the hypothalamus are activated and set up reflex responses. The body temperature may tend to rise (a) because of external conditions, (b) with intake of hot food and drink, (c) with muscular activity, during which only about 25% of the energy is used for muscle contraction and the other 75% is liberated as heat, (d) due to chemical activity for general metabolic purposes, e.g. heat release during oxidative breakdown of glucose. When a rise above normal limits is sensed by the hypothalamus, autonomic responses dilate the blood vessels of the skin so that more blood can circulate near the surface of the body to be cooled by simple radiation from the unclothed parts. At the same time the sweat glands are activated and watery secretion passes onto the skin surface where it evaporates. The heat required for this evaporation greatly increases the cooling effect. Failure of this evaporation due to tight-fitting and impermeable clothing or a water-saturated environment can result in an unpleasant and eventually dangerous rise in body temperature, similar to the effects of fever produced by toxins. Emotions can also produce hypothalamic response with effects similar to those of raised body temperature, e.g. fear—cold sweat; rage—'hot under the collar' or red with anger; embarrassment—blushing. If the body temperature tends to fall, the sweat glands cease to function and the blood vessels of the skin contract so that less blood is exposed to cooling effects. Skin from which blood is retracted may become bluish or pale with cold. Shivering may occur in an attempt to generate internal warmth. Gooseflesh is a vestigeal response by the erector pili muscles—see above. If the condition continues for too long, it may cause localised frostbite in the extremities or general hypothermia with slowing of all the body functions.

The Trunk

The trunk has a framework of bone, cartilage and muscle surrounding two cavities within which lie the viscera. The cavities are separated from one another by a muscular partition called the diaphragm—see pages 58 and 59. The upper cavity with its walls constitutes the **thorax** and the lower cavity with its walls is the **abdomen**.

THE THORAX

The walls of the thorax are formed by the thoracic vertebrae, sternum, ribs and costal cartilages and the intercostal muscles. The floor of the thorax is formed by the diaphragm.

The **thoracic cavity** is conical in shape and contains:

1. the heart and roots of the great blood vessels;
2. the lungs, bronchi and part of the trachea;
3. part of the oesophagus;
4. the thoracic duct;
5. parts of the sympathetic cords and vagus nerves.

THE ABDOMEN

The walls of the abdomen are formed by the lumbar vertebrae, parts of the lower ribs, the abdominal muscles and the pelvis. The roof of the abdomen is formed by the diaphragm.

The **abdominal cavity** is ovoid and its lower end opens into the **pelvic cavity**, which lies below the brim of the true pelvis and has the levator ani muscle for its floor. For convenience in describing the location of the abdominal viscera the abdomen is divided by four imaginary lines into nine regions.

The abdominal cavity contains:

1. the stomach, small intestine and large intestine except part of the pelvic colon and the rectum;
2. the liver and pancreas;
3. the spleen;
4. the kidneys and the greater part of each ureter;
5. the abdominal aorta and the greater part of the inferior vena cava;
6. the lower parts of the sympathetic cords.

The pelvic cavity contains:

1. part of the pelvic colon and the rectum;
2. the lower ends of the ureters, the bladder and the urethra;
3. the ovaries, Fallopian tubes, uterus and vagina of the female and the vasa deferentia and seminal vesicles of the male.

Note. During pregnancy the uterus rises into the abdominal cavity.

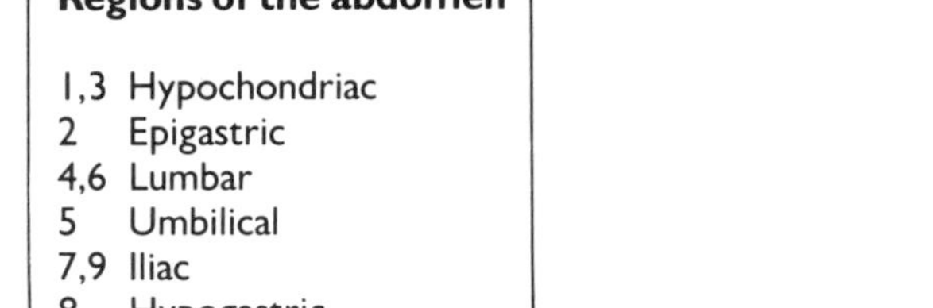

Regions of the abdomen

1,3	Hypochondriac
2	Epigastric
4,6	Lumbar
5	Umbilical
7,9	Iliac
8	Hypogastric

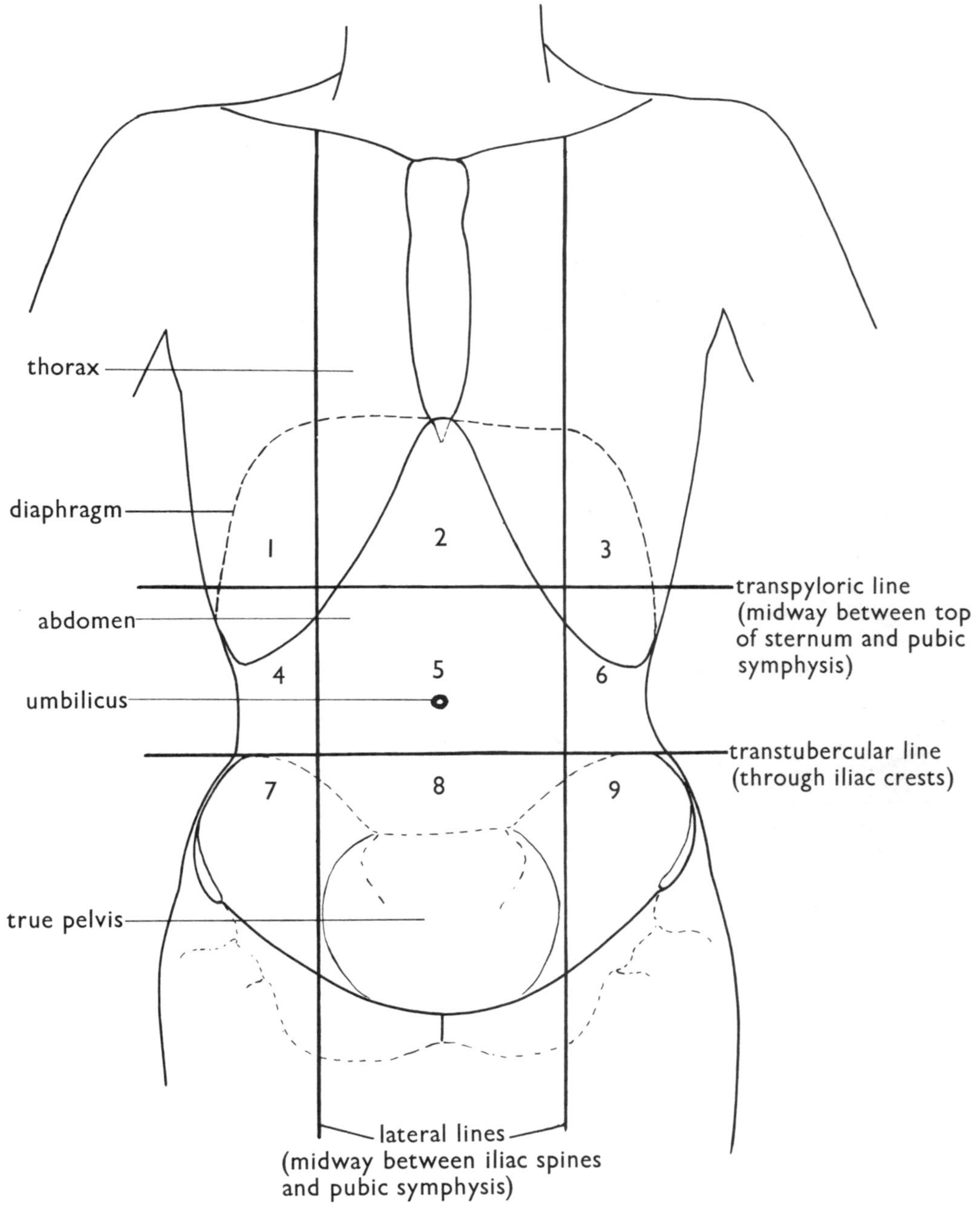

thorax
diaphragm
abdomen
umbilicus
true pelvis
1
2
3
4
5
6
7
8
9
transpyloric line (midway between top of sternum and pubic symphysis)
transtubercular line (through iliac crests)
lateral lines (midway between iliac spines and pubic symphysis)

Trunk

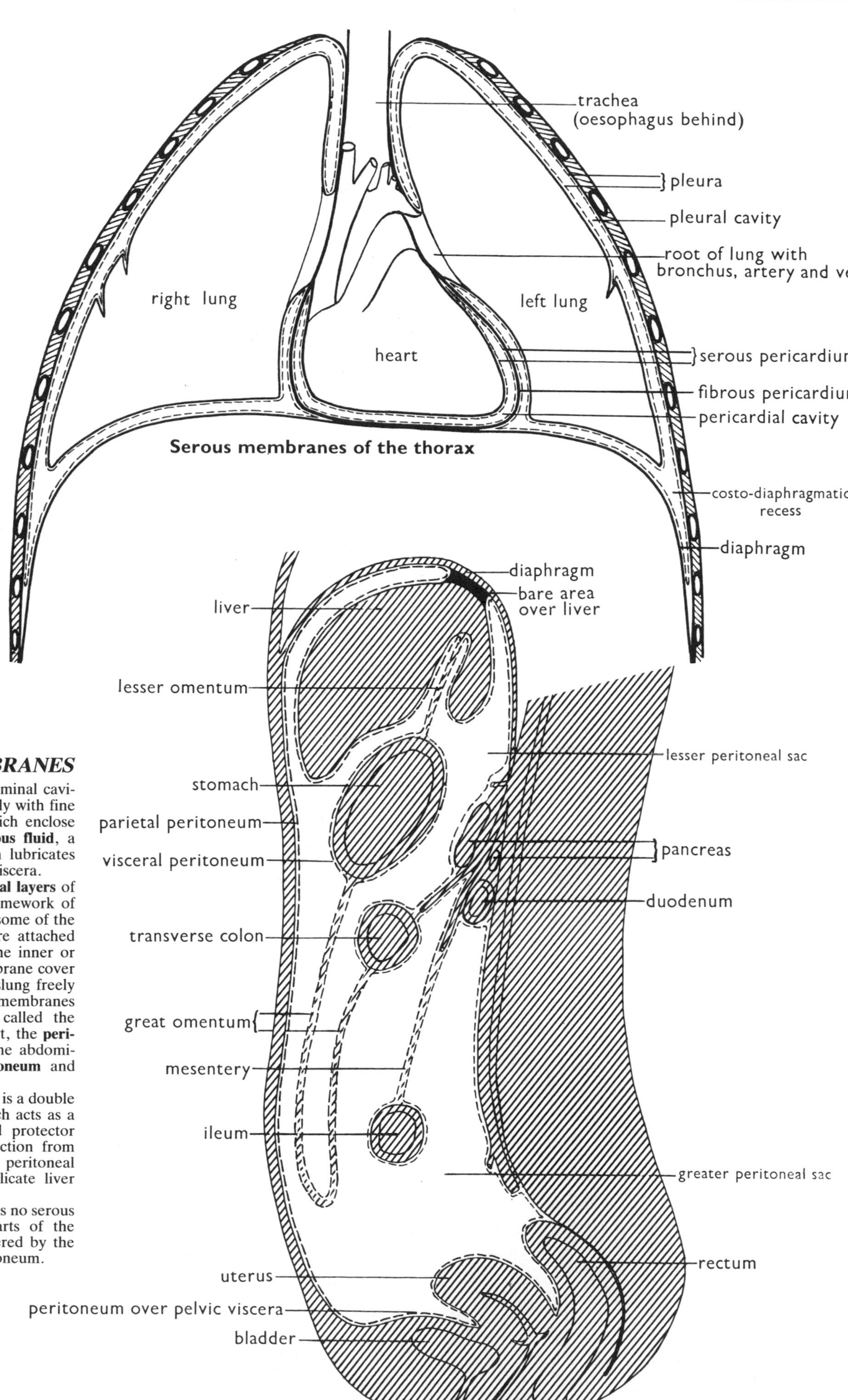

Serous membranes of the thorax

Serous membranes of the abdomen

SEROUS MEMBRANES

The thoracic and abdominal cavities are lined completely with fine serous membranes which enclose spaces filled with **serous fluid**, a form of lymph, which lubricates the movements of the viscera.

The outer or **parietal layers** of membrane line the framework of the cavities and cover some of the organs so that they are attached firmly to the walls. The inner or **visceral layers** of membrane cover the organs which are slung freely in the cavities. The membranes round the lungs are called the **pleura**, round the heart, the **pericardium**, and round the abdominal viscera, the **peritoneum** and **mesenteries**.

The **great omentum** is a double fold of mesentery which acts as a storehouse of fat and protector against spread of infection from the intestines into the peritoneal cavity, around the delicate liver and pancreas.

The pelvic cavity has no serous spaces. The upper parts of the pelvic viscera are covered by the lowest part of the peritoneum.

The Digestive System

The digestive system is concerned with nutrition, which involves ingestion, digestion, absorption and assimilation.

Ingestion of food requires opening and closing of the **mouth** by the jaw muscles aided by facial muscles which control the lips—see pages 42 and 43.

Digestion, by which food is broken down, involves two processes:

1. **mechanical digestion** whereby solid food is chewed and churned into small pieces;
2. **chemical digestion** whereby large molecules in the food are split by **hydrolysis** into smaller ones. Hydrolysis is catalysed by **digestive enzymes**, which are highly specific and very sensitive to conditions, especially pH. As a result of hydrolysis proteins are split to shorter-chain polypeptides, known as proteoses and peptones, and these in turn to dipeptides and finally amino acids. Similarly fats are split into fatty acids and glycerol, and polysaccharides, e.g. starch, are split into disaccharides and then monosaccharides (hexose sugars). For a full table of digestive enzymes see page 94.

Absorption is the process whereby dissolved substances pass from the alimentary canal (gut) into the blood stream.

Assimilation is the utilisation of the absorbed nutrients for the metabolic processes of the living cells.

The **feeding reflex** is centred in the mammillary body—see page 62. Food ingested at the mouth is digested and absorbed during passage through the alimentary canal. Undigested residues are **ejested** (pass out) through the **anus** as **faeces**.

The **alimentary canal** or **gastro-intestinal tract** is a continuous tube about 9 m long in the average adult. Most of its length is coiled up in the abdominal cavity. It is divided into the following regions:

1. buccal cavity;
2. pharynx;
3. oesophagus;
4. stomach;
5. small intestine—duodenum, jejunum and ileum;
6. large intestine—caecum, appendix, colon and rectum.

Associated with the alimentary canal, and having ducts leading into it, there are three pairs of large salivary glands, the pancreas and the liver.

THE BUCCAL CAVITY

The buccal cavity is supported by the jaws in which the teeth are set in sockets—see page 17. These separate the space inside the lips and cheeks, the **vestibule**, from the **oral cavity** proper. The muscular tongue—see page 43—lies on the floor of the oral cavity, partially held down by an under-tongue fold, the **frenulum**. The oral cavity is separated from the nasal cavities by the **palate**, the front part of which is stiffened by bone. The pendulous **uvula** hangs from the back edge of the **soft palate**.

The buccal cavity is lined with stratified epithelium continuous with that of the skin and like it capable of growth to replace sloughed-off surface cells. These cells become flattened but are not cornified. There are sensory **taste buds** on the tongue and palate—see page 74.

In the buccal cavity food is chewed by the teeth, turned over and mixed with saliva by the tongue. Certain dissolved substances may be tasted, while other flavours may be sensed from their smell both before and after entering the mouth. These sensations set up reflexes which stimulate the **salivary glands** to increase their output. **Saliva** is very slightly acid. It contains bicarbonates, which act as buffers, and the enzyme **salivary amylase**, which starts digestion of starch. It softens the food mass or **bolus** which is then swallowed.

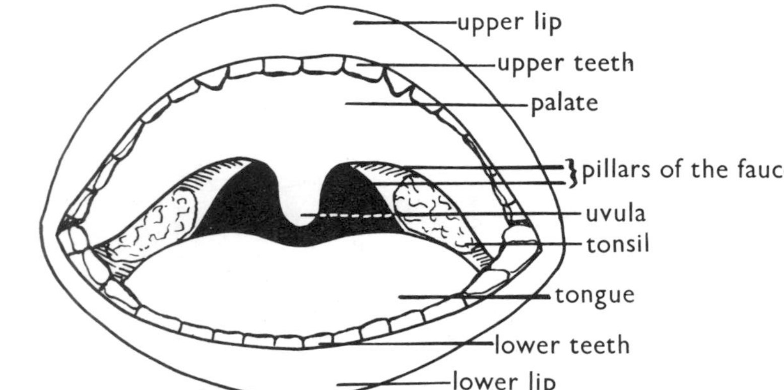

View into the mouth

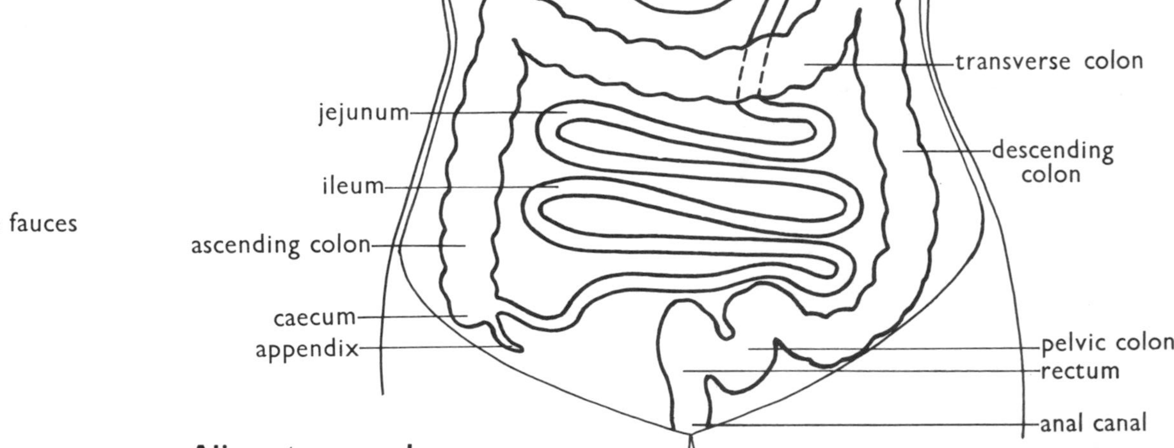

Alimentary canal

THE PHARYNX

The pharynx is divided into the **naso-pharynx** at the back of the nose and the **oro-pharynx** at the back of the oral cavity. The oro-pharynx is continuous with the oesophagus and has an opening, the **glottis**, into the **larynx**—see page 96. The glottis has a cartilage-stiffened lid, the **epiglottis**.

The **tonsils** are special large patches of lymph tissue—see page 115—which lie at the sides of the pharynx between folds called the **pillars of the fauces**. **Adenoids** are similar patches in the naso-pharynx.

Swallowing is initiated by the tongue pressing against the palate. As the bolus passes, the soft palate closes off the naso-pharynx and the larynx is raised and pulled forwards so that the epiglottis lies over the glottis and the air passages are closed. Thus food is normally prevented from going the wrong way.

From the pharynx onwards the alimentary canal is a simple tube, whose walls are made up of four coats: (a) an outer fibrous or serous coat; (b) a muscular coat for the most part made up of smooth muscle with the outer layers of fibres longitudinal and the inner layers circular; (c) submucosa of connective tissue with many blood vessels; (d) mucosa consisting chiefly of connective tissue with blood vessels, but separated from the submucosa by a muscle layer called the muscularis mucosae and having epithelium for its surface layer.

THE OESOPHAGUS

The oesophagus is a straight tube through the neck and thorax, close behind the trachea or wind pipe, but in front of the descending part of the main blood vessel, the aorta. It passes through the oesophageal opening of the diaphragm—see page 59. Its distinguishing features are:

1. stratified epithelium for its mucous membrane (continuous with that of the pharynx);
2. numerous longitudinal folds which allow stretching to accommodate the bolus;
3. relatively thick, strong muscularis mucosae;
4. striated muscles for the upper two-thirds of its length which carry through the swallowing movement initiated in the oral cavity.

Regular waves of muscular movement called **peristalsis** sweep the food downwards, involving contraction and relaxation of antagonistic groups of longitudinal and circular muscle. Occasional antiperistalsis produces vomiting, which is aided by contraction of the abdominal muscles and the diaphragm—see pages 56 to 59.

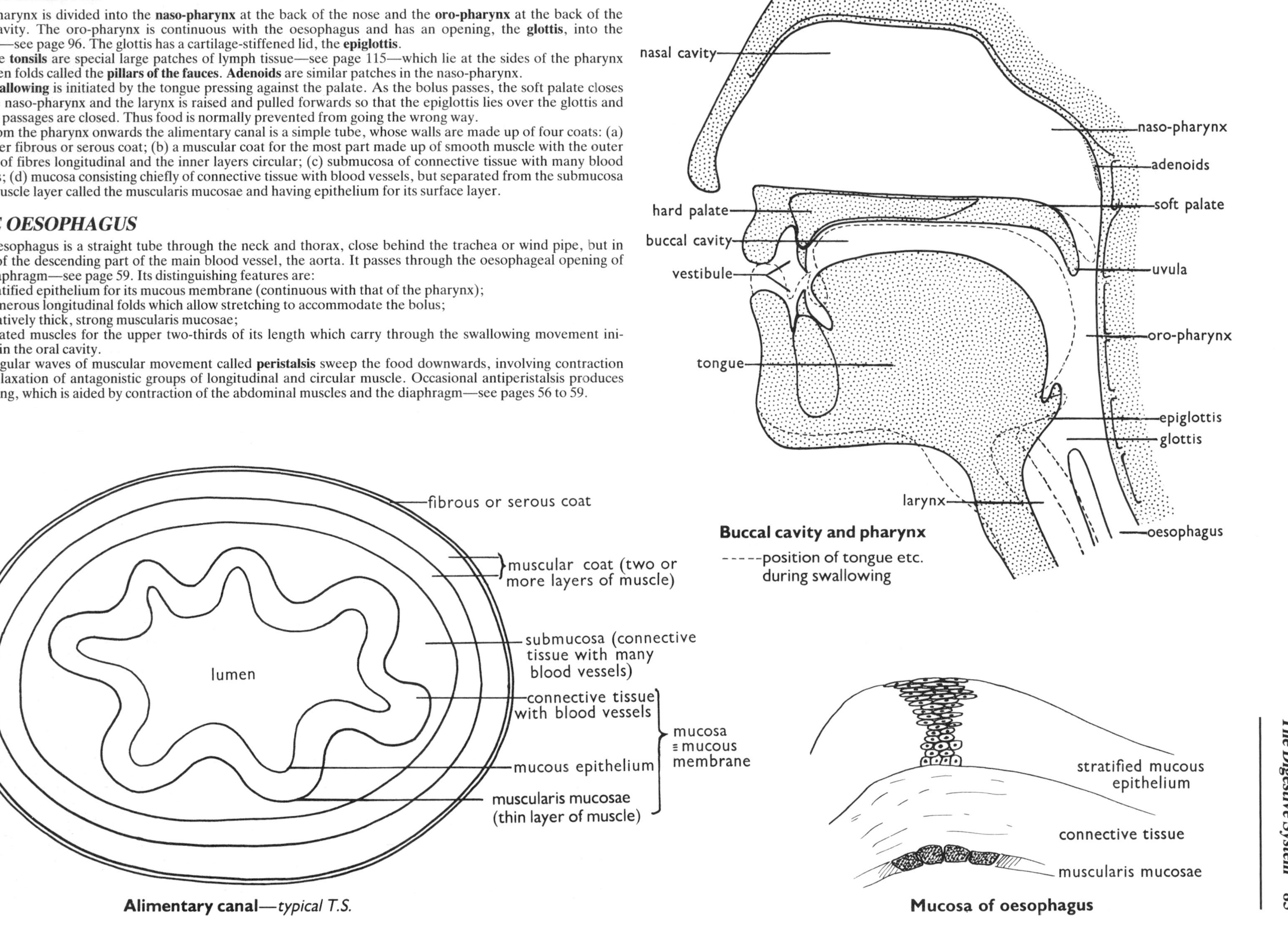

Buccal cavity and pharynx

Alimentary canal—*typical T.S.*

Mucosa of oesophagus

THE STOMACH

The stomach lies close below the diaphragm to the left of the liver. It is suspended by its mesentery, a fold of which hangs down as the great omentum—see page 83. It varies in shape and has an average capacity of about 1 litre. It is capable of considerable distension, sending stimuli to the hypothalamus—see page 62, where the **hunger centre** and the **satiety centre** control desire to eat. The opening from the oesophagus is guarded by the **cardiac sphincter** and the opening to the duodenum by the **pyloric sphincter**.

The wall of the stomach is characterised by:

1. a layer of oblique muscle in addition to the circular and longitudinal muscle;
2. temporary folds of the mucosa called **rugae**, which flatten out as the stomach fills;
3. unstratified (single-layered) lining epithelium;
4. numerous tubular **gastric glands**, which vary in detail in the different regions—see diagrams.

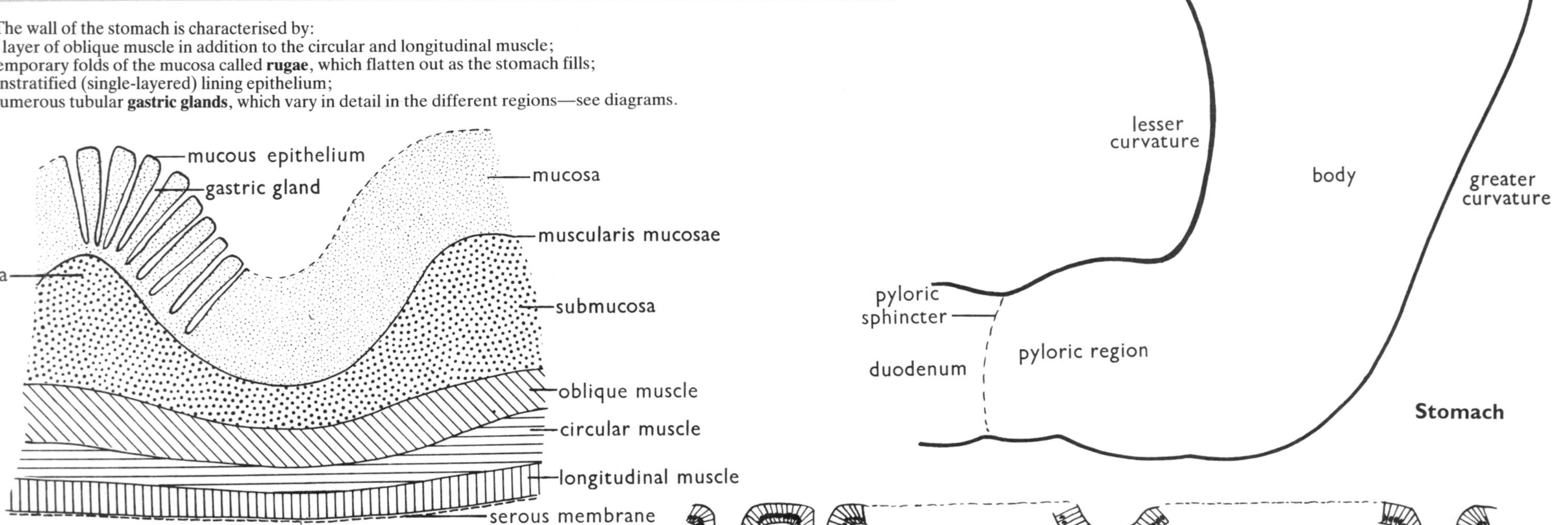

Stomach—*T.S.*

oesophagus, cardiac sphincter, cardiac region, fundus, lesser curvature, body, greater curvature, pyloric sphincter, duodenum, pyloric region

Stomach

goblet cell, tubular gland, duct, racemose gland, parietal or oxyntic cells, central or peptic cells, argentafin cell, duct, mucoid cells

(a) cardiac glands from cardiac region

(b) fundus gland from fundus or body

(c) pyloric glands from pyloric region

Diagrams of gastric glands

Functioning of the stomach

1. **Churning**. The food mass is held for a while in the fundus and passed little by little to the **body** and **pyloric region**, where rippling movements complete mechanical digestion and mix the particles thoroughly with gastric juice. The product called **chyme** is allowed through the pyloric sphincter in small spurts.

2. **Mucus** is produced by cardiac and pyloric glands and by neck cells of fundus and body glands. It lubricates passage of the food and protects the lining of the stomach from its own digestive enzymes.

3. **Gastric juice** is secreted by glands of the fundus and body. Secretion is initiated by the vagus nerve as a reflex response to sight, smell, taste and even thought of food. It is increased by further neural stimulation in response to pressure of food in the stomach and by hormonal stimulation by gastrin. Gastric juice contains:

(a) **pepsinogen**, an inactive enzyme precursor, produced by the central or peptic cells;
(b) **hydrochloric acid**, from oxyntic or parietal cells, which curdles milk and, when churned into the food mass, stops the action of salivary amylase and provides the acid conditions, pH = 2, needed for activation of pepsinogen to **pepsin**, which splits proteins to proteoses and peptones;
(c) **intrinsic factor** involved in absorption of vitamin B_{12} by the intestine

4. **Gastrin** is a hormone, produced from isolated cells in the glands of the fundus and body, as a reflex response to pressure of food in the stomach and chemical stimulation by proteoses and peptones (**secretagogues**). It circulates in the blood and stimulates the gastric glands to greater activity. It also contracts the cardiac sphincter, mildly increases movement of the alimentary canal, and relaxes the pyloric sphincter and the ileo-caecal sphincter—see next page.

5. **Serotonin** is also a hormone produced by isolated cells, **argentaffin cells**, mainly in the fundus glands. It is vasoconstrictive—see page 110.

6. Other digestive enzymes which may be active in the stomach are:
(a) **amylase** from the salivary glands, which goes on acting in the food mass until it is mixed with HCl and may complete digestion of up to 70% of the starch eaten;
(b) **gastric lipase** of unknown source, which is scarcely active in the adult at the very low pH necessary for pepsin activity, but is useful in hydrolysing fat in the stomach of the infant before full peptic digestion is established;
(c) **rennin**, which is found in the stomach of many young mammals, but there is doubt as to whether it is produced in man. In the presence of calcium, it curdles milk.

7. **Absorption** in the stomach is slight. The stomach wall is impermeable to most substances, but takes up some water, electrolytes, certain drugs and alcohol.

8. **Gastric digestion** is slowed down by high fat content of food, but the stomach normally empties in 2–6 hours, after which gastric secretion is inhibited.

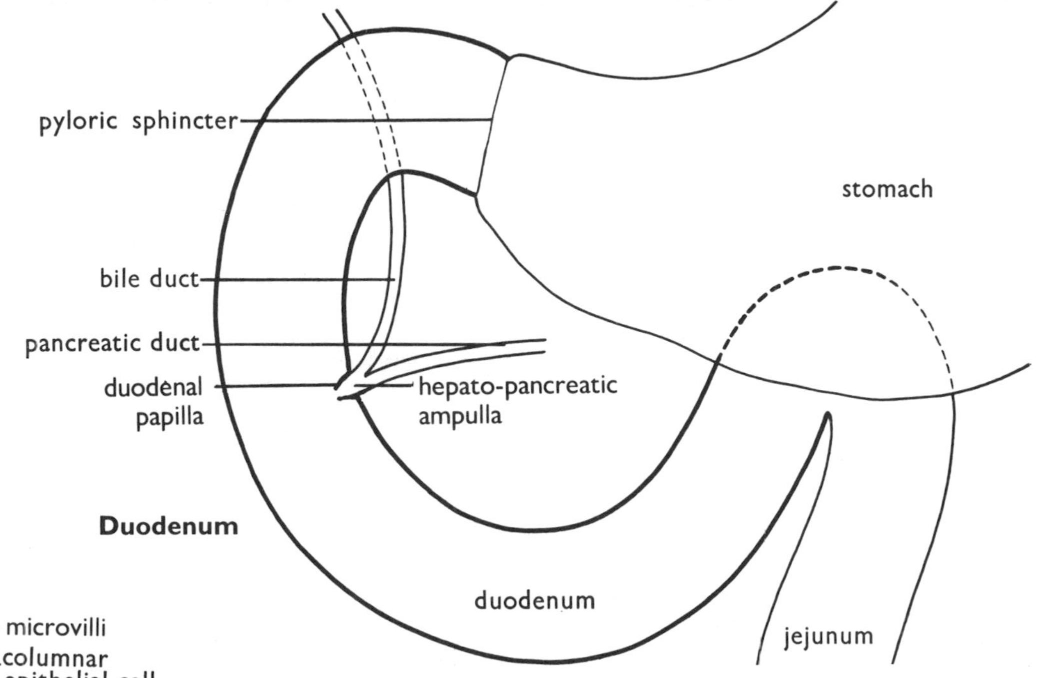

Duodenum

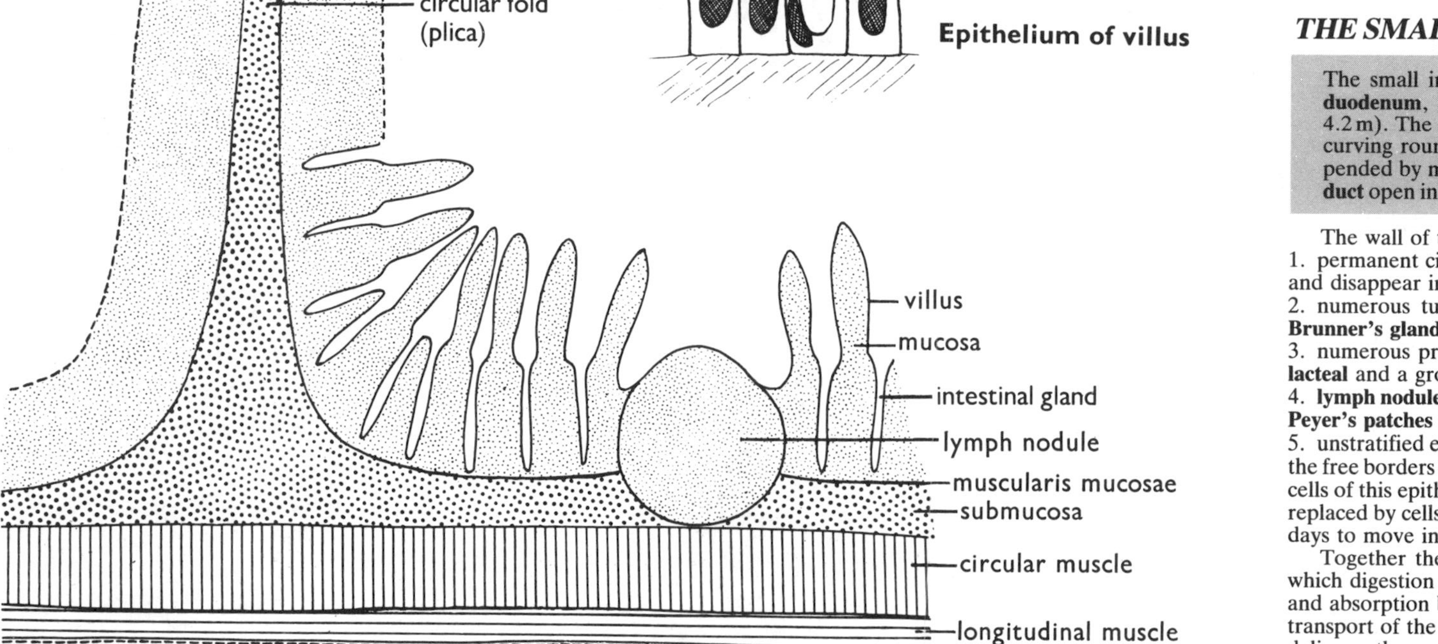

Epithelium of villus

Diagrammatic longitudinal section through part of the small intestine

THE SMALL INTESTINE

The small intestine is about 7 m long. The first 240–290 mm is **duodenum**, the rest is **jejunum** (about 2.5 m) and **ileum** (about 4.2 m). The duodenum lies against the back wall of the abdomen, curving round the pancreas, but the jejunum and ileum are suspended by mesentery—see page 83. The **bile duct** and **pancreatic duct** open into the duodenum together.

The wall of the small intestine is relatively thin. It has:
1. permanent circular folds of the mucosa called **plicae**, which decrease and disappear in the ileum;
2. numerous tubular **intestinal glands** throughout and small rounded **Brunner's glands** in the duodenum only;
3. numerous projections called **villi**, each with a lymph vessel called a **lacteal** and a group of **blood capillaries** in its core;
4. **lymph nodules**, which increase in number and become aggregated into **Peyer's patches** in the ileum;
5. unstratified epithelium with numerous **goblet cells** and columnar cells, the free borders of which have up to 1700 **microvilli**—see page 5. Exposed cells of this epithelium are short-lived. After only 1–2 days' use, they are replaced by cells from the depths of the intestinal glands, which take 2–4 days to move into place.

Together the above features form (a) a long, very flexible tube in which digestion can proceed, (b) a vast lining area for further digestion and absorption by the cells, and (c) a large lymph and blood supply for transport of the nutrient products to the rest of the body. In spite of its delicacy, the small intestine is self-protective with much mucus and rapid replacement of potentially damaged cells.

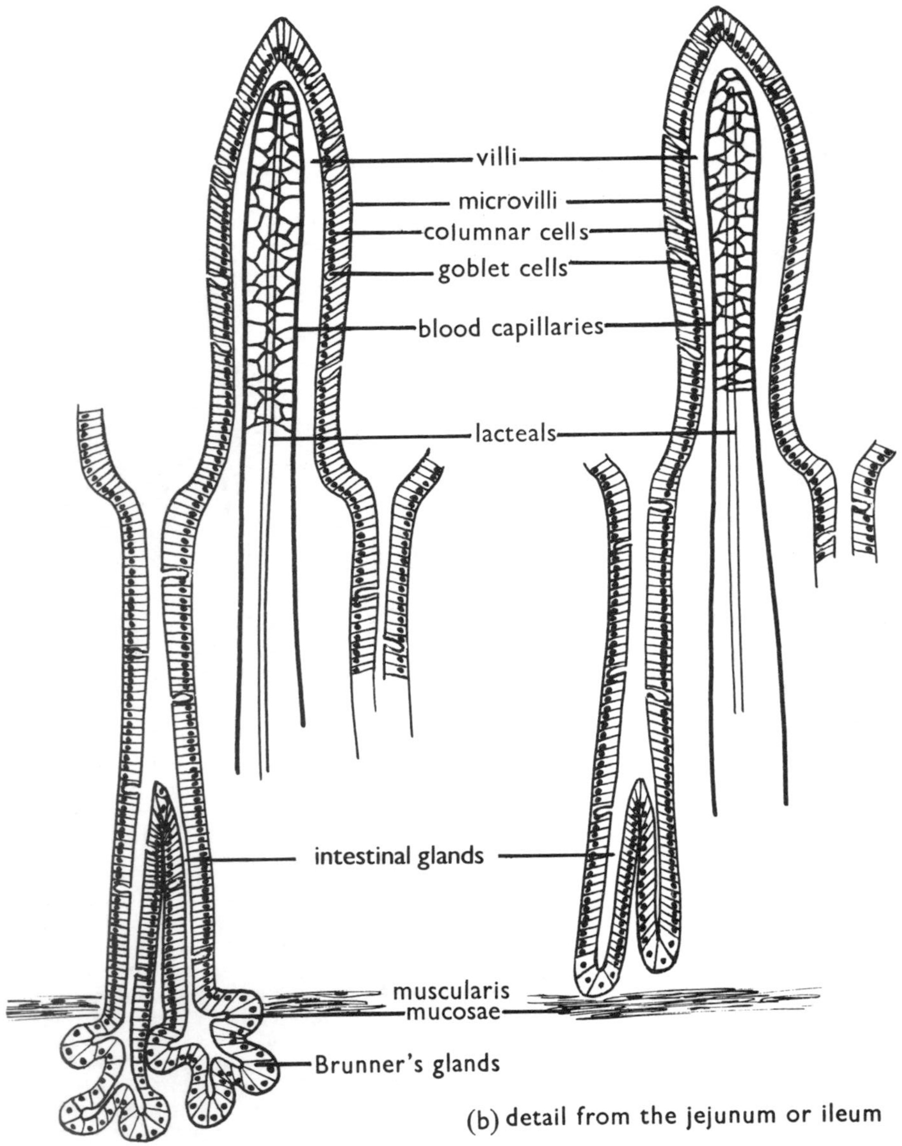

(a) detail from the duodenum

(b) detail from the jejunum or ileum

Glands and villi of the small intestine

Functioning of the small intestine

1. **Peristaltic**, **antiperistaltic** and **occlusive movements**, produced as a reflex response to the presence of food, rock the semi-fluid contents of the intestine to and fro and force it against the villi. Localised strong contractions of circular muscle produce the appearance of a string of sausages. Contractions are repeated 12–16 times a minute.

2. **Chyme** entering the duodenum from the stomach is rapidly mixed with **bile**, **pancreatic juice** and **intestinal juice**. The bicarbonate in these fluids buffers the HCl, neutralising it and bringing the pH to a level at which the intestinal enzymes can act, while at the same time stopping the action of pepsin.

3. **Intestinal juice** is produced chiefly by the glands of the duodenum, but with contributions from shed epithelial cells and intestinal bacteria. It contains:

(a) large quantities of **water**, which dilutes the food, eases movement and provides the right environment for enzyme action;
(b) large quantities of **mucus**, particularly from the Brunner's glands and goblet cells, which protects the epithelial cells from the effects of protein-splitting enzymes;
(c) **enterokinase**, which converts **trypsinogen** of the pancreatic juice into **trypsin**.

Thereafter the trypsin is autocatalytic, activating further trypsinogen and also **chymotrypsinogen** to form more trypsin and **chymotrypsin** which, with a third pancreatic enzyme, **elastase**, are capable of hydrolysing the majority of the peptide bonds, though they are individually active only at points adjacent to specific amino acid units. Much of the protein is thus split into amino acids, though some short polypeptide chains and dipeptides remain.

4. **Pancreatic lipase** can hydrolyse fat to fatty acids and glycerol. The enzyme can only act on the surface of the insoluble fat molecules, but its action is helped by the **emulsification** of fats by **bile salts**, formed from the combination of **bile acids** with some of the **lipid** substances.

5. **Pancreatic amylase** finishes the digestion of polysaccharides into disaccharides—sucrose, fructose, and maltose with less abundant trehalose and isomaltose.

6. **Nucleases**, also from the pancreas, split RNA and DNA, liberating mononucleotides.

7. **Final digestion** of disaccharides and remaining polypeptide and dipeptide units occurs on contact with the microvilli of the intestinal epithelium. Surface enzymes produced there are highly specific. **Sucrase** acts on sucrose; **lactase** on lactose; **maltase** on maltose; **trehalase** on trehalose; **isomaltase** on isomaltose; **carboxypeptidase** on the acid ends of polypeptides; **aminopeptidase** on the amino ends; and **dipeptidase** on the dipeptides.

8. The **microvilli** also provide an area up to 40 times the surface of the villi for **absorption** of the digested products, viz. monosaccharides, amino acids, nucleotides, fatty acids and glycerol, and also for water, inorganic salts, vitamins, bile salts and neutral fats, i.e. fats in finely emulsified form, closely associated with the bile salts. Much of the absorption of organic materials is by **active transport** against the concentration gradient. Most of the materials are passed rapidly into the blood vessels and are taken via the hepatic portal vein to the liver—see pages 92 and 107. Neutral fats, long-chain fatty acids and fat-soluble vitamins A, D, E and K pass into the lacteals. The milky-appearing lymph is called chyle. Absorption of vitamin B_{12} needs the presence of intrinsic factor from gastric juice.

9. Four **hormones** are known to be produced by the small intestine.

(a) **Secretin** stimulates the pancreas to release fluid and bicarbonate. It is produced in response to acidic conditions in the first part of the duodenum, i.e. the arrival of chyme.
(b) **Cholecystokinin** (pancreozymin) stimulates the pancreas to produce its enzymes and enzyme precursors (zymogens). It also stimulates contraction of the gall bladder and relaxes the sphincter at the junction of the common bile duct with the duodenum, thus allowing bile to reach the food. Together with secretin it inhibits secretion of gastric juice and decreases peristalsis so that food is not swept on before it can be digested. Cholecystokinin is produced in response to the presence of amino acids and fatty acids in the intestine.
(c) **Enterocrinin** stimulates secretion of water and mucus by the whole intestine. It is produced in response to the presence of food or acid.
(d) **Villikinin** stimulates movements of the villi so that they are exposed to fresh material. It is produced in the upper part of the intestine when food is present.

THE LARGE INTESTINE

The large intestine averages 1.5 m long and is formed of the **caecum** with the **appendix**, the **colon** and the **rectum**. It is coiled around the small intestine—see page 84. Only the transverse colon and a small part of the pelvic colon hang freely suspended by mesentery, the rest is more closely attached to the abdominal wall. The rectum passes through the pelvis.

The **caecum** is a small pouch, about 60 mm long, to which the 80 mm long **appendix** is attached and into which the ileum opens through the **ileo-colic valve** with its **ileo-caecal sphincter**. This valve is guarded by folds and contains muscle fibres which relax as a reflex response to pressure from distension of the ileum as peristalsis brings solid matter to this point. After the mass has passed, reflex response to pressure in the caecum causes contraction so that the valve is closed and there is no return.

The **colon** is sacculated owing to restriction of its longitudinal muscle to three bands, the **taeniae coli**. It has three distinct bends or **flexures**.

The **rectum** is only about 120 mm long and runs directly to the **anal canal**. The latter is guarded by an **internal sphincter** of smooth muscle and an **external sphincter** of striated muscle.

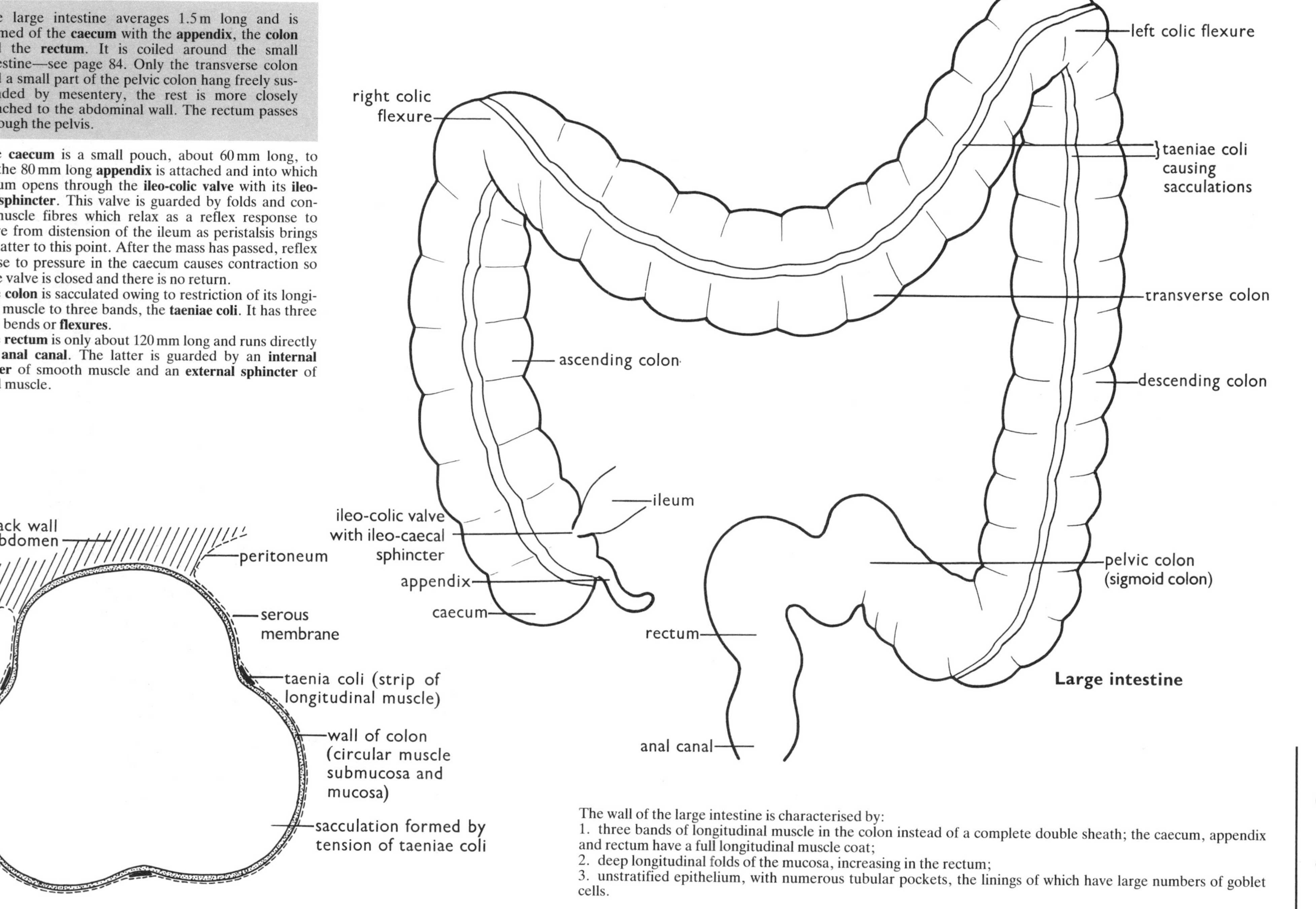

Large intestine

Colon showing taeniae coli and peritoneum—*T.S.*

The wall of the large intestine is characterised by:

1. three bands of longitudinal muscle in the colon instead of a complete double sheath; the caecum, appendix and rectum have a full longitudinal muscle coat;
2. deep longitudinal folds of the mucosa, increasing in the rectum;
3. unstratified epithelium, with numerous tubular pockets, the linings of which have large numbers of goblet cells.

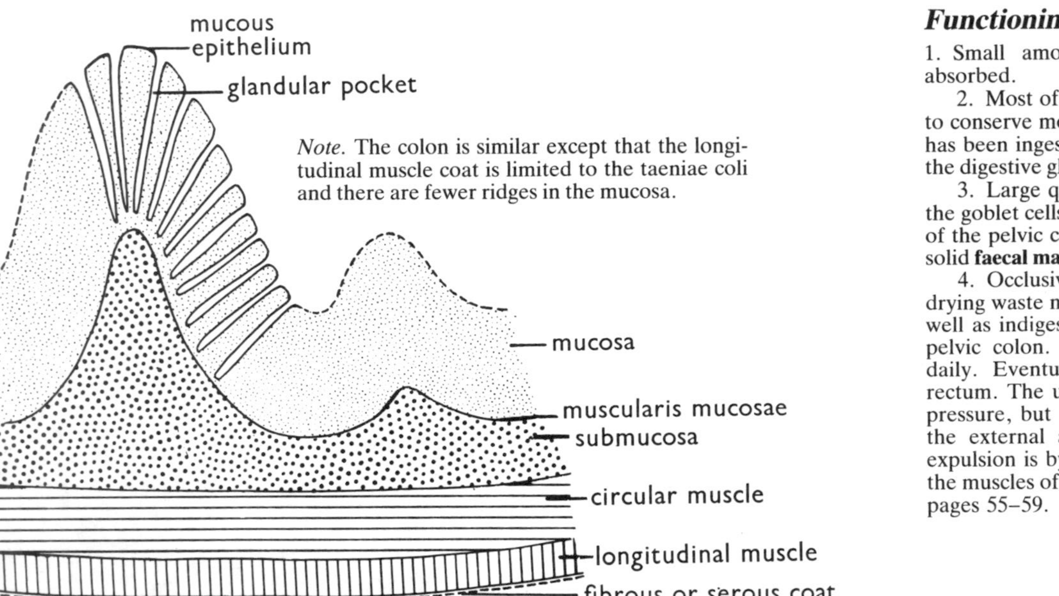

Note. The colon is similar except that the longitudinal muscle coat is limited to the taeniae coli and there are fewer ridges in the mucosa.

Rectum—*T.S.*

Functioning of the large intestine

1. Small amounts of digested food still present are absorbed.

2. Most of the water is absorbed to dry the faeces and to conserve moisture in the body. Only some of this water has been ingested. Much has come from the secretions of the digestive glands.

3. Large quantities of mucus are produced, mainly by the goblet cells, to protect the epithelium, particularly that of the pelvic colon and rectum, from friction by the semi-solid **faecal mass** and to lubricate the passage.

4. Occlusive and peristaltic movements squeeze the drying waste material, which contains sloughed-off cells as well as indigestible food, and move it slowly towards the pelvic colon. Larger mass movements occur 3–4 times daily. Eventually the waste matter is forced into the rectum. The urge to **defaecate** is a reflex response to the pressure, but actual release is under voluntary control of the external anal sphincter. When this is relaxed, the expulsion is by the rectal muscles aided by contraction of the muscles of the abdominal wall and the diaphragm—see pages 55–59.

Glandular pocket from large intestine

THE SALIVARY GLANDS

There are three pairs of large compound salivary glands as well as numerous small isolated salivary glands on the lips, cheeks and tongue. The ducts of these glands open into the buccal cavity as indicated in the diagram.

Each salivary gland is composed of secretory alveoli and their ductules (small ducts). A group of alveoli constitutes an acinus and, in the large glands, many acini are bound by connective tissue into lobules and the lobules into lobes.

Functioning of the salivary glands

The salivary glands produce **saliva**, which is a mixture of **mucus** from the mucus-secreting cells and a watery solution of salts including **bicarbonate**, with the enzymes **salivary amylase** and some **lysozyme**. Water moistens and softens the food, mucus makes it slippery for swallowing, bicarbonate buffers any acidity, so that salivary amylase can hydrolyse cooked starch into dextrin and maltose. Lysozyme attacks bacteria by dissolving their outer membranes. This reduces the amount of acid produced by growth of bacteria on food caught between the teeth.

The salivary glands may also **excrete** both organic and inorganic substances, e.g. lead in cases of lead poisoning.

Control of secretion of saliva is neural, though massage by jaw muscles can speed delivery. The motor neurones involved are parasympathetic fibres of the VII and IX cranial nerves—see page 64. These respond reflexly to the stimuli of taste, and are indirectly influenced by the smell, sight and thought of food.

Salivary glands

The pancreas lies behind the stomach, in the loop of the duodenum. It is about 125 mm long and weighs about 85 g. Like the salivary glands, it is composed of numerous **lobules**, each containing secretory **alveoli** connected to a branching system of **ducts**. Between the alveoli there are masses of cells called **islets of Langerhans**, which are derived from alveoli in the developing foetus, but have no ducts and pass their secretion as hormones directly into the blood stream. The actively secreting cells have very large nuclei, generating the RNA needed for protein synthesis.

Functioning of the pancreas

Pancreatic juice has pH 7.5–8.8. As already described—see page 88, it contains **water**, **salts** (including bicarbonate), **pancreatic lipase**, **pancreatic amylase**, **nucleases** and **zymogens** (trypsinogen, chymotrypsinogen and elastase). All these substances are involved in the digestive processes in the small intestine. The pancreas is stimulated to start releasing its secretion as soon as food distends the fundus region of the stomach. This neural stimulus is boosted by hormonal stimulation by gastrin formed when food reaches the pyloric region of the stomach and by secretin and cholecystokinin from the intestinal wall.

The **islets of Langerhans** secrete the hormones **glucagon**, **insulin** and **somatosatin** from alpha, beta and delta cells respectively. Glucagon, produced when the level of sugar in the blood tends to fall, stimulates breakdown of liver glycogen to glucose, thus raising the level of sugar again. It also promotes the conversion of other nutrients into glucose. Insulin, produced when the level of sugar in the blood tends to rise, helps many cells, including muscle, to take up glucose and use it or store it as glycogen, thus lowering the level of sugar. It also promotes protein and lipid synthesis. Together, therefore, glucagon and insulin help to transfer carbohydrate reserves from the liver to active tissues, while maintaining an almost constant concentration of sugar in the blood and minimising osmotic fluctuations. Slight rises occur for a short time after meals. The action of glucagon is similar to that of **adrenaline**—see page 122, but insulin has no counterpart. Lack of the latter deprives some tissues, particularly muscle, of nourishment and upsets homeostatic balance. This lack is diagnosed as diabetes melitus. Somatosatin is a growth-hormone inhibiting factor—see page 122.

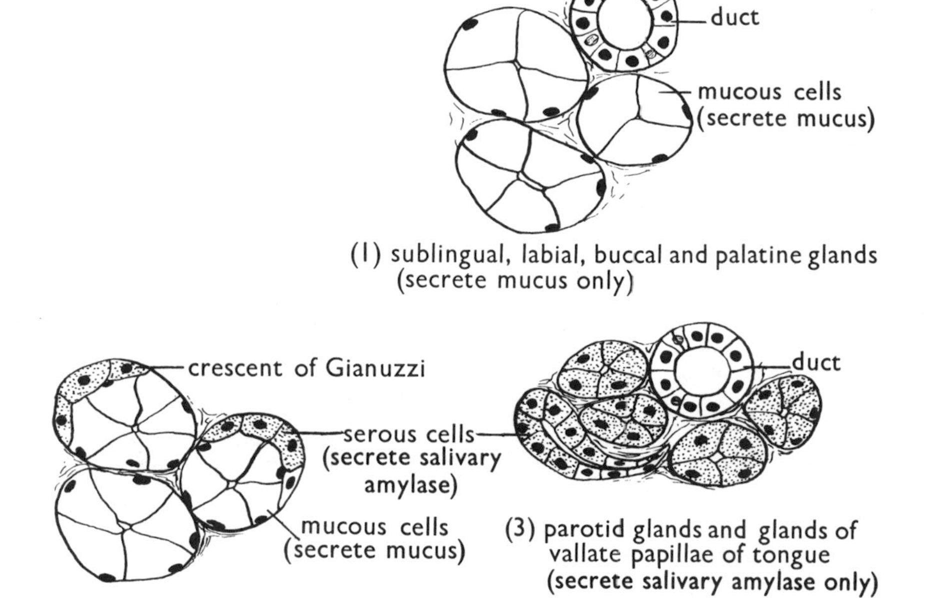

Sections of parts of various salivary glands

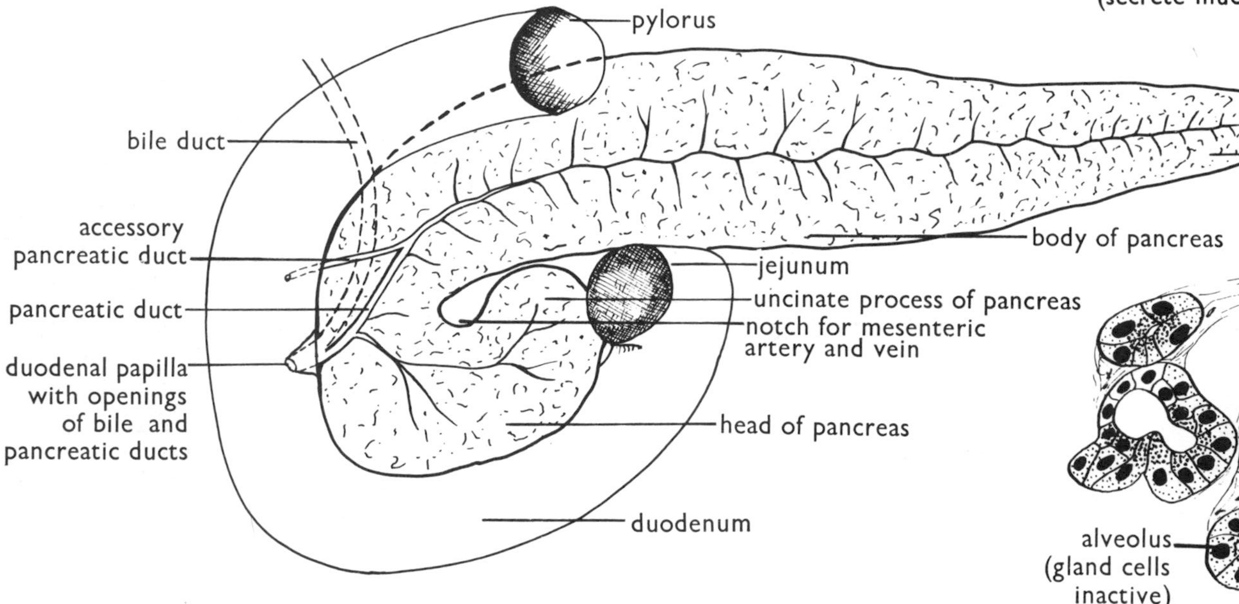

Pancreas in duodenal loop

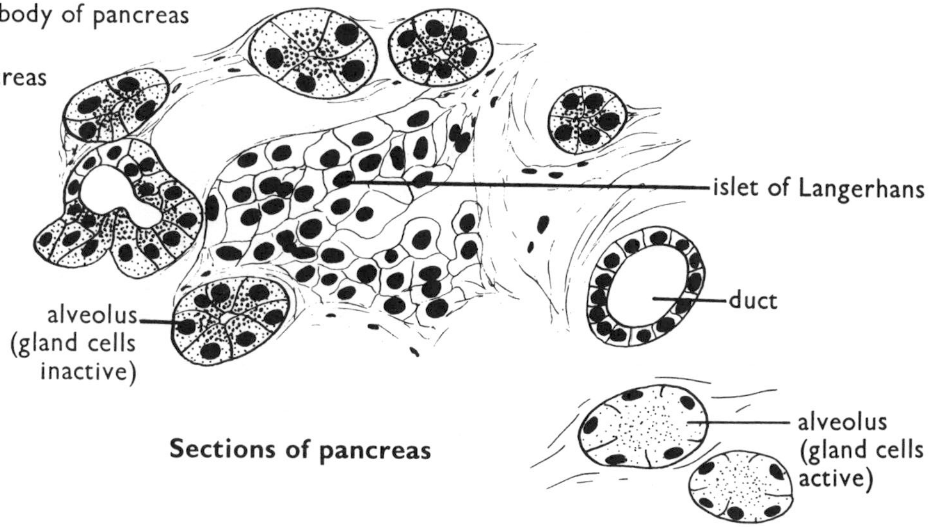

Sections of pancreas

THE LIVER

The liver lies below the diaphragm to the right of, and overlapping, the stomach. It is held in place by the **falciform** (sickle-shaped) **ligament**, the right and left **lateral ligaments** and the **round ligament**. The round ligament is formed from the umbilical vein through which the foetus was nourished during its time in the uterus—see page 130. The liver weighs about 1.3 kg and is the largest gland in the body, having large right and left lobes and small caudate and quadrate lobes.

The liver is served by five sets of vessels.

1. The **hepatic artery** branches into arterioles and brings fresh, oxygenated blood to the liver tissue at a rate of about 350 ml per minute.
2. The **hepatic portal vein** brings nutrients from the alimentary canal and breakdown products of worn-out red blood cells from the spleen, but little oxygen because this has been removed by the vital processes of the organs through which the blood has passed. The blood flows at up to 1100 ml per minute.
3. The **hepatic veins** take blood away from the liver to the inferior vena cava and thence to the heart for recirculation.
4. The **lymphatics** remove the small amounts of lymph which are formed.
5. The **hepatic ducts** carry bile secreted into them by the liver cells. The main right and left hepatic ducts join to form the common hepatic duct, which in turn joins the cystic duct to and from the gall bladder. The common bile duct leads from this junction to the duodenum, emptying into the latter along with the pancreatic duct, through the **hepatopancreatic ampulla** and the **duodenal papilla**—see diagram on page 87.

Structure of the liver

The liver appears to be made up of many tiny **lobules** which are polygonal in transverse section and thimble-shaped in vertical section. Each lobule is surrounded by a connective tissue sheath called a **Glisson's capsule**. Inside each lobule the liver cells are arranged in radiating rows called **liver cords**, between which blood flows in channels called **sinusoids**. These channels have isolated **Kupffer cells**, but otherwise the blood is in direct contact with the liver cells. The functional units of the liver centre on the strands of connective tissue known as **portal canals** in which lie (a) **arterioles**, (b) branches of the hepatic portal veins (**interlobular veins**) and (c) small tributary bile ducts receiving secretion from the liver cells through **bile canaliculi**. The **sublobular** veins are tributaries of the hepatic veins.

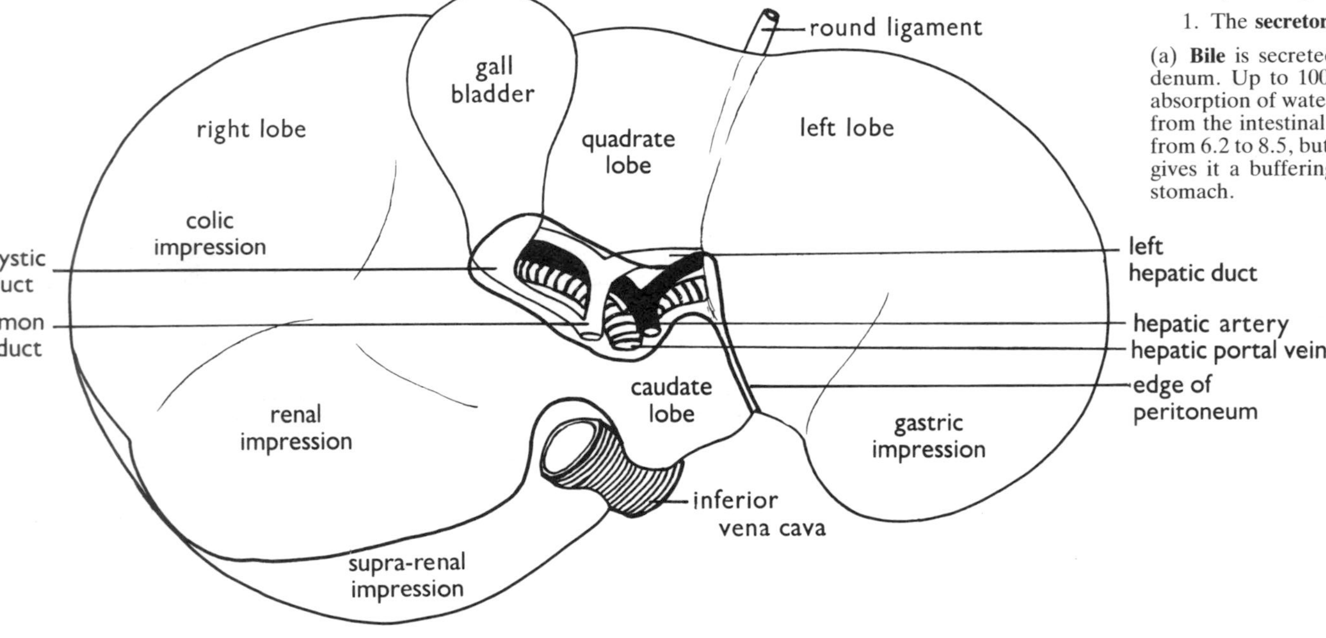

Liver viewed from beneath

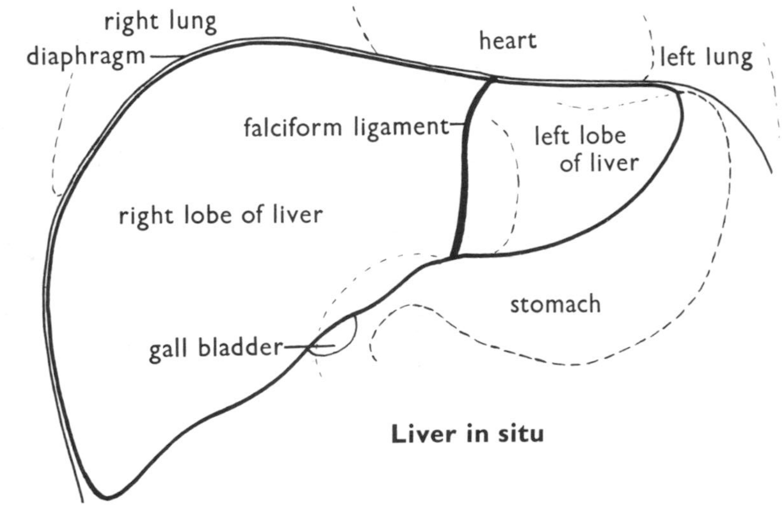

Liver in situ

Functions of the liver

With its enormous vascularity the liver acts as a reservoir for blood. It is also a primary filter. The Kupffer cells remove bacteria which may have penetrated the intestinal wall and are brought to the liver by the portal vessels before they can get into the general circulation.

The liver tissue itself has many functions which may be considered under three arbitrary headings: secretory, storage, and metabolic.

1. The **secretory functions** include the production of **bile** and **heparin**.

(a) **Bile** is secreted continuously, but released only intermittently into the duodenum. Up to 1000 ml a day can be produced. It is stored and concentrated by absorption of water in the gall bladder, from which it is released under stimulation from the intestinal hormone, cholecystokinin—see page 88. The pH of bile varies from 6.2 to 8.5, but it is usually somewhat alkaline and the presence of bicarbonates gives it a buffering action, which helps to neutralise acidity of chyme from the stomach.

Bile contains **bile acids** and **bile pigment**. The bile acids form **bile salts** with some of the lipids in the intestine. These salts emulsify fats thus aiding digestion and uptake of fat by the intestinal cells. The bile salts themselves are absorbed for re-use. The pigment **biliverdin** (green) and its reduced form **bilirubin** (red) is waste material produced by the Kupffer cells from the remains of worn-out red blood cells (erythrocytes), which have been partially destroyed in the spleen. Prior to secretion bilirubin is rendered water soluble by conjugation (joining) with **glucuronic acid**. The conjugate is oxidised by bacteria in the intestine to form **urobilinogen** (**stercobilinogen**). Some of this is absorbed and either resecreted by the liver or excreted by the kidneys, becoming **urobilin** in the urine, while the rest is oxidised by further bacterial action to **stercobilin** which colours the faeces brown.

(b) **Heparin** is an anticoagulant—see page 102.

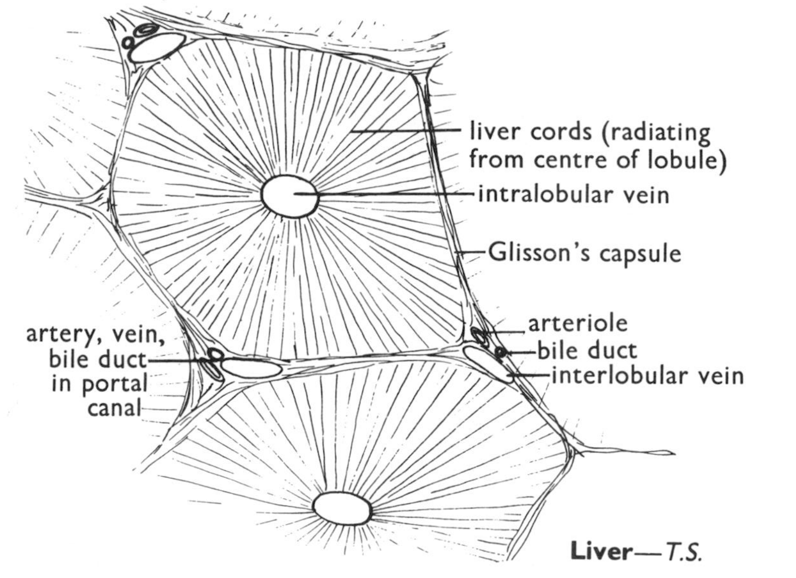

Liver—*T.S.*

2. The **storage functions** of the liver include hoarding of vitamins A, B, and D, iron and glycogen.
(a) **Vitamins** A, B, and D are rapidly removed from circulation after they have been absorbed. They are stored until required. Vitamin B_{12} (cyanocobalamin) is identical with the antianaemic factor or erythrocyte maturation factor of the liver and is essential for formation of new red blood cells. (*Note*. Vitamin C cannot be stored.)
(b) **Iron** comes from the breakdown of worn-out erythrocytes. It is stored combined with protein until required for the manufacture of the red pigment, haemoglobin, for new red blood cells.
(c) **Glycogen** is formed from glucose in the presence of specific enzymes in the liver cells, but the reaction is reversible and whether glycogen is formed or glucose liberated depends on the amount of sugar in the blood and the hormonal balance of insulin, glucagon and adrenaline—see page 91.

3. The **metabolic functions** include conversions and synthesis involving fats, proteins and carbohydrates and also detoxification processes.
(a) Many **fats** taken in with food are unsuitable for assimilation and have to be desaturated before they can be used. Fatty acid oxidation breaks down the higher fatty acids so that the products can be stored and used by other tissues.
(b) Excess and unsuitable **amino acids** are deaminated by removal of amino groups. The organic radicals or ketoacids are converted into carbohydrates and used as such. Some of the amino acid groups are stored and can be used to reaminate ketoacids, but most of them are converted into urea to prevent accumulation of free ammonia, which is poisonous to tissues.
(c) Unwanted **nucleo-proteins** are broken into uric acid.
(d) All the **proteins** of the blood fluid (plasma), except globulins, are synthesised in the liver. Amongst these are the blood-clotting factors, prothrombin and fibrinogen—see page 102.
(e) **Carbohydrate** in excess of what can be stored as glycogen is converted to fat for storage elsewhere. The otherwise unusable monosaccharide, galactose, is converted to glucose.
(f) Many **drugs** and other poisonous substances are destroyed or neutralised. **Detoxification** often involves conjugation with other substances.
(g) The liver of the foetus produces **red blood cells**.
Note. With so much metabolic activity, the liver releases a great deal of heat, and is the warmest organ of the body.

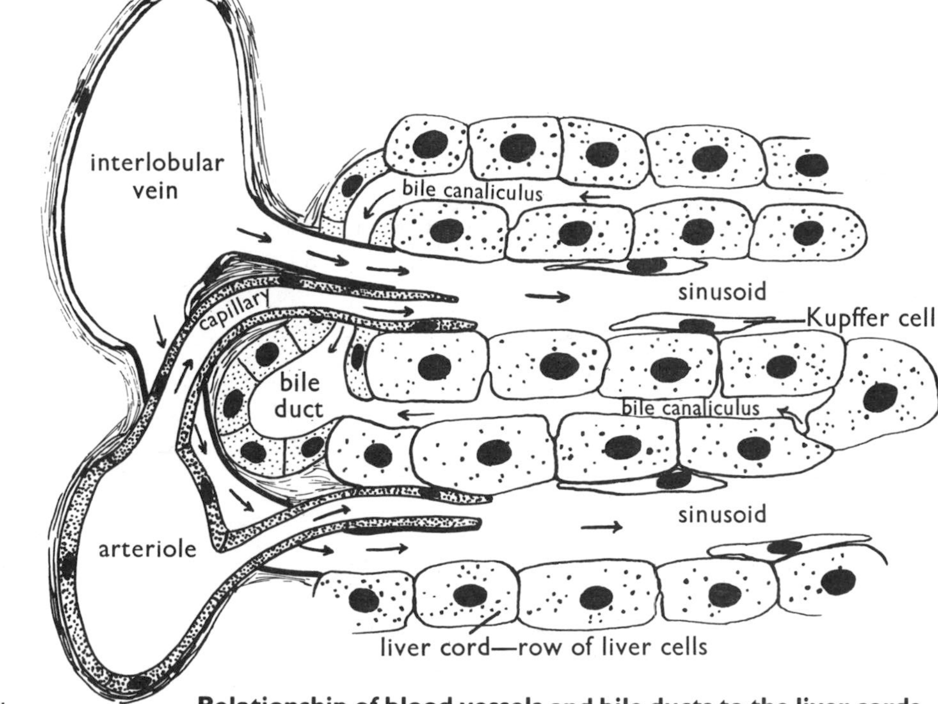

Relationship of blood vessels and bile ducts to the liver cords

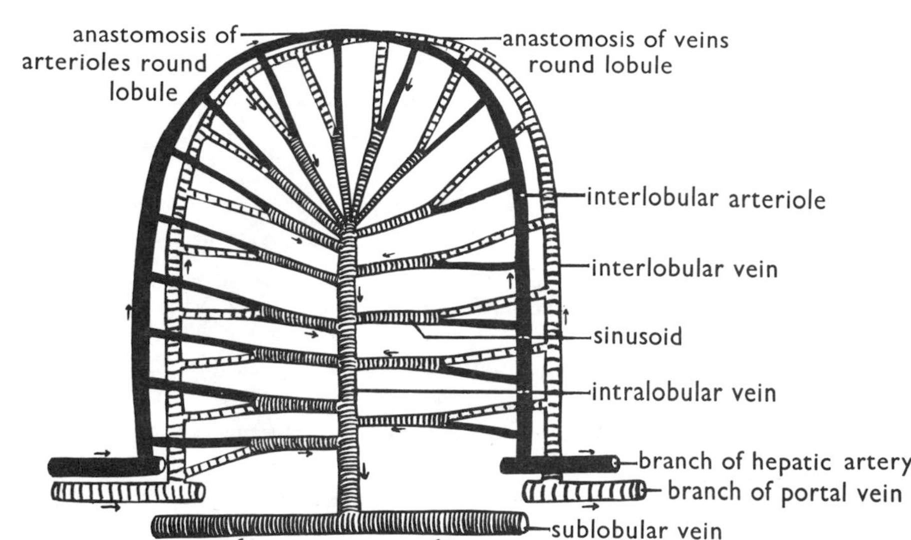

Blood supply to a liver lobule

DIGESTIVE JUICES

Secretion	Where produced	Where effective	Principal components	Action of components
Saliva	Salivary glands	Mouth (and temporarily in stomach)	Water	Softens food
			Mucus	Makes food slippery
			Bicarbonate	Buffer action provides almost neutral medium for action of salivary amylase
			Lysozyme	Attacks bacteria
			Salivary amylase	Splits cooked starch into dextrin and maltose
Gastric juice	Gastric glands	Stomach	Water	Further softens food
			Mucus	Prevents gastric juice from damaging the stomach wall
			Hydrochloric acid	Stops the action of salivary amylase and allows pepsin to work; kills many germs; curdles milk
			Pepsin (secreted as pepsinogen)	Splits certain proteins into proteoses and peptones, i.e. shorter chain polypeptides; curdles milk in adults (when rennin scarce or absent and in any case ineffective)
			Rennin	Curdles milk in many young mammals; presence in man doubtful
			Gastric lipase	Splits fats into fatty acids and glycerol
Bile	Liver (stored in the gall bladder)	Small intestine	Water	
			Bile pigment	Waste material—excreted with faeces or absorbed and re-excreted later
			Bicarbonate	Buffer action stops pepsin, helps intestinal enzymes
			Bile acids	Form bile salts with lipids. These emulsify fats and help fat absorption
Pancreatic juice	Pancreas	Small intestine	Water	
			Bicarbonate	Buffer action as above
			Pancreatic lipase	Splits fats into fatty acids and glycerol (acts more effectively than gastric lipase as fat is emulsified)
			Pancreatic amylase	Splits all forms of starch and dextrin into maltose
			Trypsin (secreted as trypsinogen) **Chymotrypsin** (secreted as chymotrypsinogen) **Elastin**	Split certain proteins, proteoses and peptones into shorter polypeptide chains and liberate some amino acids
Intestinal juice	Duodenal glands	Small intestine	Water	
			Mucus	Protects intestinal mucosa
			Enterokinase	Activates trypsinogen forming trypsin; trypsin then activates more trypsinogen and chymotrypsinogen
	Surface of villi	Microvilli	**Peptidases**	
			Carboxypeptidase Aminopeptidase	Split amino acids, one at a time, from the acid and amino ends, respectively, of the polypeptide chains
			Dipeptidase	Splits the final dipeptide residues
			Maltase	Splits maltose into glucose
			Sucrase	Splits sucrose into glucose and fructose
			Lactase	Splits lactose into glucose and galactose

ASSIMILATION

Assimilation is the process whereby nutrients are metabolised. It includes **anabolic** or constructive processes and **catabolic** or destructive ones. The former usually use energy derived from the latter. A good balanced diet should contain all the raw materials needed for this metabolism: carbohydrates, fats, proteins, vitamins, inorganic salts and water, and also enough indigestible roughage (fibre) to stimulate movements of the alimentary canal and prevent constipation.

Carbohydrates, **fats** and **proteins** are used for activity, growth and repair of tissue, including production of the non-living intracellular organic compounds, e.g. collagen. Carbohydrates are the principal source of energy, but fats and proteins may also be used—see pages 3 and 4. Fats are a convenient, compact storage material and their derivatives are used for special purposes, e.g. cell membranes and some hormones. Proteins provide the amino acids which are the building blocks for the body's own proteins, including enzymes.

Vitamins are substances which, though essential in very small quantities, cannot be synthesised in the human body. They vary in chemical structure and use—see page 99.

Inorganic salts are needed to maintain the delicate balance between living cells and their environment and also for a number of specific purposes—see page 99.

Water is needed for replacement of inevitable losses. It is an essential component of all living substance and body fluids.

The Respiratory System

Respiration is the process whereby oxygen is obtained and used for the oxidation of food materials to liberate energy and to produce carbon dioxide and water as waste materials.

INTERNAL OR TISSUE RESPIRATION

Internal or tissue respiration is the chain of chemical processes which take place in every living cell to free energy needed for its vital activities. A number of reactions, each with its own catalytic enzymes, are involved and the energy is liberated in small quantities at a number of different stages in the chain. When carbohydrate is the raw material oxidised—see page 3—the total effect is:

$$C_6H_{12}O_6 + 6O_2 = 6CO_2 + 6H_2O + \text{energy}$$
glucose + oxygen = carbon dioxide + water + energy

In this case the respiratory quotient (amount CO_2/amount O_2) = 1, but if fat is oxidised, relatively more oxygen is required and the respiratory quotient is about 0.7. The normal average quotient due to the oxidation of a mixture of food substances is about 0.85.

EXTERNAL RESPIRATION

External respiration is the means by which oxygen is obtained from the environment and carbon dioxide is released into it, and is therefore sometimes called **gaseous exchange**. It takes place in the **lungs**, from which the oxygen is carried and to which CO_2 is brought by the **blood stream**. Air reaches the lungs through the **respiratory passages** and is changed regularly by the **breathing movements**. The complete system dealing with external respiration (the **gaseous exchange system**) involves nasal passages, **pharynx** (shared with the alimentary canal—see page 85), **larynx**, **trachea**, **bronchi** and lungs, and also the muscles involved in making the breathing movements—see page 58.

The nasal passages and pharynx

The nasal passages are separated from one another by the **nasal septum** and are lined with ciliated mucous membrane whose surface is vastly increased by the **conchae**. The **nasal mucosa** is continuous with the linings of the frontal, ethmoid, sphenoid and maxillary sinuses—see page 16.

Air passing through the nose is warmed to body temperature and saturated with moisture by contact with the warm, damp mucous membrane. Dust and germs in the air are filtered out partly by the hairs round the nostrils and partly by adhesion to the slimy **mucus** which is produced by glands in the mucosa. **Cilia** move this mucus gradually into the pharynx, whence it is swallowed. The slightly salty exudation from the mucosa and lacrimal secretion from the eyes—see page 77—have a mild disinfectant action. Smells in the air are detected by **olfactory nerve endings** in the mucosa of the upper parts of the nasal cavities—see page 74.

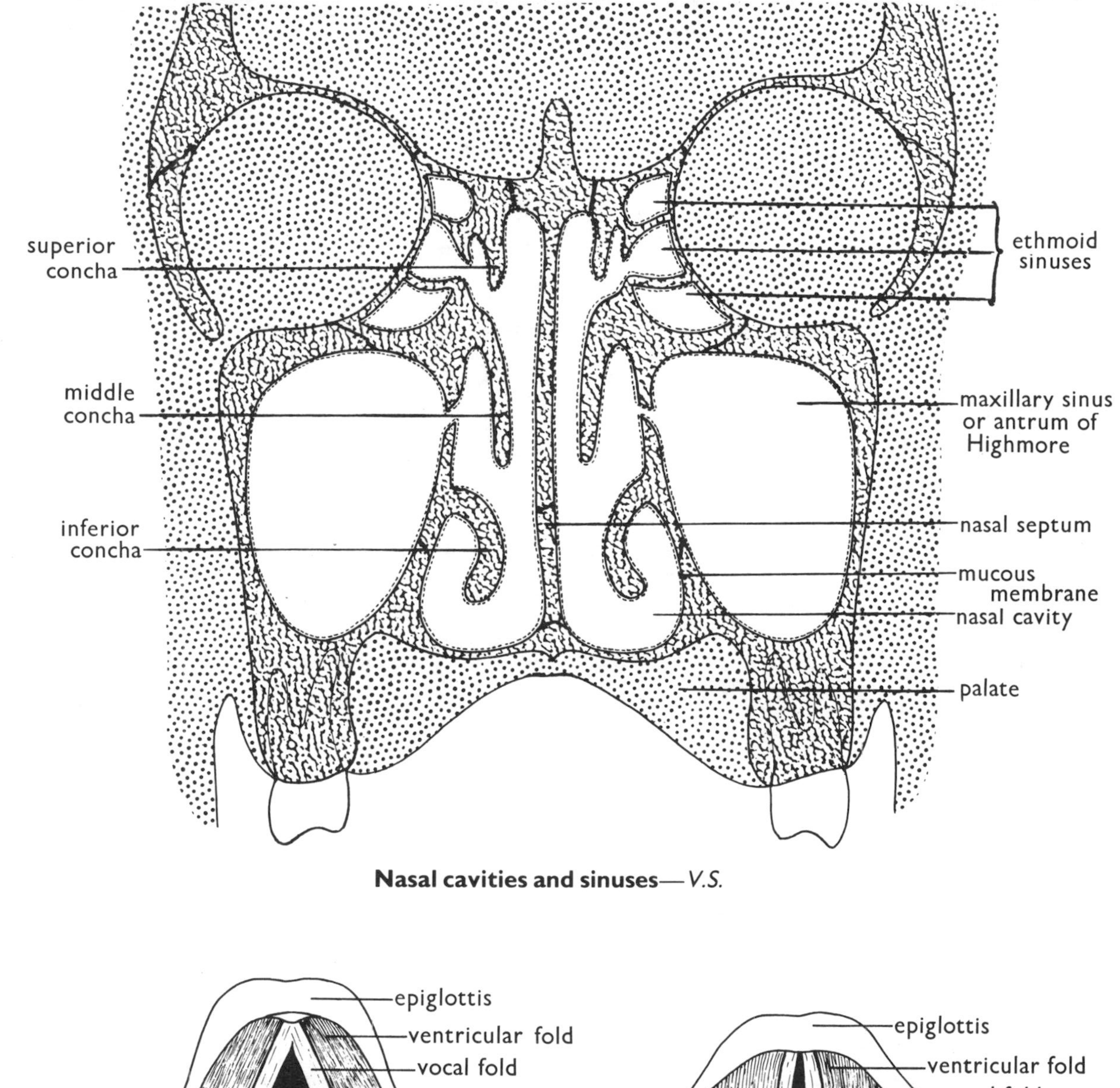

Nasal cavities and sinuses— *V.S.*

epiglottis
ventricular fold
vocal fold
rima

Vocal folds with rima of the glottis open, as when breathing quietly

epiglottis
ventricular fold
vocal fold
rima

Vocal folds with rima of the glottis closed, as when speaking

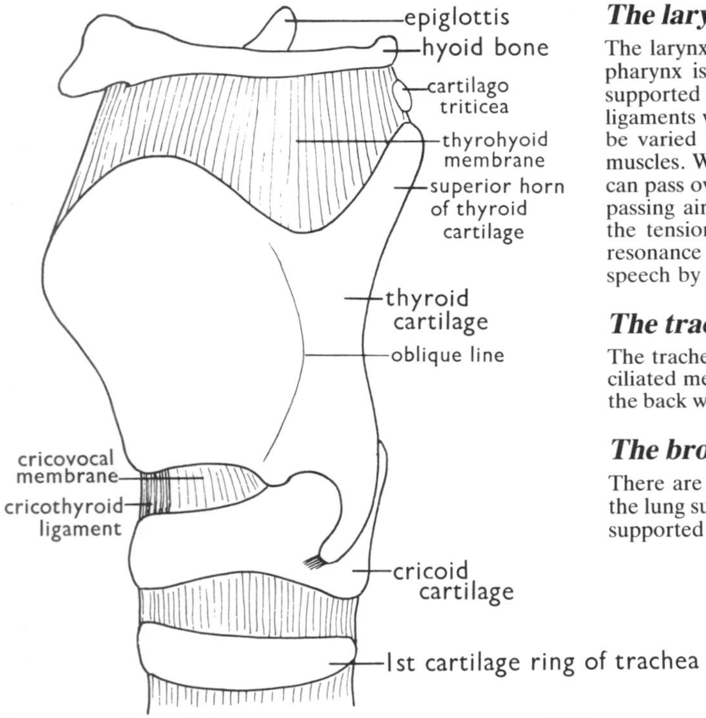

Larynx—*lateral view*

The larynx

The larynx or voice box lies below the hyoid bone. The opening into it from the pharynx is called the **glottis** and can be closed by the **epiglottis**. The larynx is supported by flexible elastic cartilages and contains the **vocal folds**. The tension of ligaments within these folds and the width of the slit between them (the **rima**) can be varied by movements of the arytenoid cartilages brought about by the vocal muscles. When the ligaments are slack and the slit wide, the air breathed in and out can pass over the folds soundlessly, but when they are taut and the slit narrow, the passing air causes the folds to vibrate, producing sounds whose pitch varies with the tension. The volume of the sounds is increased and the quality affected by resonance in the mouth, nose and sinuses. The sounds may be converted into speech by the lips, tongue and teeth.

The trachea

The trachea or wind pipe stretches through the neck to the thorax. It is lined with ciliated membrane and is supported by bands of cartilage which are interrupted at the back where it is adjacent to the **oesophagus**.

The bronchi

There are two main bronchi, one to each lung. These divide into finer bronchi in the lung substance, eventually giving off very fine **bronchioles**. The larger tubes are supported by cartilage and all are lined with ciliated membrane.

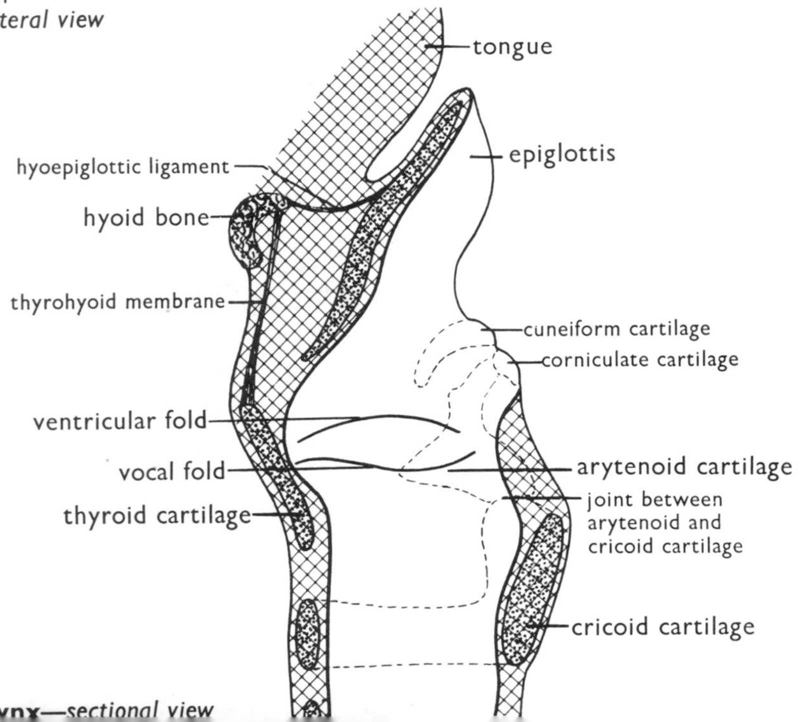

Larynx—*sectional view*

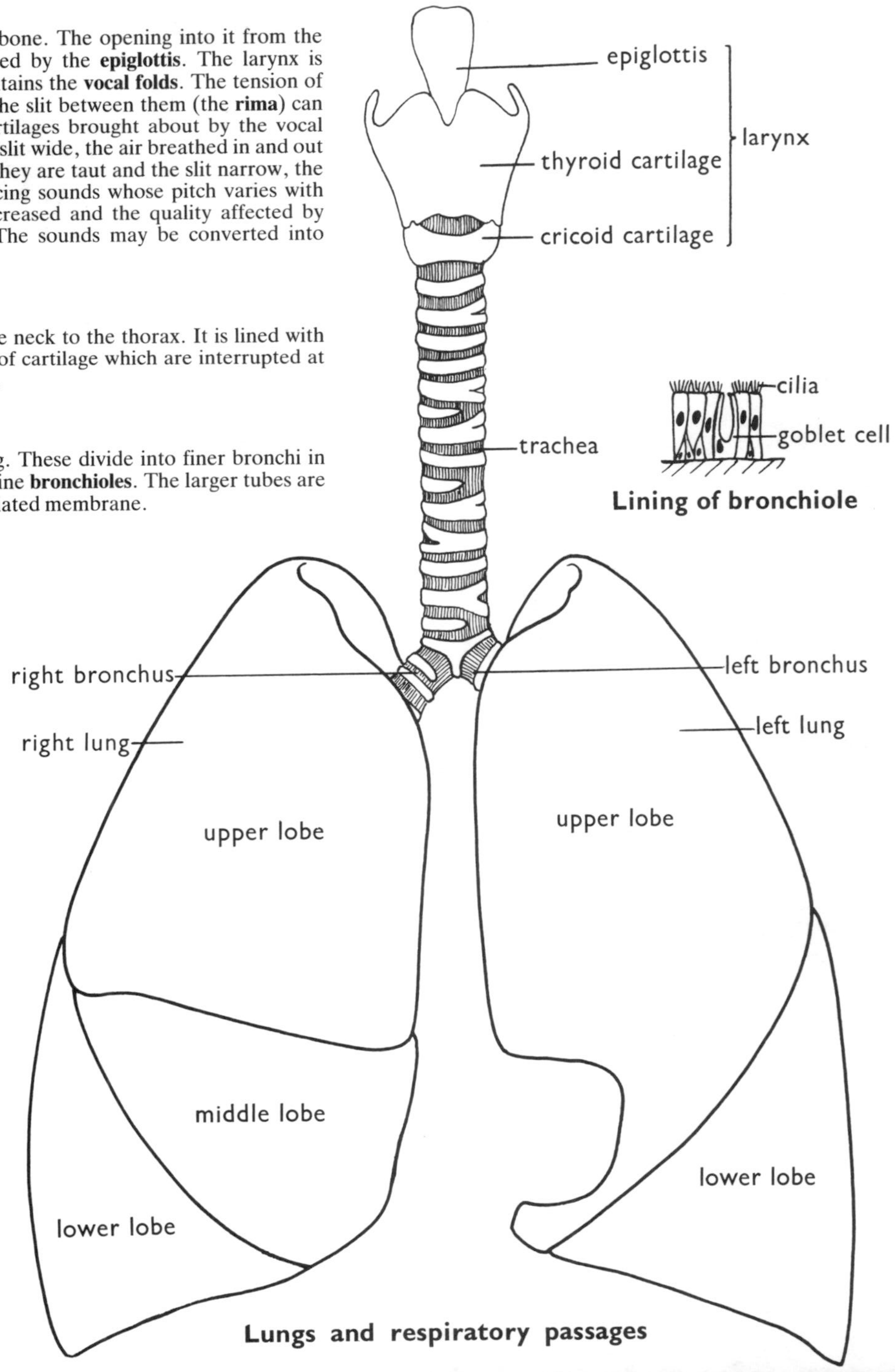

Lining of bronchiole

Lungs and respiratory passages

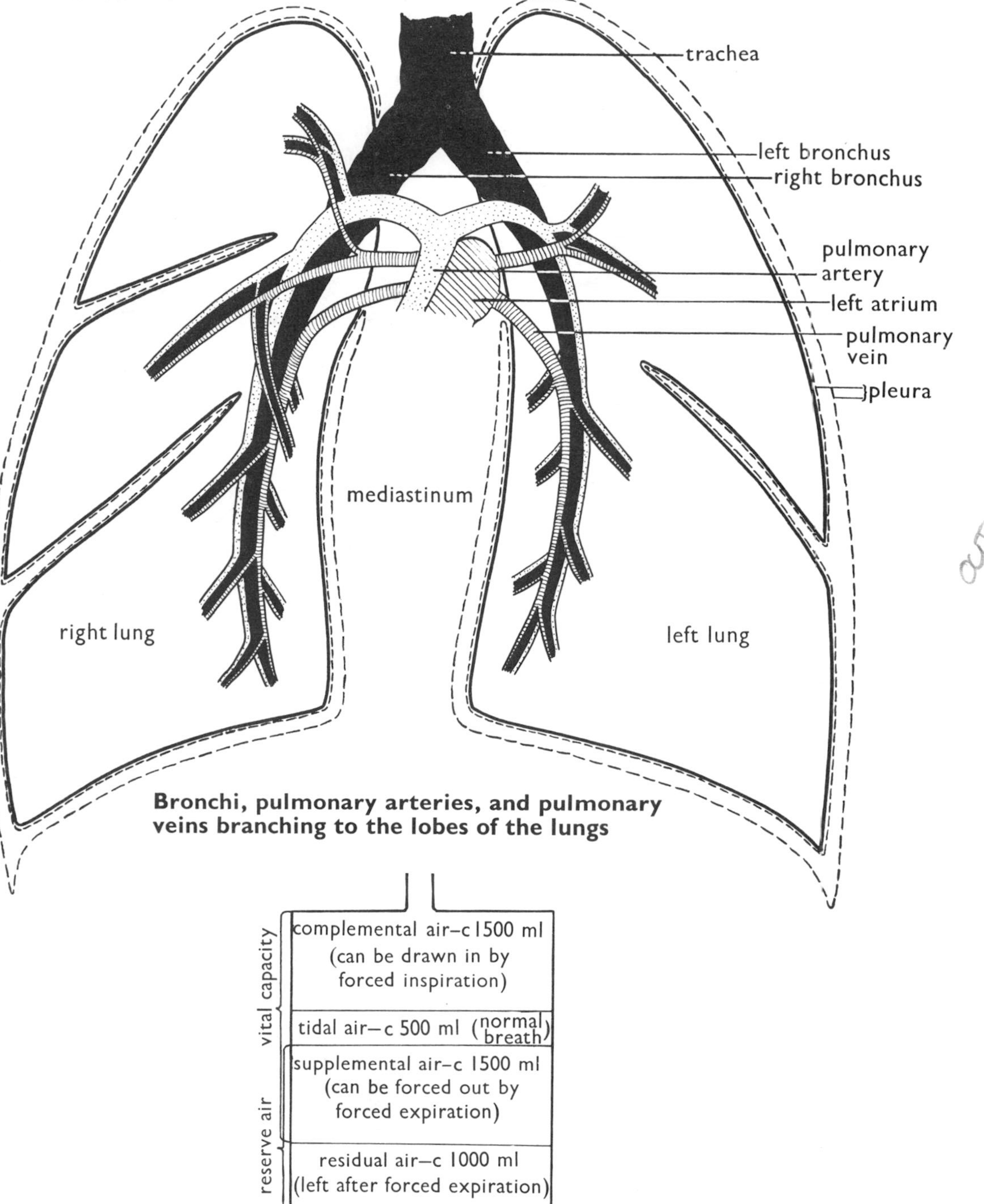

Bronchi, pulmonary arteries, and pulmonary veins branching to the lobes of the lungs

Capacity of the lungs *Note.* 500ml = approx. 1pint

The lungs

The lungs lie in the thorax, invested in serous membranes called **pleura** and separated from one another by the mediastinum, in which lie the heart and the great blood vessels—see page 83.

The right lung has three lobes and the left lung two lobes. Each lobe is made up of numerous lobules bound together by loose connective tissue. Each lobule consists of a group of air chambers attached to a terminal bronchiole. These end sacs or **infundibula** bear small pouches called **alveoli**, lined with pavement epithelium. Close to this epithelium is a dense network of very fine blood capillaries which link the pulmonary arteries to the pulmonary veins—see diagram on page 98.

Functioning of the lungs

Oxygen from the air in the lungs dissolves in the thin film of moisture on the cells lining the alveoli, and then diffuses through these cells and through the walls of the capillaries into the plasma of the blood. From the plasma it diffuses into the red blood cells (erythrocytes)—see page 100, and combines with **haemoglobin** to form **oxyhaemoglobin**. In this way the blood can carry about 70 times more oxygen than would be possible in a simple solution.

In the other parts of the body the utilisation of oxygen produces an oxygen gradient in the opposite direction. The oxyhaemoglobin breaks down and oxygen diffuses out of the blood, while carbonic acid from dissolved carbon dioxide diffuses in. Carbonic acid is carried in the blood as bicarbonate in the plasma and cells and also in combination with haemoglobin as **carbaminohaemoglobin**.

When blood returns to the lungs, the carbaminohaemoglobin and some of the bicarbonate break down to liberate carbonic acid, which in turn liberates carbon dioxide in conditions of low carbon dioxide concentration. If the carbon dioxide concentration in the lungs rises, it interferes with this breakdown, and the consequent slight increase in acidity of the blood (decrease in pH) stimulates the respiratory centres of the brain to increase the breathing movements.

Note. The lung tissue itself uses some oxygen and has its own blood supply.

Breathing

Though breathing can, to a certain extent, be controlled voluntarily, it is normally a reflex action whose rate varies with body activity, i.e. with carbon dioxide production.

Inspiration, or breathing in, is brought about by (a) contraction of the diaphragm, which increases the depth of the thorax, and (b) contraction of the intercostal muscles which swing the ribs outwards and upwards and thus increase the diameter of the thorax. Both movements combine to increase the capacity of the thorax and thus suck air into the lungs through the respiratory passages. The air drawn in mixes with the air already present; therefore the air in the lungs always contains less oxygen and more carbon dioxide than the air actually drawn in.

Expiration, or breathing out, is brought about by elastic recoil when the muscles relax.

Normally about 500 ml (1 pint approx.) of air is changed at each breath (tidal air), but additional air can be inspired (complemental air), and more air can be forced out (supplemental air), by the use of the pectoral and abdominal muscles in addition to the normal muscles of respiration.

Note. For a more detailed description of the breathing movements see page 58.

Average composition of air

Inspired		Expired	
Nitrogen	78%	Nitrogen	78%
Inert gases	1%	Inert gases	1%
Oxygen	21%	Oxygen	17%
Carbon dioxide—negligible		Carbon dioxide	4%
Water vapour—variable		Water vapour—to saturation at body temp.	

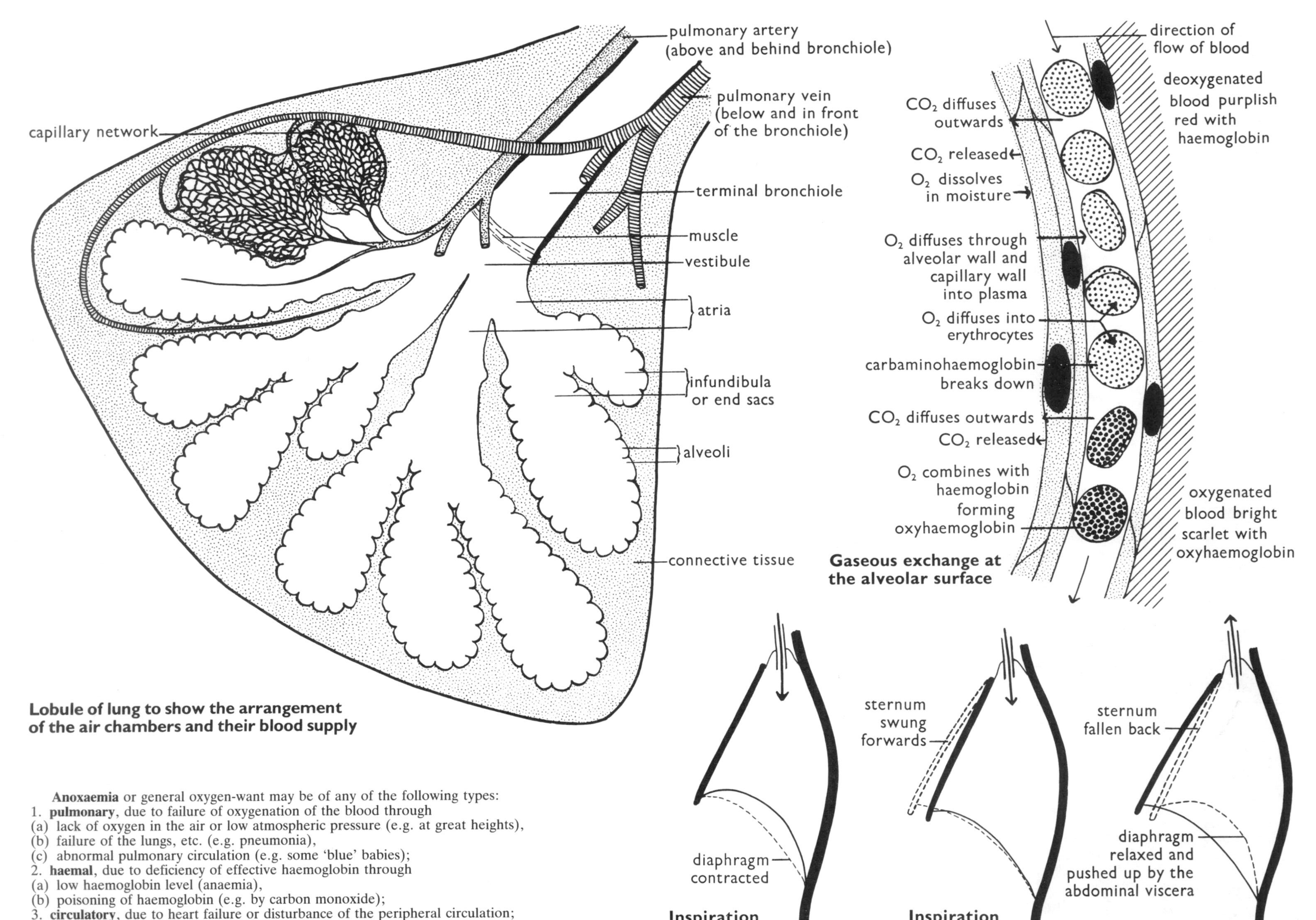

Lobule of lung to show the arrangement of the air chambers and their blood supply

Gaseous exchange at the alveolar surface

Inspiration (1) *diaphragmatic breathing*

Inspiration (2) *costal breathing*

Expiration

Anoxaemia or general oxygen-want may be of any of the following types:

1. **pulmonary**, due to failure of oxygenation of the blood through
   (a) lack of oxygen in the air or low atmospheric pressure (e.g. at great heights),
   (b) failure of the lungs, etc. (e.g. pneumonia),
   (c) abnormal pulmonary circulation (e.g. some 'blue' babies);
2. **haemal**, due to deficiency of effective haemoglobin through
   (a) low haemoglobin level (anaemia),
   (b) poisoning of haemoglobin (e.g. by carbon monoxide);
3. **circulatory**, due to heart failure or disturbance of the peripheral circulation;
4. **tissue**, due to interference with the action of the enzymes in the cells (e.g. the effect of cyanide, narcotics and anaesthetics).

The Blood Vascular System

Blood is a fluid connective tissue contained within a closed system of vessels, the arteries, veins and capillaries, through which it is made to circulate by the pumping action of the heart. It is the chief transport system of the body.

Blood forms about 7–9% of the total body weight. It appears to the naked eye as a viscous red fluid, but when examined more carefully it is found to consist of a yellow liquid portion, the **plasma**, and large numbers of cells (**corpuscles**) and cell fragments (**platelets**). About 55% of the volume of blood is plasma and the other 45% corpuscles and platelets (formed elements). When blood is centrifuged, the latter separate out. The rate at which they do so is known as the **sedimentation rate**. It is influenced by shape of the cells and also by conditions in the plasma.

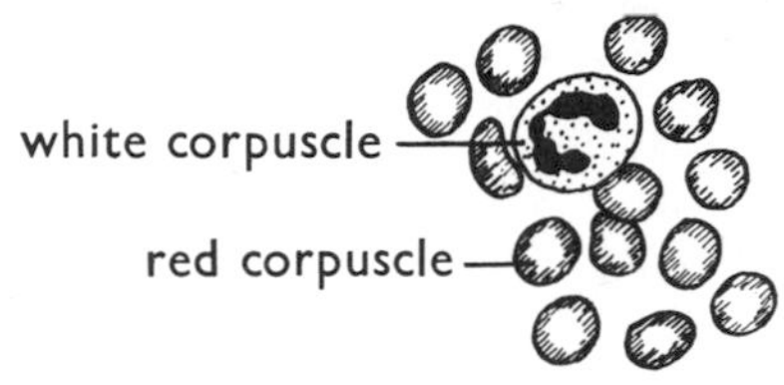

Blood smear

BLOOD PLASMA

Plasma is a clear, slightly alkaline, yellow fluid. It is less viscous than whole blood and consists largely of **water**—90–93%—in which the following substances are dissolved.

1. **Blood proteins**—7–9%. These are responsible for the viscosity of the plasma. They include serum **albumins**; alpha (α), beta (β) and gamma (γ) **globulins**; **prothrombin** and **fibrinogen**. The albumins and most of the globulins produce colloidal osmotic pressure helping to maintain a stable osmotic environment for the cells of the blood itself and also binding certain substances for transport, e.g. albumin with thyroxine; α globulins with lipids and thryoxine; β globulins with iron and cholesterol. Many γ globulins are antibodies and as such part of the body's immune system—see page 101. Prothrombin and fibrinogen are involved in blood clotting—see page 102. **Serum** is plasma minus these clotting proteins.
2. **Inorganic salts**—0.9%. These are in solution and largely ionised. The commonest ions are Na^+ and Cl^-, responsible for isosmotic conditions between the blood and tissue fluids and between the latter and the contents of living cells. Homeostasis of salinity is controlled by the kidneys—see page 120. Bicarbonate and phosphate ions have a buffering action, helping to maintain the pH at 7.0–7.8. Many other ions are present, being transported to the tissues where they are specifically required—see table.
3. **Food substances**—0.2–0.4%. These are chiefly amino acids, glucose, fats and fatty acids.
4. **Waste materials**—0.02–0.04%. These are chiefly urea, but with some uric acid and creatinine and minute amounts of urobilinogen—see page 92.
5. **Gases**: (a) oxygen, to the limit of its solubility; (b) nitrogen, to the limit of its solubility; (c) carbon dioxide, only small amounts in simple solution, the rest buffered by bicarbonate and phosphate ions, blood proteins and red corpuscles.

   Note. The solubility of gases depends not only on temperature, but on the pressure in the gaseous environment. It is therefore affected by height, e.g. on mountains and in non-pressurised aircraft, and by depth, e.g. when diving with breathing apparatus.
6. **Enzymes**. Many different enzymes are present, catalysing chemical reactions within the blood itself. The abnormal presence of some enzymes may be symptomatic of disease.
7. **Hormones**. These are secretions from the endocrine organs—see page 121, and are distributed by the blood to the tissues and organs where they take effect.
8. **Antimicrobial substances** (other than immunoglobulins) include interferon and complement—see page 101.
9. **Vitamins**. These were once known by the letters of the alphabet only, but are now identified by their chemical names. They are distributed by the blood. Their involvement is highly specific—see table.

Plasma and tissue fluid

Exchanges between the plasma and tissue fluids take place through the thin walls of the capillaries, especially under pressure. Small molecules of gases and inorganic salts pass easily, but proteins and other large molecules do so with more difficulty and, by their osmotic effects, promote return of fluid as the capillary pressure becomes less towards the veins—see also lymph, page 114.

The special capillaries of the brain have denser walls and are surrounded by neuroglial cells, which makes them less permeable and creates the blood–brain barrier—see page 65, by which the delicate neural substance is protected from the effects of potentially harmful substances. While glucose, oxygen and certain ions pass readily, urea, chloride, sucrose and insulin do so only slowly. Proteins and most antibiotics are totally excluded.

out

Special uses of some ions

Potassium	Nerve functioning
Calcium	Bone and tooth structure Blood clotting Muscle contraction Release of neurotransmitters
Fluorine	Bone hardening
Magnesium	Bone structure Co-factor to many enzymes
Iron	Part of haemoglobin —carrying oxygen Part of cytochromes —carrying electrons during electron transfer —see page 4.
Phosphates	Buffering Structurally involved in phospholipids of cell membranes, in DNA, RNA and energy transfer.
Zinc	Part of many enzymes
Copper Cobalt Manganese	Needed for haemoglobin synthesis
Iodine	Involved in structure of thyroid hormones

Uses of some vitamins

Thyamine (B_1) Riboflavin (B_2) Pantothenic acid Niacin	Precursors of co-enzymes and carriers in the Krebs cycle and electron transport—essential for cell respiration
Choline Inositol	Help fat metabolism
Peridoxine (B_6)	Co-enzyme is fatty acid and amino acid metabolism
Biotin	Helps transamination and protein synthesis
Naphthoquinone (K)	Helps prothrombin synthesis
Cyanocobalamin (B_{12}) Folic acid	Synthesis of nucleoproteins Formation of red blood cells
Tocopherol (E)	Maintenance of red blood cells In experimental animals helps pregnancy
Ascorbic acid (C)	Essential for collagen and ground substance Lack causes scurvy
Carotene (A)	Forms visual purple —rhodopsin Maintains health of epithelia
Calciferol (D)	Helps calcium absorption, thus hardening of bones and teeth

Note. Where more than one vitamin is involved they help different parts of the metabolic process.

BLOOD CELLS AND PLATELETS

All the solid components of the blood develop from undifferentiated mesenchymal cells called **stem cells**. **Mesenchyme** is embryonic connective tissue and widely spread in the foetus so that stem cells occur in the yolk-sac—see page 128, liver, spleen, lymph nodes and marrow of developing bones. From shortly after birth, stem cells are limited to red bone marrow (**myeloid tissue**) of the skull bones, vertebrae, sternum and ribs and the ends of long bones and to the **lymphoid** tissue of spleen, lymph nodes and tonsils.

The stem cells form:

1. **rubriblasts**, which give rise to **erythrocytes** (red blood cells);
2. **myeloblasts**, which give rise to **polymorphonuclear leucocytes** (granular white cells);
3. **monoblasts**, which give rise to **monocytes** (large non-granular white cells);
4. **lymphoblasts**, which give rise to **lymphocytes** (smaller non-granular white cells);
5. **megakaryocytes**, which do not enter the blood stream whole but release numerous small fragments called **platelets**.

Erythrocytes

Erythrocytes—red cells (red corpuscles)—are biconcave discs without nuclei, i.e. they are **enucleate**. There are about 5 million per cubic mm of blood (rather more in men than women). Singly they appear pale orange but in masses they give the blood its red colour.

Each erythrocyte is about 8 μm in diameter and 2 μm thick at the edges. It has an **envelope** surrounding a spongy mass called **stroma**. The stroma contains the red pigment, **haemoglobin**, which is capable of combining reversibly with oxygen—see page 97. In this way the corpuscles carry a great deal of oxygen from the lungs to the other parts of the body. They also carry carbon dioxide away, the haemoglobin acting as a buffer.

Haemoglobin is a complex molecule formed of **haem** and four polypeptide chains, the **globin**. Haem has **iron** attached to **porphyrin**. The oxygen-carrying power depends on easy change of the iron from the ferrous (Fe^{2+}) to the ferric (Fe^{3+}) state. In each haemoglobin molecule, four haem molecules are anchored by globin, which keeps them within the corpuscle. The exact structure of the polypeptide chain of the globin is very important. It is genetically controlled and any slight deviation causes defects in appearance and efficiency of the erythrocyte, e.g. sickle-cell anaemia.

Red corpuscle—*surface view*

Red corpuscle—*semilateral view*

Red corpuscle—*sectioned*

Because the erythrocytes have no nuclei they are short lived in the circulation. The average functional life is 100–120 days, after which they die and are broken down, chiefly in the spleen and liver. Iron from the haemoglobin is stored in the liver until required for new corpuscles, while the residue of the haem forms bile pigments and is removed from the body—see page 92. Replacement erythrocytes develop from the rubriblasts in the red bone marrow. During development haemoglobin is added and, just before liberation into the blood, the nucleus is shed.

Stimulation of erythrocyte production is by **erythropoietin**, produced by conjugation of erythropoietic factor from the kidneys—see page 120, with a globulin from the liver. The kidneys and liver are stimulated to produce these substances in response to oxygen shortage from whatever cause, thus normally balancing replacement with loss. Besides the essential raw materials for cell structure, e.g. lipid, protein, amino acids and nucleotides, certain factors must be present. **Cyanocobalamin (B_{12})** and **folic acid** promote DNA synthesis and thus help nuclear divisions amongst the marrow cells. **Pyridoxine**, by stimulating fatty acid and amino acid metabolism, increase rate of cell division. **Copper**, **cobalt** and **manganese** aid synthesis of haemoglobin. **Iron** is essential for incorporation into the haem. Much of the iron in the body is recycled, but additions from the diet are needed when there is blood loss, e.g. in menstruating women or through wounds. This iron is absorbed as Fe^{2+} which is rapidly combined with a globulin and carried as transferin to the liver where it joins iron already there, and is deposited as ferritin.

Anaemia, with deficiency of oxygen-carrying power, can be acute following bleeding or chronic due to failure of any part of the replacement system. Healthy red bone marrow, with adequate supplies of raw materials and essential vitamins and minerals, can generate about 3.5 million erythrocytes per kg body weight per day.

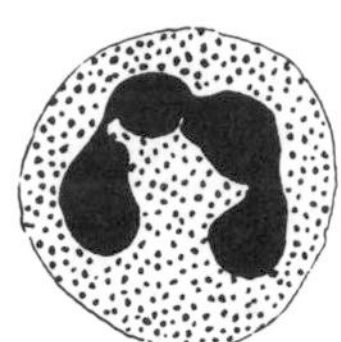
Granulocyte

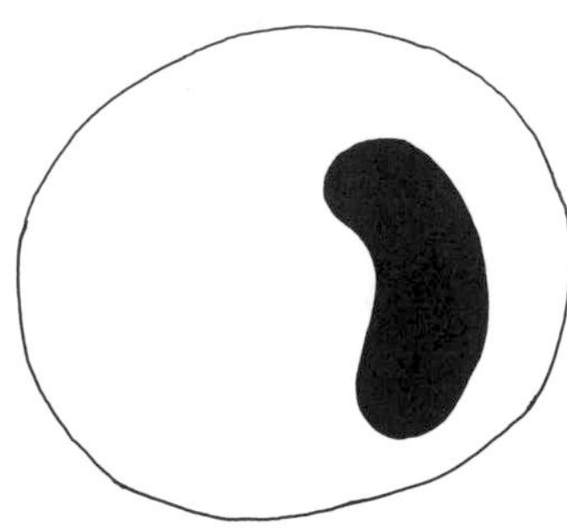
Monocyte

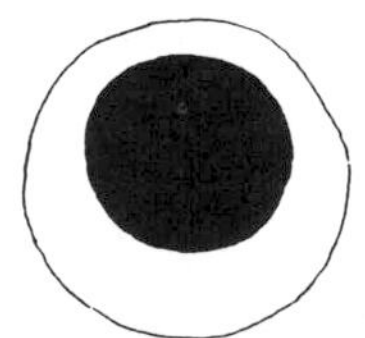
Large lymphocyte

Small lymphocyte

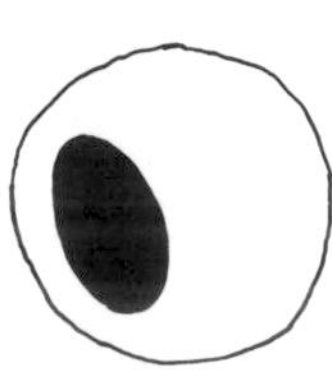
Plasma cell

Polymorphonuclear leucocytes (granulocytes)

Granulocytes make up 65–75% of the total leucocyte population of 5000–10 000 per cubic mm of blood. They develop in myeloid tissue (red bone marrow), taking 10–14 days to mature and living freely in the blood for 12 hours to 3 days before being destroyed in the lungs, intestines, liver and spleen if they have not passed into the peripheral tissues from which they cannot return. Each granulocyte has a large lobulated nucleus, the number of lobules increasing with age, and numerous granules in the cytoplasm which stain characteristically and have specific functions—see table.

Type of granulocyte	% of total	Size	Granules	Main functions
Neutrophil	60–70	12–14 μm	Fine— stain with acid or basic dyes	Actively phagocytic Produce lysozyme which destroys some bacteria and damaged tissue cells
Eosinophil	2–4	10–12 μm	Shiny— stain with acid dyes	Have some phagocytic action Granules produce destructive enzymes Increase in number during allergic reactions, auto-immune states and parasitic invasions
Basophil	0.15	9 μm	Dull— stain with basic dyes	The mast cells of tissues Release serotonin → temporary vasoconstriction histamine → vasodilation heparin, an anticoagulant

Monocytes

Monocytes make up 2–4% of the normal leucocyte count. With diameter 20–25 μm, they are much larger than the other white cells and are formed in the lymphoid tissue. Each cell has a single large round or slightly indented nucleus, few fine granules, and lives in the circulation for 100–300 days. Monocytes frequently leave the blood and wander in other connective tissues where they are recognised as macrophages. They are actively **phagocytic**. Along with the neutrophils, they ingest and destroy germs and clean up debris of damaged cells. Both are attracted to the site of invasion by release of toxins by the invaders or by chemicals released by invaded and inflamed tissue itself. The neutrophils usually arrive first, but the macrophages accumulate in greater numbers and are better scavengers. During the scavenging process many of the leucocytes die. **Pus** is a thick fluid containing dead and living white cells as well as tissue debris.

Lymphocytes

Lymphocytes make up 20–25% of the normal leucocyte count. They have rounded nuclei and clear cytoplasm. In appearance there are three types: (a) **large lymphocytes** (12–14 μm), (b) **small lymphocytes** (9 μm), (c) **plasma cells** (up to 15 μm), but functionally differentiation goes back to a very early stage in life when lymphoblasts from the foetal liver and bone marrow become precommitted. Some known as **B cells** go directly to the lymphoid tissue while others known as **T cells** go to the thymus. Before or shortly after birth the T cells are processed, then they too go to the lymphoid tissue. Through a process not yet understood both T cells and B cells come to have an infinite variety of minute differences which give them a high degree of **specificity**. These **immunologically competent** cells have the ability to clone, i.e. divide repeatedly to give groups of cells all descended from the same parental cell and having the specific characteristics of that cell. The purpose of this is to defend the body against a wide variety of foreign substances known as **antigens**. Most of these antigens are associated with the surfaces of disease-producing organisms or are their toxic products, but other foreign substances can provoke immune response. The trigger substances are small molecules called **haptens**, attached to larger molecules (**carriers**), mainly proteins. The haptens contain key regions known as **epitopes** with a small number (3–10) of amino acids, the sequence of which is characteristic of a given antigen. The lymphocytes are coded to recognise the differences between epitopes.

The antigens stimulate the appropriately coded T cells and B cells. In the presence of hormones from the thymus, these clone. T cells and macrophages are also involved in the maturation process of B cells. In most cases the antigens must first contact the immature lymphocytes in the lymphoid tissue. The mature descendants are liberated into the blood stream later, where the T cells give **cell-mediated immunity**, i.e. they bind to antigens on invading cells, while B cells give **humoral immunity** by secreting **antibodies** which link with the antigen molecules.

T cells include (a) **killer-T** (T_C) which produce cytotoxic substances which destroy the antigen and the cell which bears it and also recruit additional lymphocytes and macrophages, (b) **helper-T** (T_A) which amplify the response and help descendants of B cells to produce antibodies, (c) **suppressor-T** (T_S) which inhibit killer-T and B cells thus putting a brake on the system, (d) **memory-T** (T_D) which recognise the original invading antigen and initiate a far quicker response to a subsequent invasion. Memory-T clones are long-lived so that immunity, once established, may be retained for life. T lymphocytes have a slow rate of production and considerable recirculation from the lymph nodes through the lymph vessels to the blood and back so that they keep guard against rogue cells including cancer cells, which are defective body cells acting as antigens.

When **B cells** are presented with the appropriate antigen by T cells and macrophages, they clone actively to produce **plasma cells** with highly developed endoplasmic reticulum, which can release antibody molecules at a rate of up to 2000/s/cell for the 4–5 days of their active life. Though a given plasma cell may exist for only a few days some of its descendants remain as **memory-B cells** with capacity to respond to future invasion by the same antigen. The antibodies are γ-globulins known as **immunoglobulins** (IgM, IgG, IgA, IgE). These may act as (a) **antitoxins**, neutralising toxic effects, (b) **precipitins**, precipitating antigens out of the system, (c) **agglutinins**, causing clumping of bacteria and other invading organisms so that their growth and multiplication is inhibited, (d) **opsonins** which prepare invading cells for further destructive treatment by phagocytosis, **IgE** also causes mast cells to release histamine and heparin.

After the antibody has become attached to a surface antigen, a series of 11 proteins produced by the liver and collectively called **complement**, react with the antibody and each other to produce enzymes capable of damaging the foreign cell wall so that it becomes more permeable to water and salts. The cell undergoes **lysis**, i.e. blows up and bursts. Other forms of complement increase opsonisation, and are chemotactic, attracting leucocytes, or stimulate mast cells to release histamine so that there is increased permeability of blood capillaries and easier access of leucocytes to the site of invasion.

The body may have 10^5–10^8 clones, each for a single or closely related group of antigens. The potential to respond is present even if never stimulated and is passed from one generation to the next, accounting for the susceptibility differences between members of different families and races. After the first encounter with an antigen, the response activates an immunological memory so that response to subsequent invasion is more rapid.

Immunological response is prevented from getting out of hand by:

1. suppressor antibodies and suppressor-T cells, of which there are normally half as many as killer-T cells;
2. limited life of antibody-producing cells;
3. loss of stimulatory effect once an antigen is bound to an antibody, or coated or destroyed.

Normally the body does not react against the proteins of its own tissues, but the high degree of specificity of these is responsible for tissue rejection after transplant operations and for the problems of blood grouping.

Some definitions

Immunity—protection against invasions.

Natural immunity—total immunity due to the body's own defences;
(a) **tissue immunity**—antibodies retained in plasma cells—long-lasting;
(b) **humoral immunity**—antibodies secreted—does not last as long;
(c) **cell-mediated immunity**—T-cells bind to antigens of invading cells;
(d) **active immunity**—body's response to invading antigens;
(e) **passive immunity**—antibody transferred from another individual.

Immunisation—giving of substances which boost the body's defences without producing clinical symptoms:
(a) by live, weakened (attenuated) organisms which induce defence;
(b) by killed organisms which induce defence;
(c) by toxoids (modified toxins) which induce defence;
(d) by human globulin from serum of ex-patients which contains appropriate antibodies;
(e) by antitoxins which are extracts of antibodies to specific toxins.

Interferon—a non-specific antiviral substance which prevents synthesis of new viral particles and reduces ability of the virus to attach to membranes of uninfected cells.

Transfer factor—an antibody-like substance, which can be isolated from individuals who have developed immunity and can be transferred to possibly help sufferers.

Allergy—incomplete or abnormal reaction to an antigen, which often involves liberation of histamine.

Anaphylactic shock—rapid, disastrous, allergic response.

Serum sickness—a milder allergic response, which comes on slowly.

Atopic allergies—localised responses, e.g. in the respiratory system (hay fever or asthma) or skin (dermatitis and rash).

Autoimmunity—the body treats one of its own proteins as an antigen.

Note. The disease **AIDS** (acquired immune deficiency syndrome) is caused by a virus which destroys the killer-T cells.

Platelets

Platelets or **thrombocytes** are minute rounded discs 2–4 μm in diameter. There are about 300 000 per cubic mm of blood. They are not nucleated, but have a central granular portion (**chromomere**) surrounded by clear cytoplasm (**halomere**). They are shed by break-up of very large cells called **megakaryocytes**, which originate in myeloid tissue but do not themselves enter the blood stream. The platelets have a life span of 5–9 days. They are sticky and can become involved with collagen at the site of an injury, where they adhere to torn surfaces to form **platelet plugs**. They release ADP as they adhere, which stimulates further adhesion till the damage is sealed. Their disintegration also liberates substances which help clotting.

HAEMOSTASIS

Haemostasis is the minimisation of loss of blood from damaged blood vessels. It involves:
1. **vasoconstriction**, reflexly produced by neural response to pain of wounding and locally produced by serotonin release from mast cells and damaged tissue cells, thus temporarily reducing blood flow to the wounded area;
2. **platelet plug** formation which can effectively seal small holes in blood vessels;
3. **blood clotting** which seals larger holes.

Blood clotting

This is a complex chain of events with three main stages:
(a) formation of the enzyme **thrombokinase** (thromboplastin);
(b) use of thrombokinase, in the presence of Ca^{2+}, to activate **prothrombin** to the enzyme **thrombin**;
(c) action of thrombin to convert **fibrinogen** to **fibrin**, which entangles blood corpuscles and cell debris to form the **clot**.

Thrombokinase is derived from lipoprotein released from damaged tissue and blood vessels (extrinsic source) or from disintegrating platelets (thrombocytes) adhering to a roughened surface (intrinsic source). Conversion to the active state requires a number of **plasma-coagulating factors**, normally synthesised by the liver. Deficiency of these results in loss of clotting power, e.g. in haemophylia.

Normally danger of clotting while in circulation is reduced by:
(a) the anticoagulant, **antiprothrombin**, more usually known as **heparin**, which is produced by many tissues, including the liver and linings of blood vessels;
(b) **smoothness** of the blood vessel walls to which the platelets cannot adhere.

While heparin decreases thrombokinase production naturally, certain drugs can be used to decrease it artificially. Removal of Ca^{2+} by precipitation from a blood sample will keep it in a fluid condition, by preventing conversion of prothrombin to thrombin.

A **thrombus** is a clot within a blood vessel.

An **embolus** is a clot floating in the blood stream, which may jam in a small blood vessel and cause an embolism.

REACTION TO WOUNDING

A series of events occurs when the body surface is penetrated:
1. transient, localised vasoconstriction—see above;
2. platelet plug and clot formation to seal the wound—see above;
3. vasodilation, chemically stimulated by products of cell damage and histamine from mast cells, resulting in reddening and warmth round the damaged area;
4. chemical attraction of leucocytes which start phagocytosis, having passed through the capillary walls by amoeboid action called diapedesis;
5. initiation of immune response;
6. swelling due to fluid entering tissue spaces;
7. formation of pus in some cases;
8. release of lysozyme from damaged cells, helping to digest the products of cell death and of clotting, and clearing the way for healing;
9. healing, involving mitosis—see page 124—of remaining cells and fibroblasts;

If replacement of the original type of cells is impossible a scar is left.

SHOCK

Shock most commonly occurs when there is loss of fluid in circulation either by external or internal haemorrhage (bleeding) or by exudation of plasma from burns. It may also occur during heart disturbances or as a result of toxins.

The characteristic signs of shock are:
1. low blood pressure through loss of fluid;
2. pallor of the skin due to vasoconstriction, which helps to redirect remaining fluid to more essential organs;
3. coldness of the extremities in which blood is not circulating freely;
4. depression of the central nervous system through shortage of oxygen, with resultant depression of muscle activity, reduced tone and decreased body temperature;
5. rapid heart beat in a reflex response to lowered blood pressure—see page 110;
6. reduced cardiac (heart) output and weak pulse through poor oxygen supply to the heart muscle.

If only 5% of the blood volume is lost there will be little evidence of any of these symptoms. Up to 15% loss, compensation can easily be achieved by the body's reflex responses. Over 15% loss, compensation is more difficult and over 25% loss there is usually death, because the reflex shunting of blood to heart and brain deprives other essential organs, especially the kidneys, so that they cease to function and the body is poisoned by its own waste products.

Giving of serum will replace volume loss and rapidly remove the symptoms of shock, but leave the blood to make its own replacement erythrocytes. Transfusion of whole blood may be necessary to restore oxygen-carrying power if there has been massive haemorrhage.

In the case of burns there is the added complication that the burned tissues release toxins and are extremely susceptible to entry of infective organisms.

BLOOD GROUPS

Blood groups are genetically controlled. The major grouping is into AB, A, B and O. This depends on two allelic genes—see page 124. These genes are responsible for the potential to form the antigens known as **agglutinogen A** and **agglutinogen B** on the surfaces of the erythrocytes. The same genes control formation of **agglutinins** α and β in the blood serum. Agglutinin α causes agglutination of those corpuscles carrying agglutinogen A and similarly β those with B. However, the gene which promotes formation of A prevents formation of α in the same individual, and similarly formation of B prevents formation of β—see table. Thus the serum does not agglutinate its own corpuscles.

If, however, a blood transfusion is given, it is important to ensure that the donor blood does not contain cells liable to be agglutinated by the recipient's blood.

There are other potentially incompatible factors, including the **rhesus factor**. This is under the control of three pairs of genes but the combined effect of these is to produce either Rh positive individuals with antigen but no antibody or Rh negative individuals with neither antigen nor antibody. The latter, however, have the potential to produce antibody if invaded by cells carrying the antigen. Such antibody production is in effect an immune response and produces erythrocyte destruction. It can occur during transfusion with incompatible blood or when an Rh positive foetus is being carried by an Rh negative mother. In the latter a few Rh antigen-carrying corpuscles escaping from the foetus into the maternal blood stream are enough to stimulate her to produce antibody which can diffuse back to affect the infant. The amount of antibody builds up from one pregnancy to the next like any other immune response and may necessitate complete changing of the baby's blood.

The HLA complex is a group of genes which have been identified on chromosome 6. They play a major part in the identification of foreign cells and immune responses.

Blood group	AB	A	B	O
Agglutinogen present	A and B	A	B	O
Agglutinin present	None	β	α	α and β
Can receive blood types	AB, A, B and O	A and O	B and O	O only

SUMMARY OF THE FUNCTIONS OF THE BLOOD

The functions of the blood as a whole are the combined functions of its parts.

(a) **Transport** of

1. food from the alimentary canal to the tissues;
2. oxygen from the lungs to the other tissues;
3. waste materials from the tissues to the excretory surfaces;
4. hormones from the endocrine glands to the other organs or tissues whose metabolism they affect;
5. non-hormonal products, e.g. complement from the liver;
6. leucocytes to the site of infection;
7. heat from more active tissues to less active ones and to the skin for removal—see page 81.

(b) **Protection** by

1. salts which provide a suitable environment for the life of cells by maintaining osmotic pressure and buffering;
2. clotting which stops bleeding from wounds and helps to keep out invading organisms;
3. leucocytes which combat invading organisms and their products, and help healing of wounds by scavenging debris.

THE CIRCULATION

Blood can carry out its functions only when it is kept circulating in blood vessels. The maintenance of **heart beat** and unimpeded flow in the vessels is essential to the life of the tissues. When the heart stops beating the whole body soon dies, while if the circulation to a part only is stopped, as in frostbite or prolonged pressure with a tourniquet, that part alone may die. The most urgent and continuous need is for oxygen, therefore the blood passes to the lungs and to the other organs alternately. The heart is completely divided to keep the oxygenated and deoxygenated blood separate. The blood is carried from the heart in **arteries**, through the tissues in **capillaries** and back to the heart in **veins**. Many of the smaller arteries and veins form linked loops called **anastomoses**, which are safeguards against blockage.

1. **The pulmonary circulation**. A single pulmonary artery from the heart divides into two vessels, one to each lung, where further branching takes place to supply the lobes and lobules—see page 97. The finer arterioles supply the capillary network around the alveoli, while venules collect the blood and join to leave the lungs as four pulmonary veins, which open directly into the left side of the heart.

2. **The systemic circulation**. The **aorta** is the largest artery of the body. From it are given off arteries to the various organs. These branch into **arterioles** which anastomose frequently and supply capillary networks. Blood from the capillaries is collected into **venules** and then into veins. The veins are of two kinds: (a) deep veins usually found following the corresponding arteries and known as **venae commitantes**, and (b) superficial veins draining **superficial venous networks** and opening into the larger deep veins. The blood eventually reaches the right side of the heart through two **venae cavae** and the coronary sinus. The superior vena cava drains the upper part of the body, the inferior vena cava the lower part of the body, and the coronary sinus the heart itself—see page 111. Blood from the stomach, pancreas and spleen is collected in the **hepatic portal vein** and taken to the liver where it mingles with the arterial blood—see page 93.

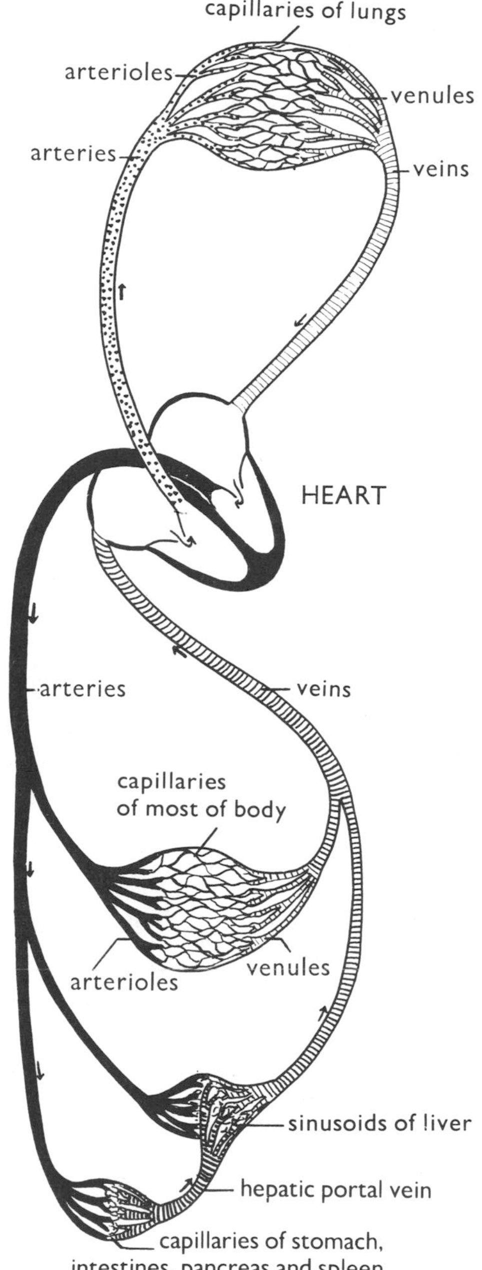

The course of the circulation

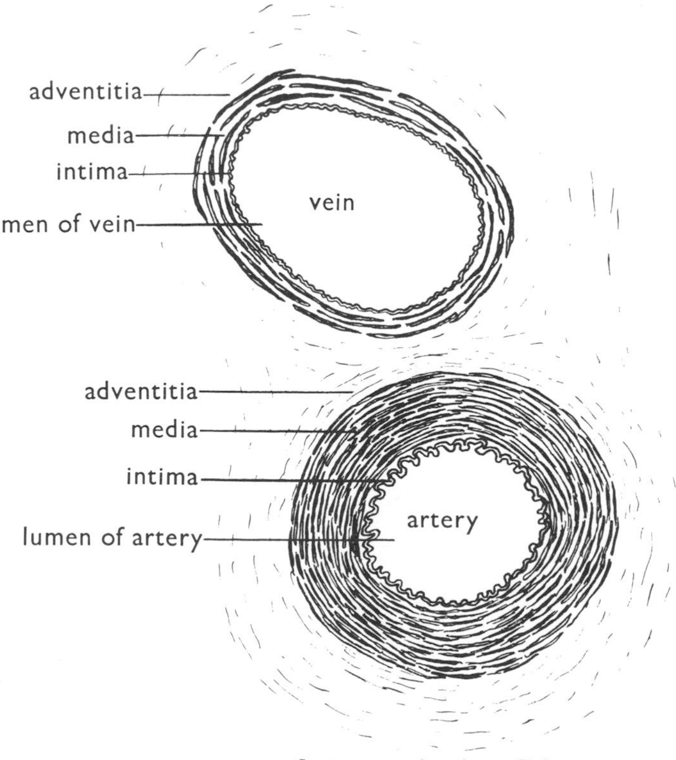

Artery and vein—*T.S.*

	Arteries	Veins
Adventitia	Common tough connective tissue sheath round the artery and its vena commitans	
Media	Strong, elastic and muscular layer—contractile	Comparatively little elasticity or muscle—not contractile
Intima	Smooth pavement endothelium—in folds when artery contracts	Smooth pavement endothelium with paired pocket-shaped valves
Blood carried	Oxygenated except in pulmonary arteries	Deoxygenated except in pulmonary veins

Note. **Endothelium** is the equivalent of epithelium, but lines spaces with no connection with the outside of the body. Most capillaries have walls of endothelium only, but those of the brain have additional layers—see page 99.

low

of flow in the largest artery, the aorta, is 100–140 cm/s. As the arteries branch repeatedly the total cross-sectional area increases, allowing pressure to decrease. At the same time the increased area of wall in contact with the blood increases friction and decreases rate of flow.

At the ends of the arterioles there are small sphincters of smooth muscle which control flow into the capillaries and also into direct connections between the arterioles and venules known as **arteriovenous shunts**. By the time the blood enters the capillary network it is flowing at 0.8–2 cm/s, but still has enough hydrostatic pressure to force water bearing dissolved food and minerals out of the blood into the tissues. Most of the proteins are held back so that when, in the exit vessels of the capillary network, the hydrostatic pressure has fallen further, the colloidal osmotic potential is able to draw fluid bearing waste materials from the tissues into the blood stream. Gaseous exchanges also take place through capillary walls in the direction of the concentration gradient—see page 98.

The exit capillary pressure forces blood on into the venules, from which it passes to the veins. The speed of flow gradually increases as the ratio of wall to cross-sectional area decreases and there is less friction. Flow in veins is maintained by:

1. flow of blood from the capillaries;
2. pressure from surrounding organs, especially skeletal muscles;
3. pairs of valves which prevent backflow;
4. suction as the thorax expands during inhalation of breath.

The better the venous return, the greater the filling of the heart during its period of relaxation and consequently the greater the stretching of the cardiac muscle. This stretching increases contraction—see page 58—and thus increases the pressure and volume of blood forced out of the heart at each stroke.

Stroke volume × heart rate = cardiac output

During rest a healthy adult has a stroke volume of about 70 ml and a heart rate of about 72 beats/min giving a cardiac output of about 5 litres/min. During exercise this volume may increase four times and in trained athletes six times. The distribution of the blood may be varied at the same time to match physiological needs—see pages 70 and 81.

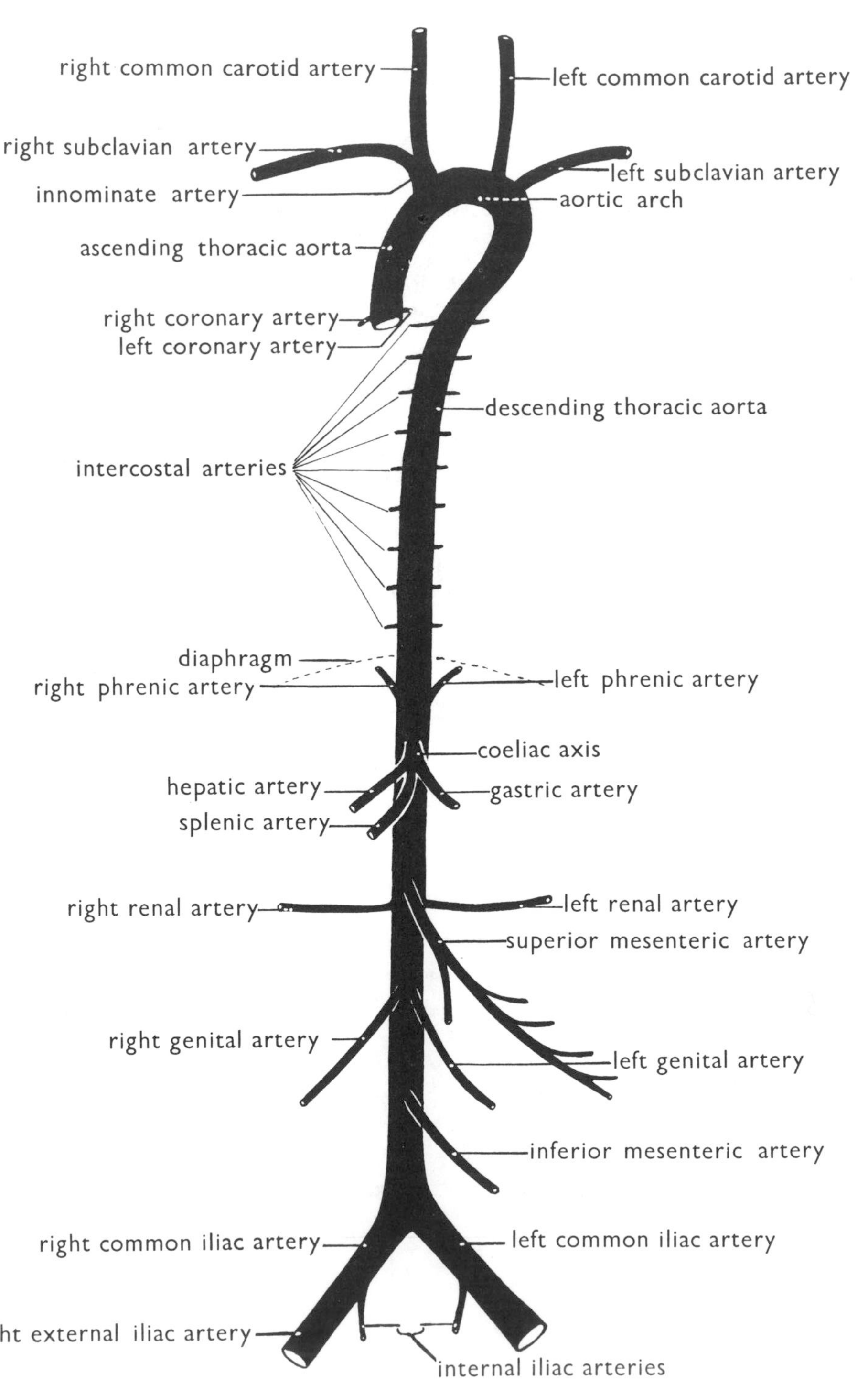

Aorta and its branches

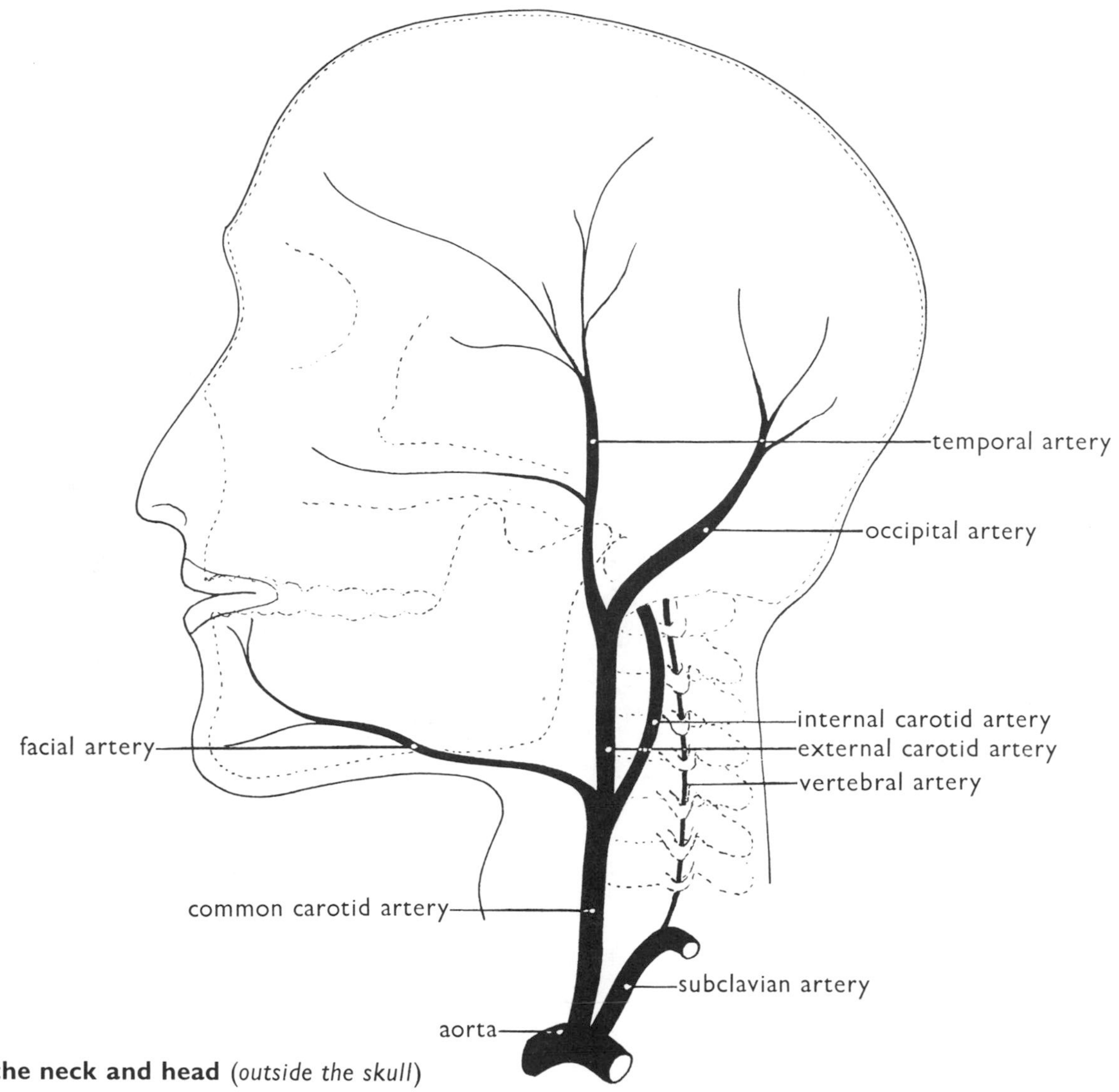

temporal artery
occipital artery
internal carotid artery
external carotid artery
vertebral artery
facial artery
common carotid artery
subclavian artery
aorta

Arteries of the neck and head *(outside the skull)*

anterior cerebral artery
internal carotid artery
middle cerebral artery
circle of Willis
posterior cerebral artery
basilar artery
cerebellar artery
vertebral artery

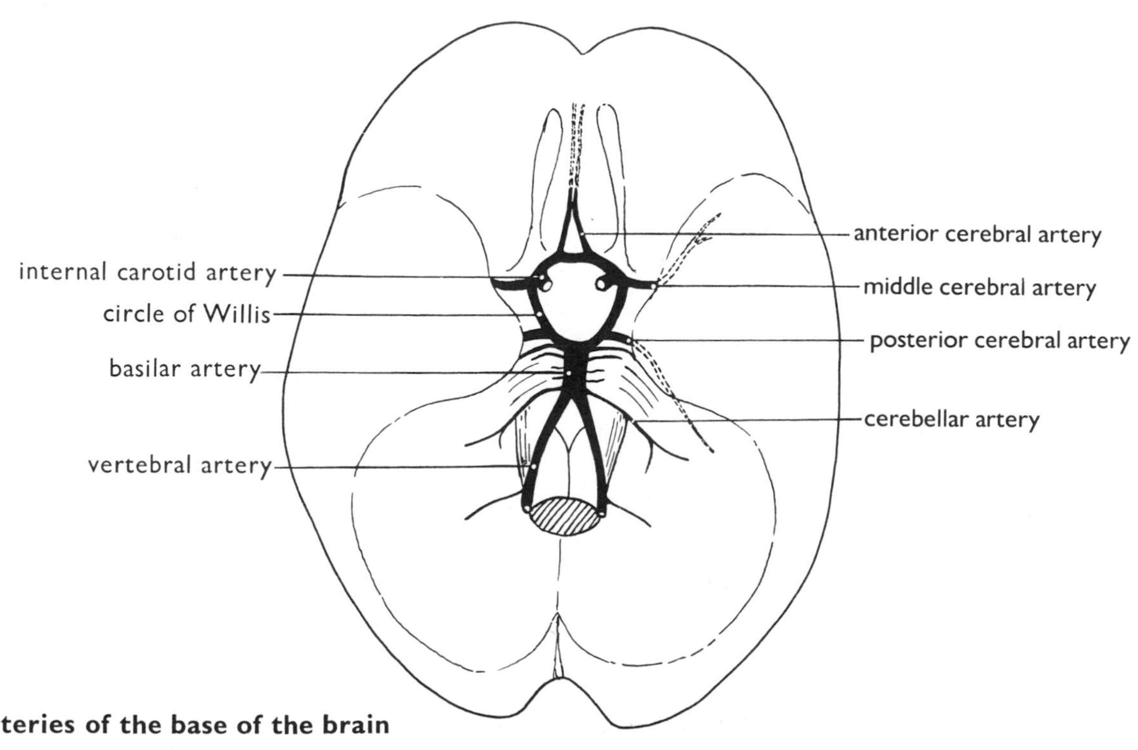

Arteries of the base of the brain

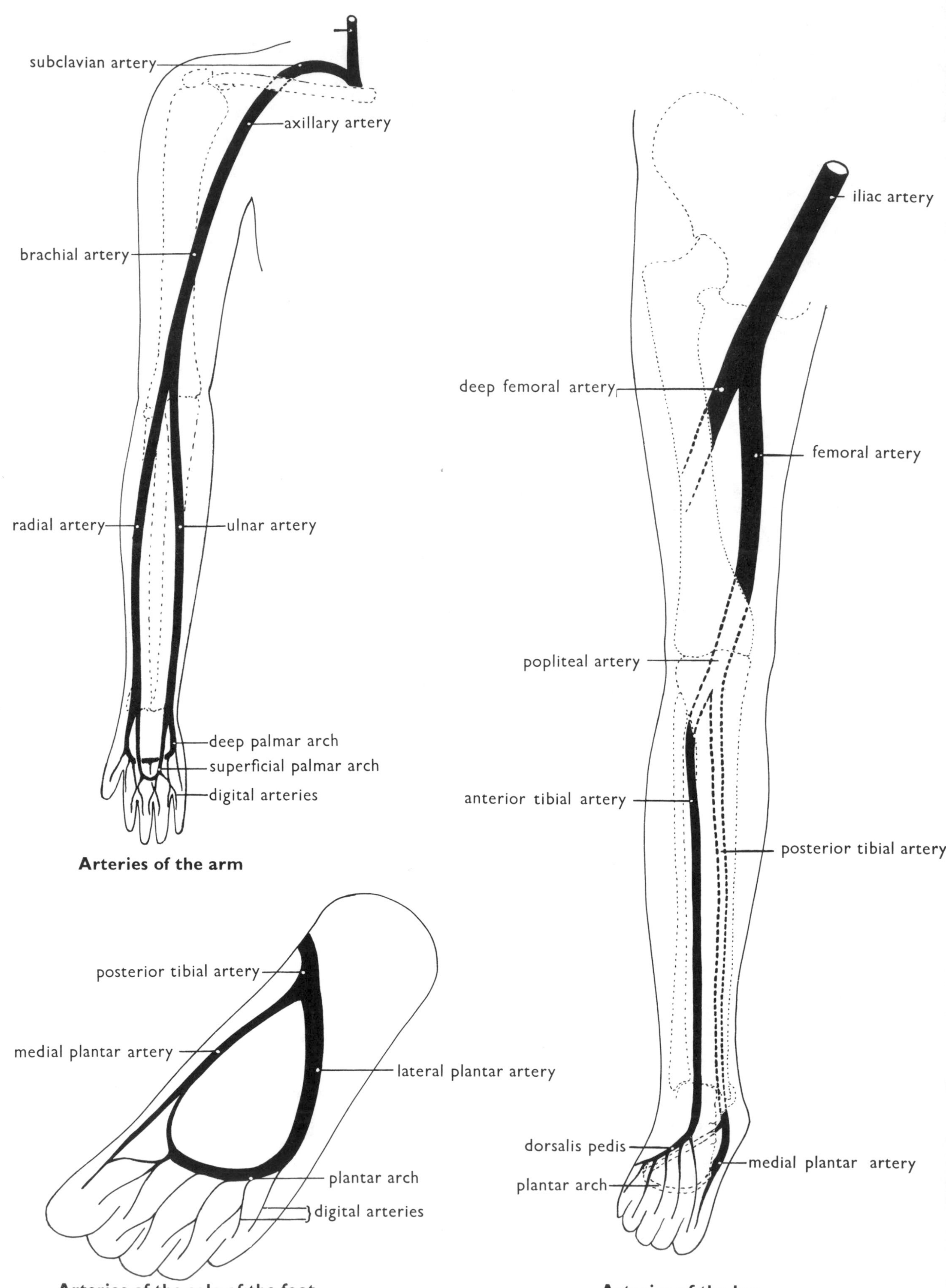

Arteries of the arm

Arteries of the sole of the foot

Arteries of the leg

The blood from the stomach, intestines, spleen and pancreas does not pass directly into the posterior vena cava but is collected in the **hepatic portal vein** and taken to the liver, where it passes through spaces in the liver tissue called **sinusoids**—see page 93—and mixes with the blood from the hepatic artery before reaching the hepatic veins.

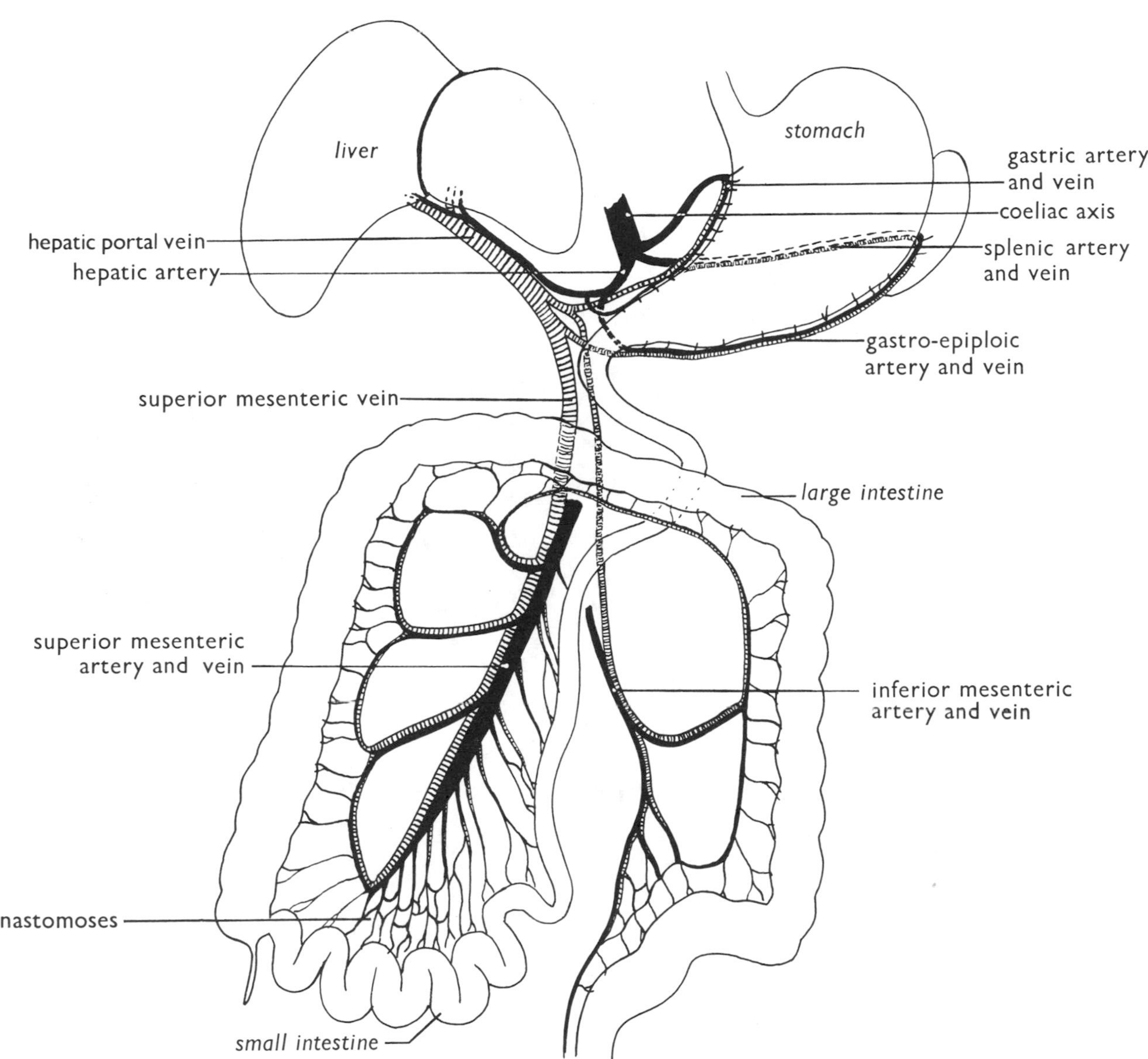

Arteries and veins of the alimentary canal

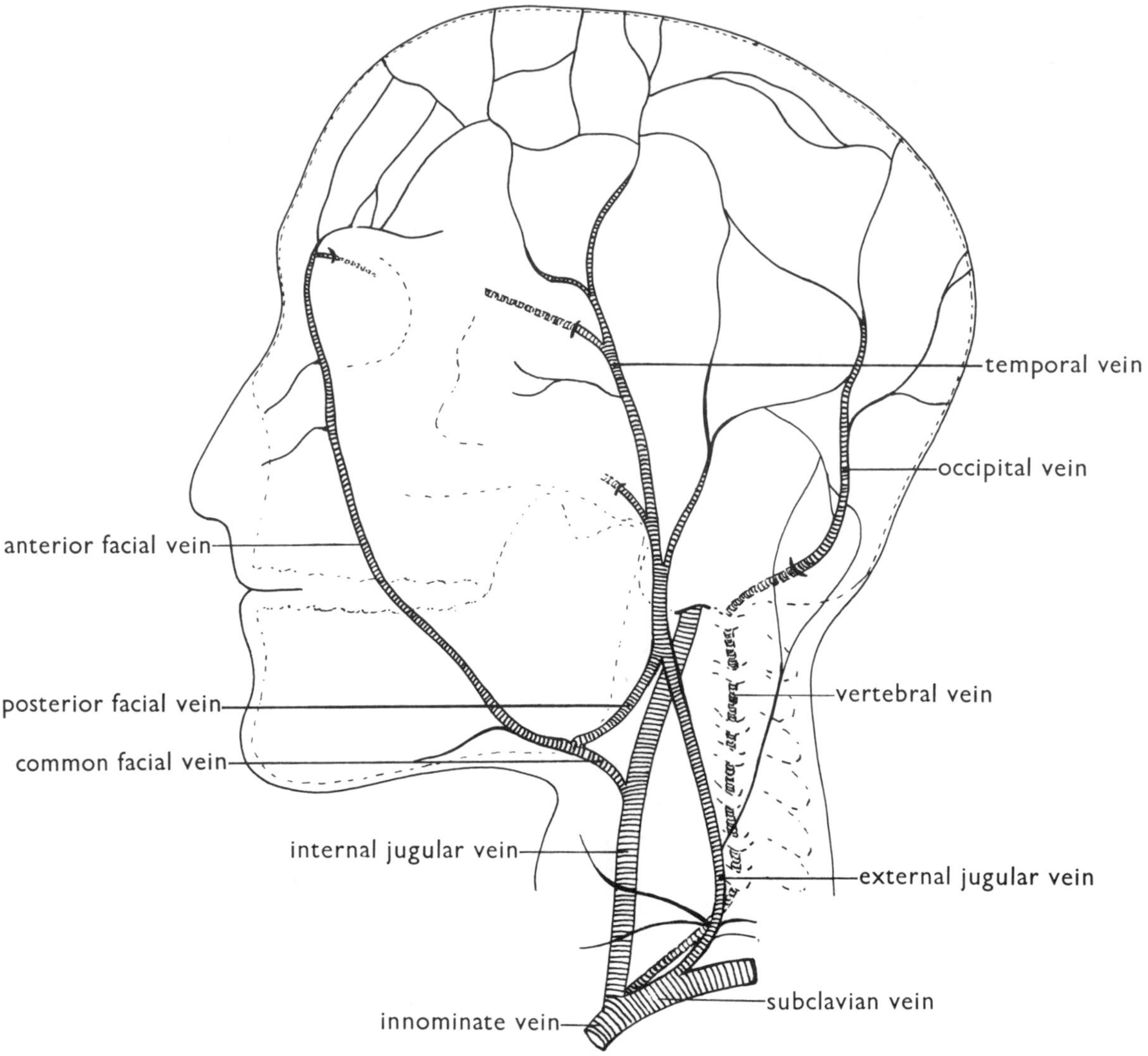

temporal vein
occipital vein
anterior facial vein
vertebral vein
posterior facial vein
common facial vein
internal jugular vein
external jugular vein
subclavian vein
innominate vein

Veins of the neck and head *(outside the skull)*

superior sagittal sinus
(in dura mater)
inferior sagittal sinus
(between cerebral hemispheres)
cerebral vein
straight sinus
right transverse sinus
left transverse sinus
cavernous sinus
meeting of sinuses
petrosal sinus
occipital sinus
internal jugular vein

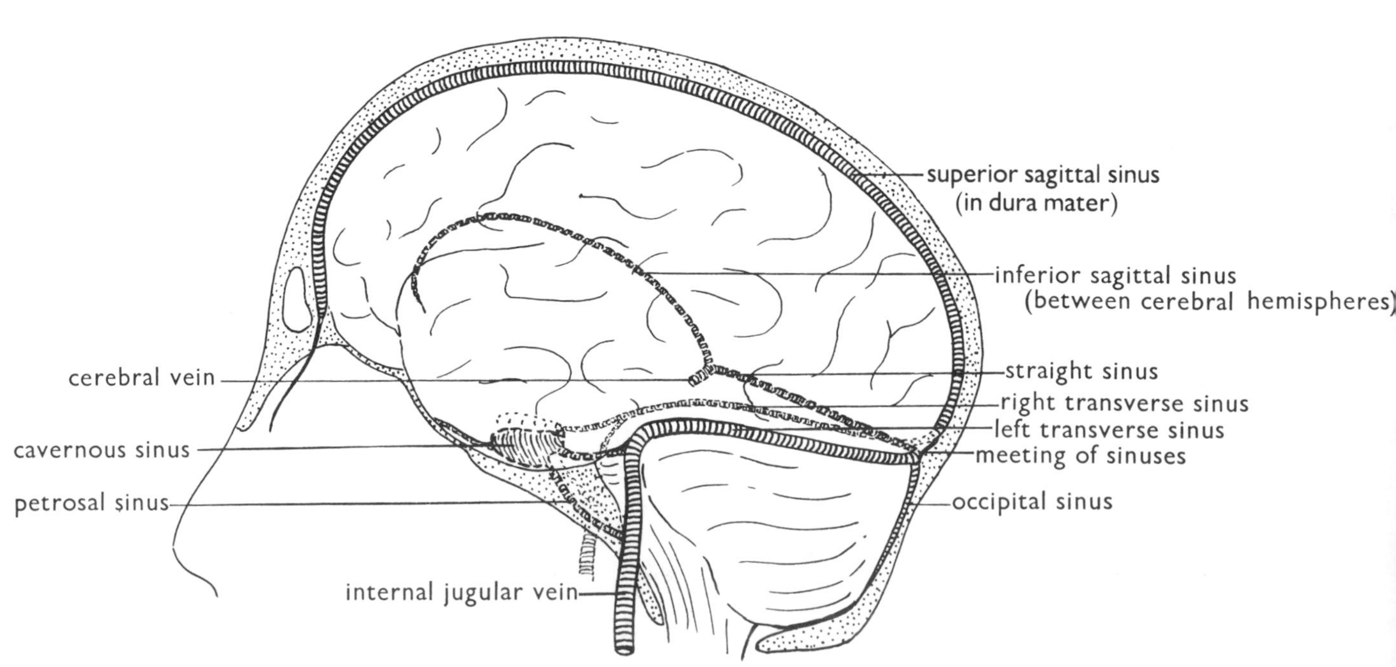

Venous sinuses *(inside the skull)*

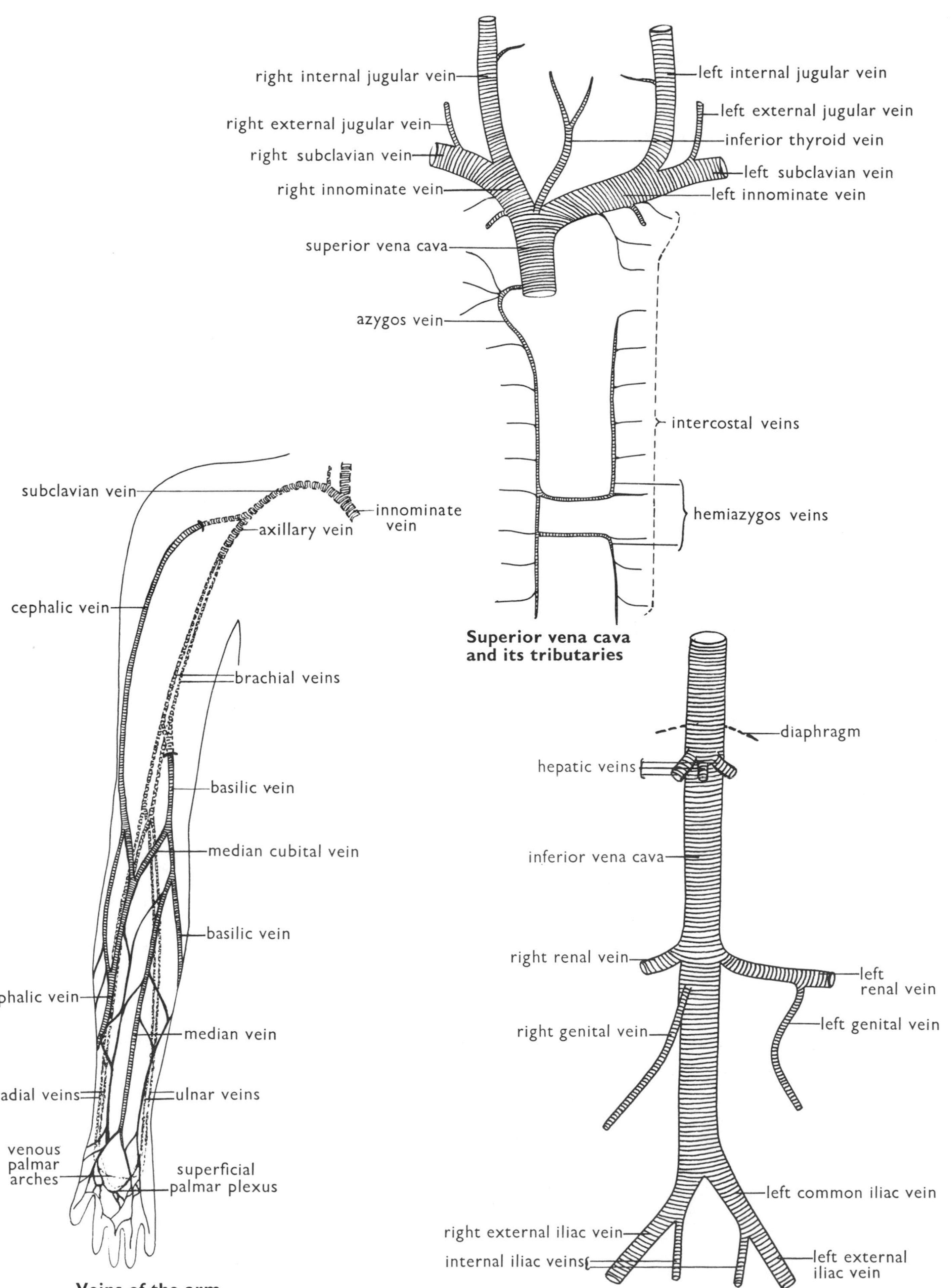

Superior vena cava and its tributaries

Veins of the arm

Inferior vena cava and its tributaries

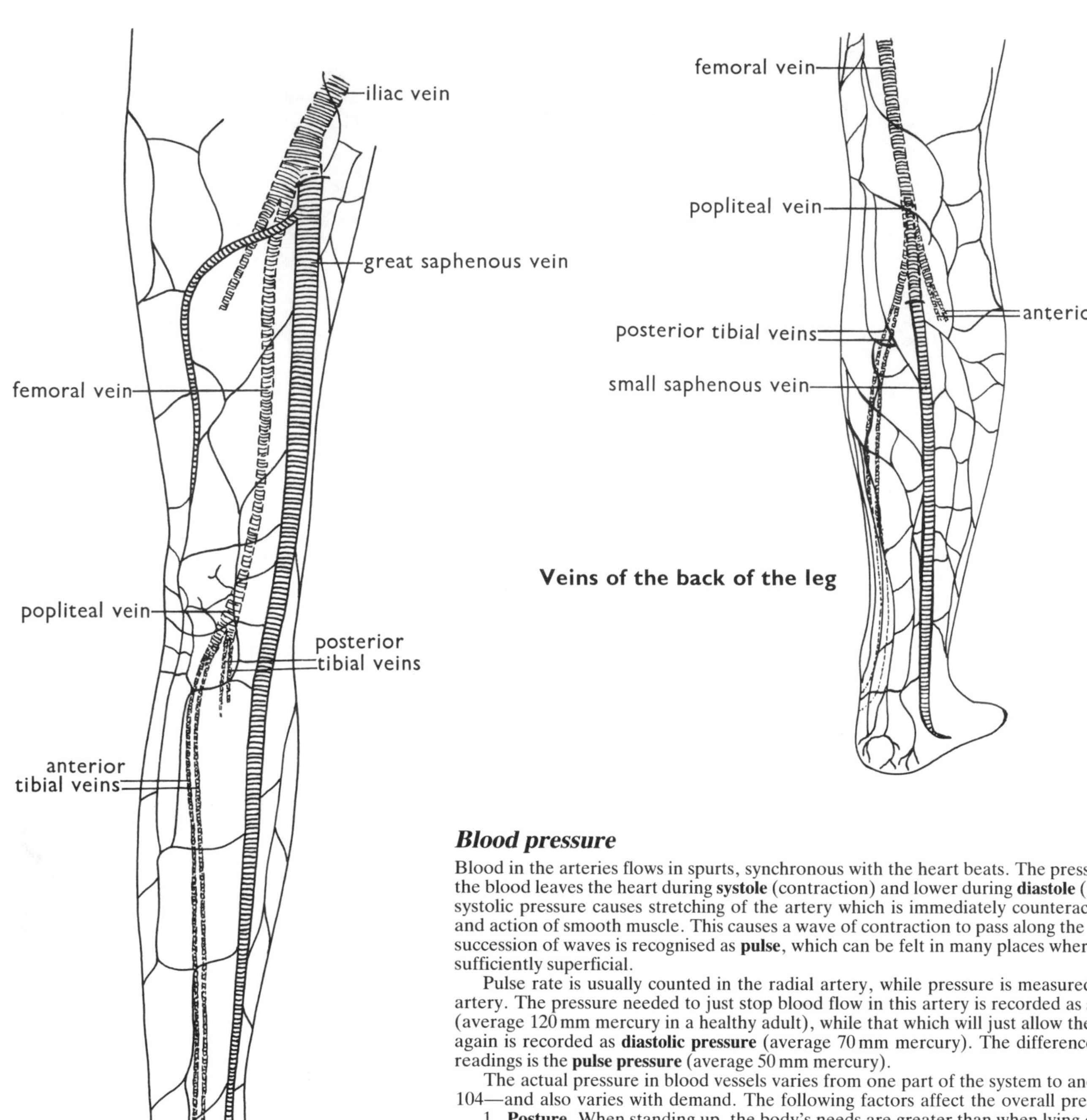

Veins of the front of the leg

Veins of the back of the leg

Chemical effects on vascular muscle

Adrenaline	Reinforces sympathetic
Noradrenaline	General constriction
Angiotensin	General constriction
Serotonin	Local constriction
Histamine	Local dilation
Bradykinin	Local dilation
CO_2	Local dilation

Blood pressure

Blood in the arteries flows in spurts, synchronous with the heart beats. The pressure is highest as the blood leaves the heart during **systole** (contraction) and lower during **diastole** (relaxation). The systolic pressure causes stretching of the artery which is immediately counteracted by elasticity and action of smooth muscle. This causes a wave of contraction to pass along the artery wall. The succession of waves is recognised as **pulse**, which can be felt in many places where the vessels are sufficiently superficial.

Pulse rate is usually counted in the radial artery, while pressure is measured in the brachial artery. The pressure needed to just stop blood flow in this artery is recorded as **systolic pressure** (average 120 mm mercury in a healthy adult), while that which will just allow the pulse to be felt again is recorded as **diastolic pressure** (average 70 mm mercury). The difference between these readings is the **pulse pressure** (average 50 mm mercury).

The actual pressure in blood vessels varies from one part of the system to another—see page 104—and also varies with demand. The following factors affect the overall pressure.

1. **Posture**. When standing up, the body's needs are greater than when lying down. The force of gravity pulling blood away from the head and retarding its return from the lower part of the body has to be overcome. Standing up suddenly without time for adjustment may cause fainting due to shortage of blood to the brain.

2. **Exercise**. When active the skeletal muscles need more oxygen and glucose. Cardiac output and blood pressure rise and there is redeployment in the capillary circulation to ensure adequate supplies.

3. **Emotional arousal**. The response to this is regulated by the autonomic nervous system—see page 70, with effects similar to those of exercise.

4. **Regulating mechanisms**. These prevent over-reaction to demand.

(a) **Baroreceptors**—see page 73—in the aortic arch, subclavian and carotid arteries and in the lungs, when stimulated by rise in pressure stretching the arterial walls, initiate reflexes which are relayed through the **cardiac inhibitory centre** in the medulla oblongata—see page 62. Parasympathetic nerve fibres then produce **inhibition of vasoconstriction** and **reduction in heart rate**, with resultant **fall** in blood pressure. Reduction in pressure removes the stimuli from the baroreceptors and allows domination of the **cardiac accelerator centre**, also in the medulla but controlled by the hypothalamus. This centre activates sympathetic nerve fibres.

(b) **Baroreceptors** in the heart itself respond to fall in pressure of the blood entering the heart from the venae cavae and initiate **cardiac acceleration** and **vasoconstriction**, which **raise** blood pressure.

(c) **Hormones** and other chemicals affect blood pressure mainly through their effects on smooth muscle of the blood vessel walls—see table.

(d) **Rise in body temperature** stimulates rise in heart rate directly, while locally applied heat or cold cause local vasodilation possibly through release of histamine.

Taken together, these responses keep the pressure in the blood vessels within the range they can tolerate without rupturing. The system is normally autoregulatory, but certain conditions, e.g. hardening (loss of elasticity) of the arterial walls, furring of linings of blood vessels, excessive number of blood corpuscles, excess weight, fluid retention, smoking and stress, tend to keep blood pressure abnormally high and put strain on the heart and blood vessel walls, while shock—see page 102—and heart failure may reduce blood pressure catastrophically.

THE HEART

The heart is a hollow, muscular organ about the size of a closed fist. It lies in the thorax, between the lungs and above the diaphragm, with its apex to the left. It is formed of three layers.
1. **Endocardium** lines the heart cavities and is continuous with the linings of the blood vessels.
2. **Myocardium** is a strong layer of cardiac muscle making up the bulk of the heart. It is supplied with blood from the **coronary arteries** and drained by the **coronary veins** and **coronary sinus**.
3. **Pericardium** is an outer covering of connective tissue and serous epithelium—see page 83. It forms a double-layered bag enclosing a cavity filled with **pericardial fluid**, which reduces friction as the heart moves.

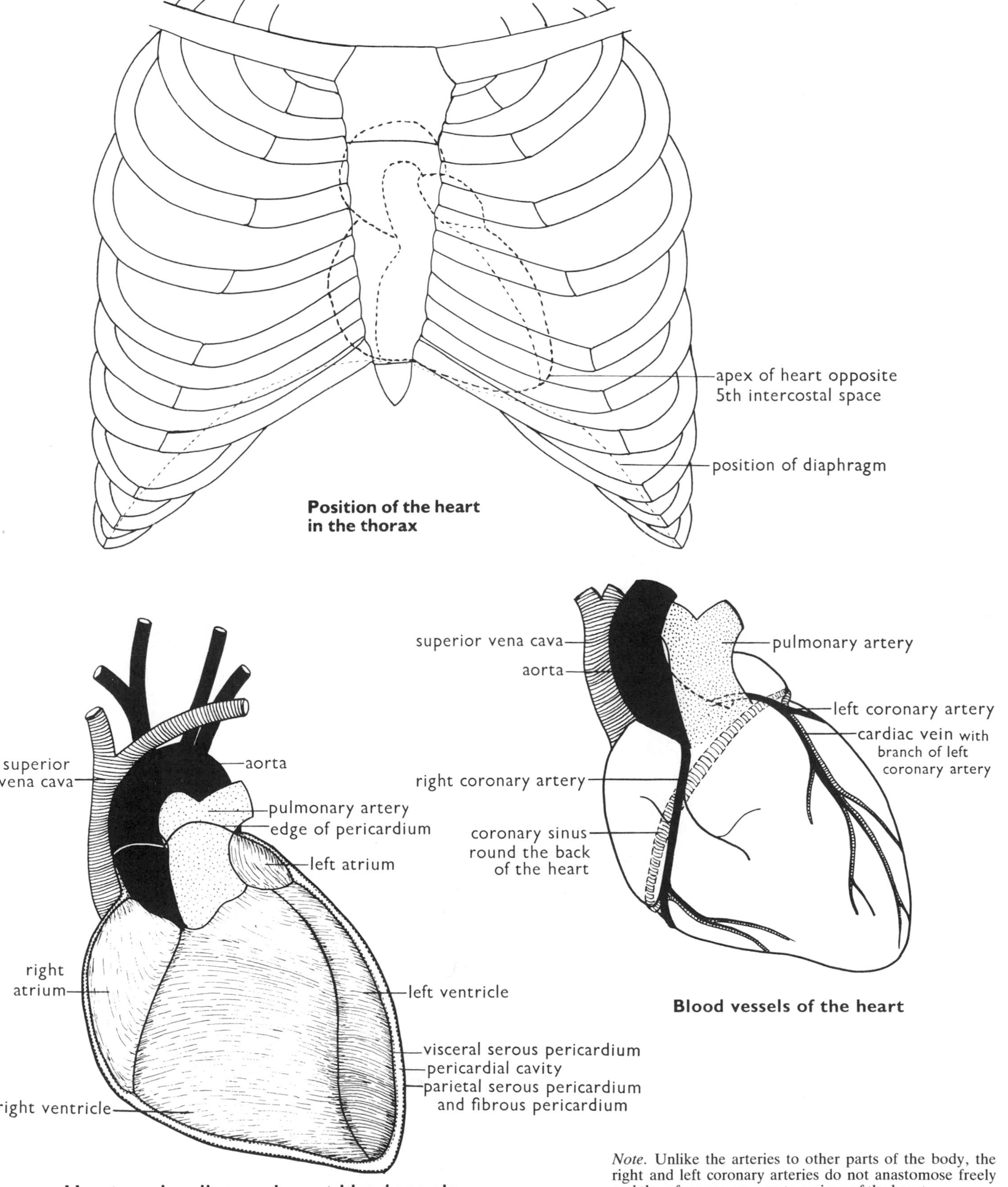

Position of the heart in the thorax

Heart, pericardium and great blood vessels

Blood vessels of the heart

Note. Unlike the arteries to other parts of the body, the right and left coronary arteries do not anastomose freely and therefore serve separate regions of the heart.

THE HEART—continued

The cavity of the heart is divided completely by a vertical medium partition called the **septum**. Each side is further divided into a thin-walled **atrium** above and a thick-walled **ventricle** below.

Between each atrium and the corresponding ventricle there is an **atrio-ventricular aperture**. In situ in the thorax, the right atrium and ventricle lie in front of the left atrium and ventricle. Blood vessels are connected to each of the four chambers of the heart.

The right **atrium** receives **deoxygenated blood** from the **superior vena cava**, the **inferior vena cava** and the **coronary sinus**.

The **right ventricle** sends **deoxygenated blood** into the **pulmonary artery**.

The **left atrium** receives **oxygenated blood** from the four **pulmonary veins**.

The **left ventricle** sends **oxygenated blood** into the **aorta**.

The deoxygenated and oxygenated blood are kept separate by the septum except during foetal life when it is perforated by the foramen ovale—see page 130.

Valves of the heart

The direction of flow of blood is maintained by valves.

The openings of the inferior vena cava and coronary sinus each have a single **pocket-shaped valve**, while the superior vena cava and pulmonary veins can be closed by **sphincter action** of the atrial wall.

Between the right atrium and right ventricle is the **tricuspid valve** with three flaps, each supported at its free edge by tendinous strands, the **chordae tendineae**, the other ends of which are attached to the ventricular wall through **papillary muscles**. Between the left atrium and the left ventricle is the **bicuspid** or **mitral valve**, similar to the tricuspid valve but with only two flaps. When the ventricles contract these valves are forced upwards to close the atrio-ventricular openings. The papillary muscles automatically adjust the length of the supporting strands so that the valves are not forced too far.

At the mouth of the pulmonary artery and of the aorta there are three pocket-shaped **semilunar valves**, which hold the blood under pressure in the arteries while the ventricles relax.

Failure of any of these valves impairs heart function.

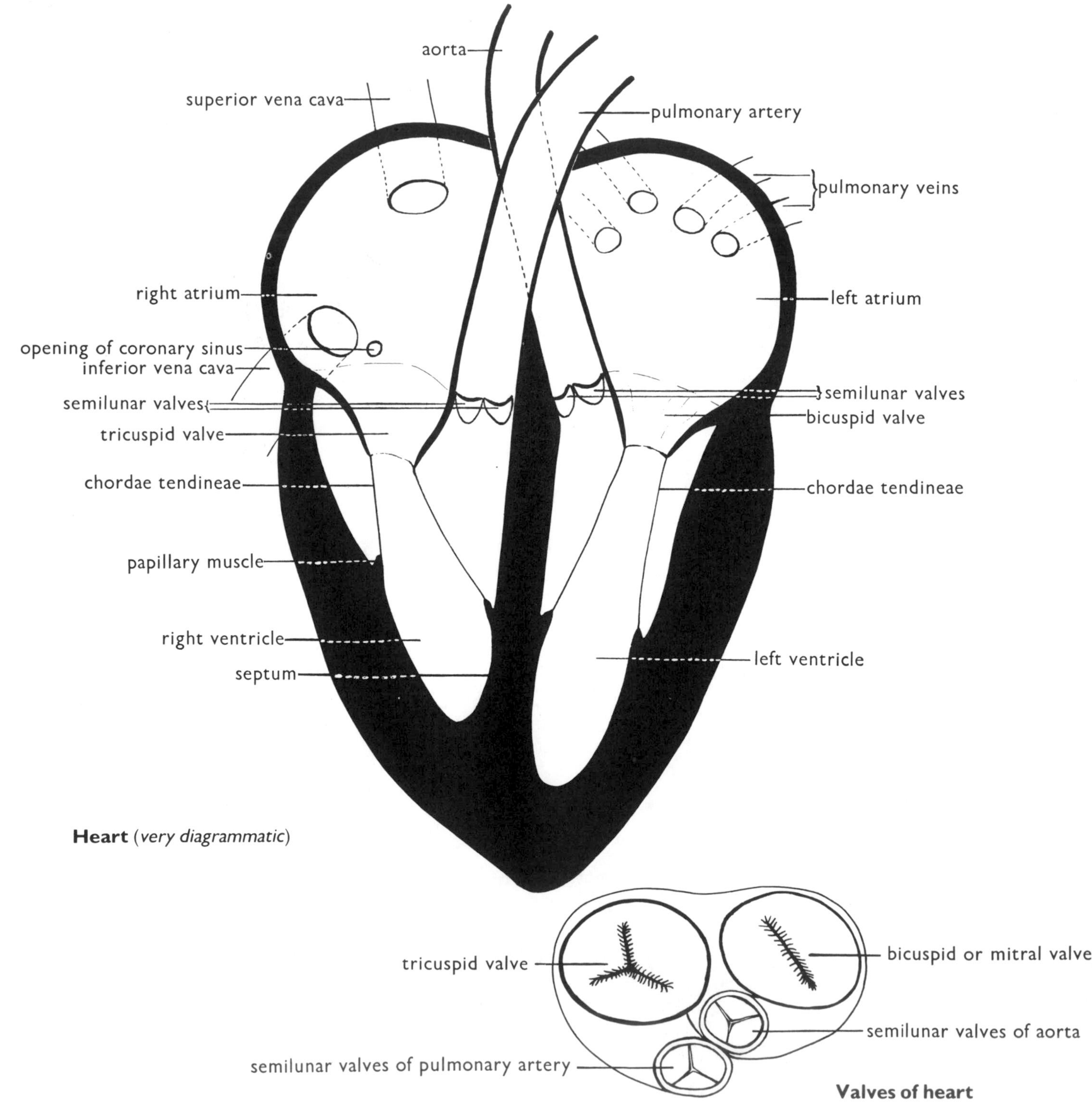

Heart (*very diagrammatic*)

Valves of heart

Heart beat

The heart muscle contracts rhythmically with a period of relaxation and rest between each contraction. The contraction period is called **systole**, the dilation period **diastole**, and the whole repetitive process is the **cardiac cycle**.

Atrial systole. The atria contract and force blood into the ventricles so that they are distended and filled with blood under pressure. The contraction affects the regions round the mouths of the veins first so that these vessels are closed and blood cannot flow back into them.

Ventricular systole. The ventricles contract and force blood, under pressure, past the semilunar valves into the arteries. Backflow of blood into the atria is prevented by the tricuspid and bicuspid valves respectively. The papillary muscles keep the chordae tendineae taut so that the valves are held in place. At the same time the muscles of the atria relax and blood begins to flow into them from the veins.

Diastole. The muscles of both atria and ventricles relax. Blood flows from the veins through the atria into the ventricles till all the cavities of the heart are filled. Thus the filling of the ventricles is mainly a passive process. Backflow of blood from the arteries is prevented by the closed semilunar valves.

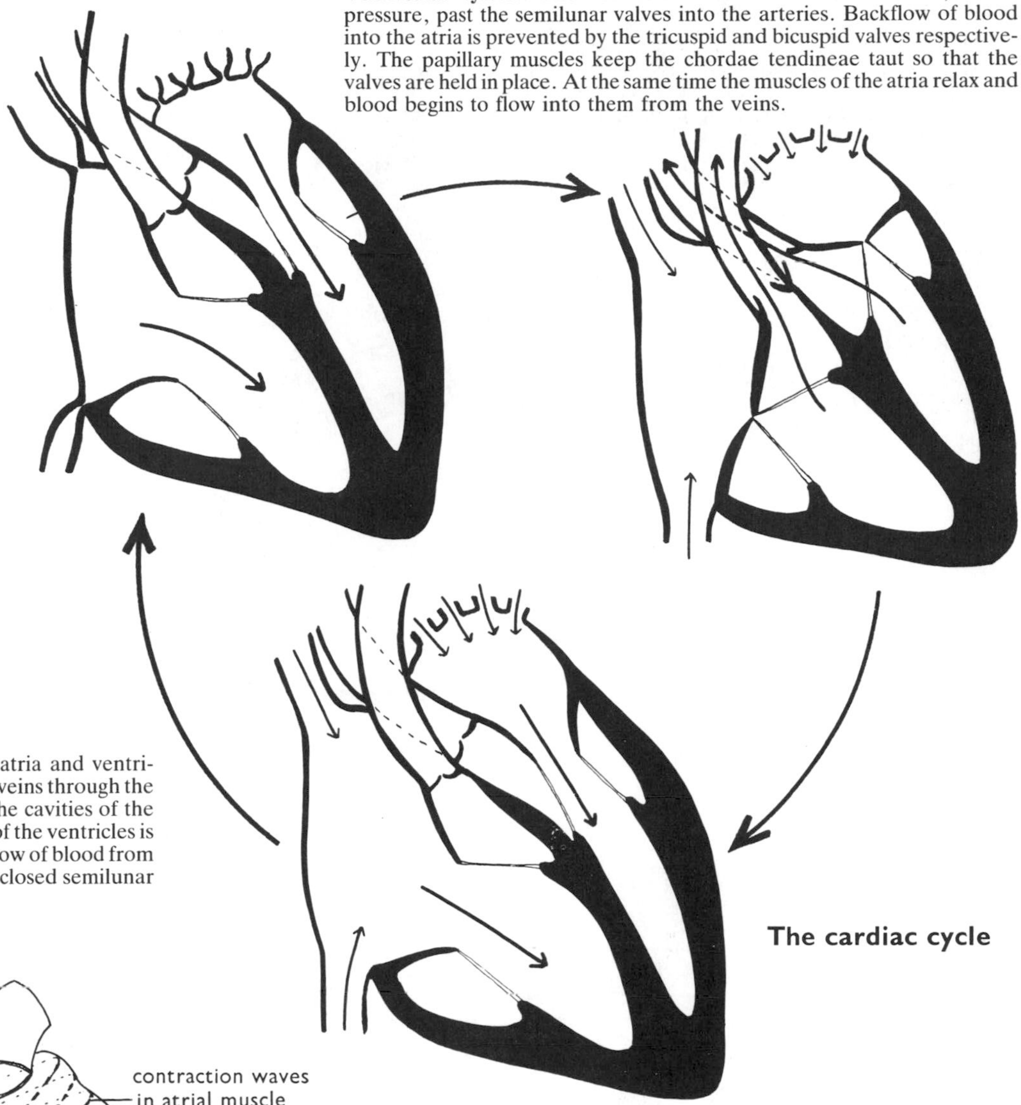

The cardiac cycle

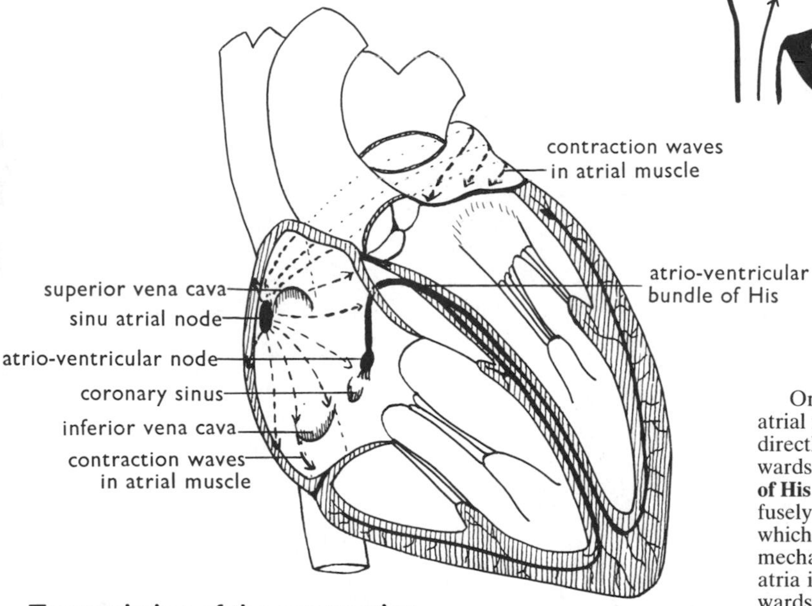

contraction waves in atrial muscle
atrio-ventricular bundle of His
superior vena cava
sinu atrial node
atrio-ventricular node
coronary sinus
inferior vena cava
contraction waves in atrial muscle

Transmission of the contraction wave through the heart muscle

Rate of heart beat

Cardiac muscle contracts rhythmically without nervous stimulation, but the pace of contraction is normally controlled by the autonomic nerves, which supply a group of special cells close to the opening of the superior vena cava, called the **sinu-atrial node**. Parasympathetic fibres from the vagus nerves and sympathetic fibres from the cardiac plexus both end here and, when stimulated, release their neurotransmitters, the balance of which overrides the inherent cardiac rhythm. **Parasympathetic** stimulation **slows** heart rate, while **sympathetic** stimulation **speeds** it up—see pages 69, 70 and 110.

Once initiated, waves of contraction pass from fibre to fibre in the atrial muscle but are prevented from reaching the muscle of the ventricle directly by a connective tissue ring. The impulses are transmitted onwards through the **atrio-ventricular node** and the **atrio-ventricular bundle of His**. The latter bifurcates as it runs down the septum and branches profusely in the ventricular muscle, each branch ending in Purkinje fibres, which directly contact the contractile cells of the myocardium. This mechanism ensures that the ventricle contracts after contraction of the atria is complete and that it starts from the apex, thus forcing blood towards the arteries.

The average rate of heart beat is 72 per minute. Thus each complete cardiac cycle takes 0.8 s. This is divided into systole and diastole as follows:

Atrial systole 0.1 s	Atrial diastole 0.7 s
Ventricular systole 0.3 s	Ventricular diastole 0.5 s
Complete systole 0.4 s	Complete diastole 0.4 s

When heart rate increases, the period of complete diastole is shortened. In extreme cases there may not be time for refilling of the ventricles before the next systole, resulting in reduced power and decreased cardiac output.

Normally there is an increase in rate of heart beat during activity, inspiration and in response to some drugs, e.g. caffeine, atropine and epinephrine, and decrease during rest, sleep, expiration and some drugs, e.g. amylnitrate and alcohol in small quantities.

Note. The relationship between heart rate and blood pressure has been dealt with already—see page 110.

The Lymphatic System

The **lymphatic system** is really part of the vascular system.

Blood is contained in a closed system of vessels and kept circulating by the heart, but some of the fluid, the plasma, escapes from the capillaries by diffusion and filtration and bathes the tissues directly. This **tissue fluid** becomes **lymph** when it is collected up in the **lymph vessels**, through which it is returned to the blood stream. The composition of lymph is similar to that of blood plasma, but with less protein, less food materials and more waste materials. It has no erythrocytes but relatively more lymphocytes.

The lymph vessels start as fine, blind-ended, **lymph capillaries**. These join to form **lymphatics** which are similar to veins in general course and structure, but are more numerous and finer and have many more **semilunar** (pocket) **valves**.

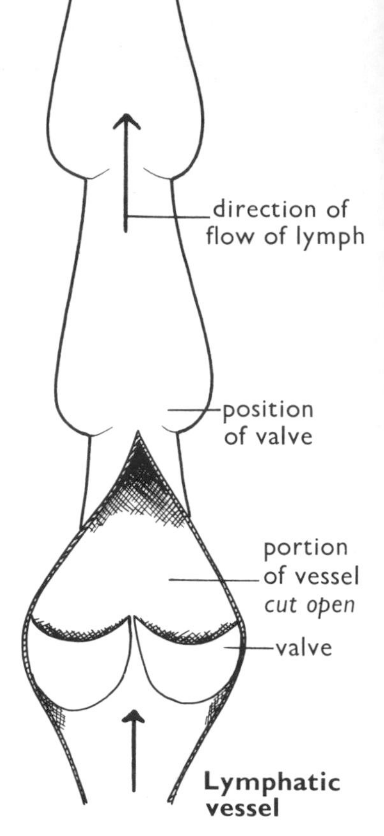

Lymphatic vessel

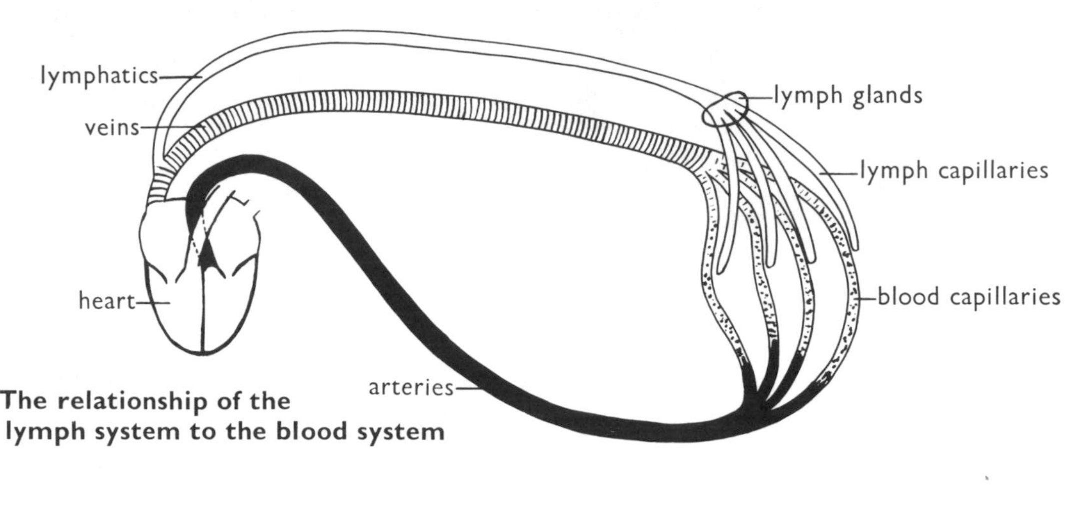

The relationship of the lymph system to the blood system

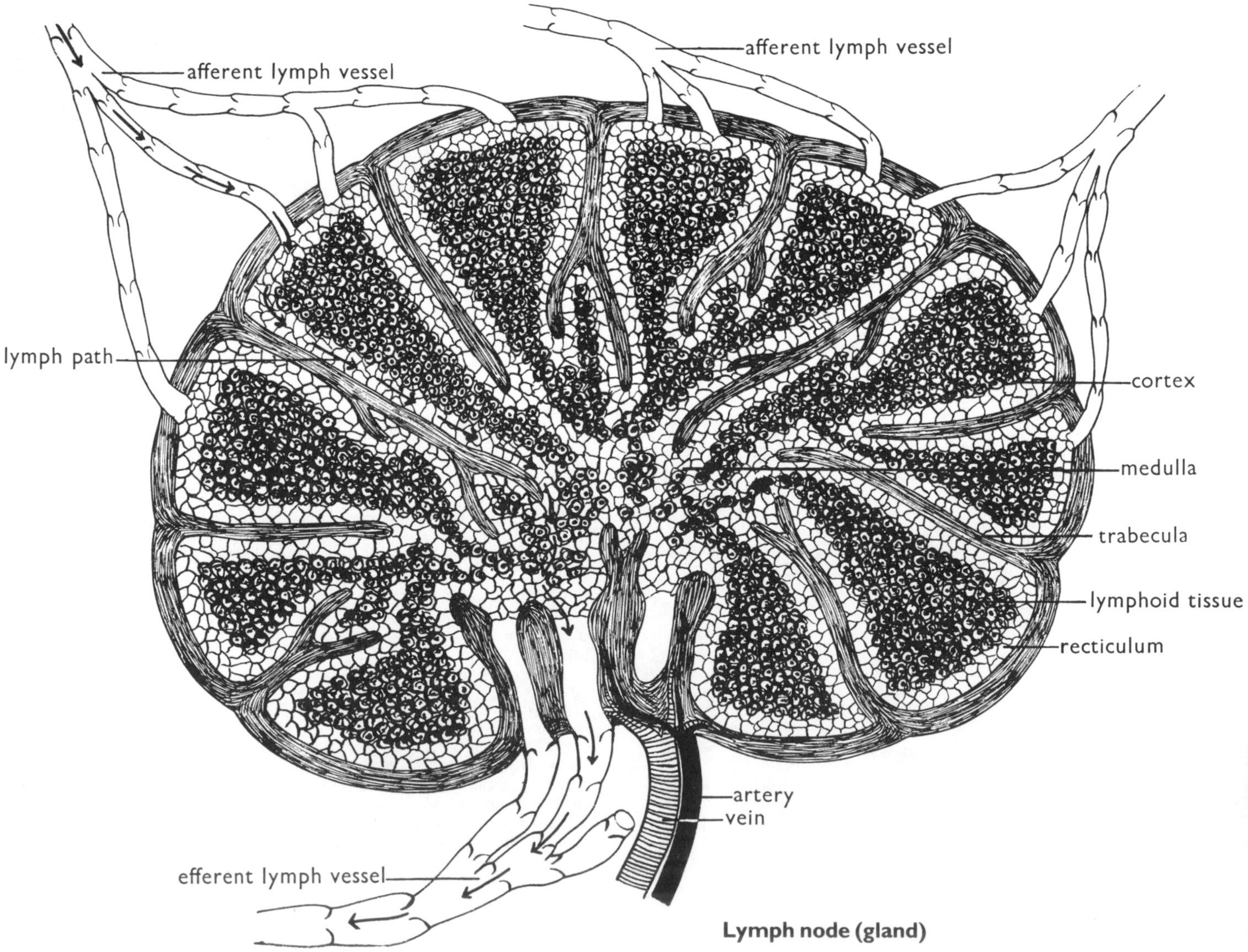

afferent lymph vessel
afferent lymph vessel
lymph path
cortex
medulla
trabecula
lymphoid tissue
recticulum
artery
vein
efferent lymph vessel

Lymph node (gland)

LYMPH DRAINAGE

Collection of tissue fluid into the lymphatics is aided by pumping action of muscle pressing on small lymphatics in the fascial connective tissue. In the absence of muscle activity, fluid may accumulate, e.g. swollen ankles when standing. The small lymphatics concerned have valve-like pores which permit entry of the whole tissue fluid including cell debris or germs from wounds, but before the lymph is returned to the blood stream it is filtered through one or more **lymph nodes**. After passing through these it is collected into the main lymph vessels. Lymph from the right side of the head and thorax and the right arm drains into the **right lymphatic duct**, while lymph from the rest of the body drains into the **thoracic duct**. These vessels open into the right and left **innominate veins** respectively.

The flow in the lymph vessels is maintained by:
1. the greater pressure in the lymph capillaries than in the lymphatics;
2. the muscular movements around the vessels which compress them, while the valves prevent backflow;
3. the inspiratory movements which suck lymph into the thoracic duct and expiratory movements which force it on into the innominate veins.

LYMPH NODES

Though non-secretory, lymph nodes are often called lymph glands. They are massed in groups on the lymph drainage system. Each node has a **fibrous coat**, extended inwards to form partitions called **trabeculae**. It contains **reticular** (net-like) **connective tissue** with spaces called **sinuses** and masses of **lymphoid tissue**. The coat and trabeculae are supplied with blood vessels, while the lymph enters the node by a number of afferent lymphatics and leaves by fewer efferent lymphatics after passing through the reticular sinuses. The reticular cells are **fixed macrophages** or **histiocytes**, part of the so-called **reticulo-endothelial system**, which is also found in the liver (Kupffer cells), lungs (alveolar macrophages), brain (microglia) and in the subcutaneous tissue, bone marrow and spleen. Like the wandering macrophages formed from the monocytes of the blood—see page 101, fixed macrophages are **phagocytic** and important in removing particulate matter and germs.

The lymphoid tissue contains cells capable of giving rise to **lymphocytes** and **plasma cells**—see page 101. These may be shed into the lymph stream or remain in the node while performing their function of antibody production. A lymph gland which is actively producing such cells may become inflamed, swollen and painful.

Besides the characteristic lymph nodes, there are masses of lymph tissue forming the **tonsils** and **adenoids**—see page 85, the **solitary glands** and **Peyer's patches** of the intestine—see page 87, and **nodules** in the spleen—see next page. The **thymus** also shares in lymphocyte production.

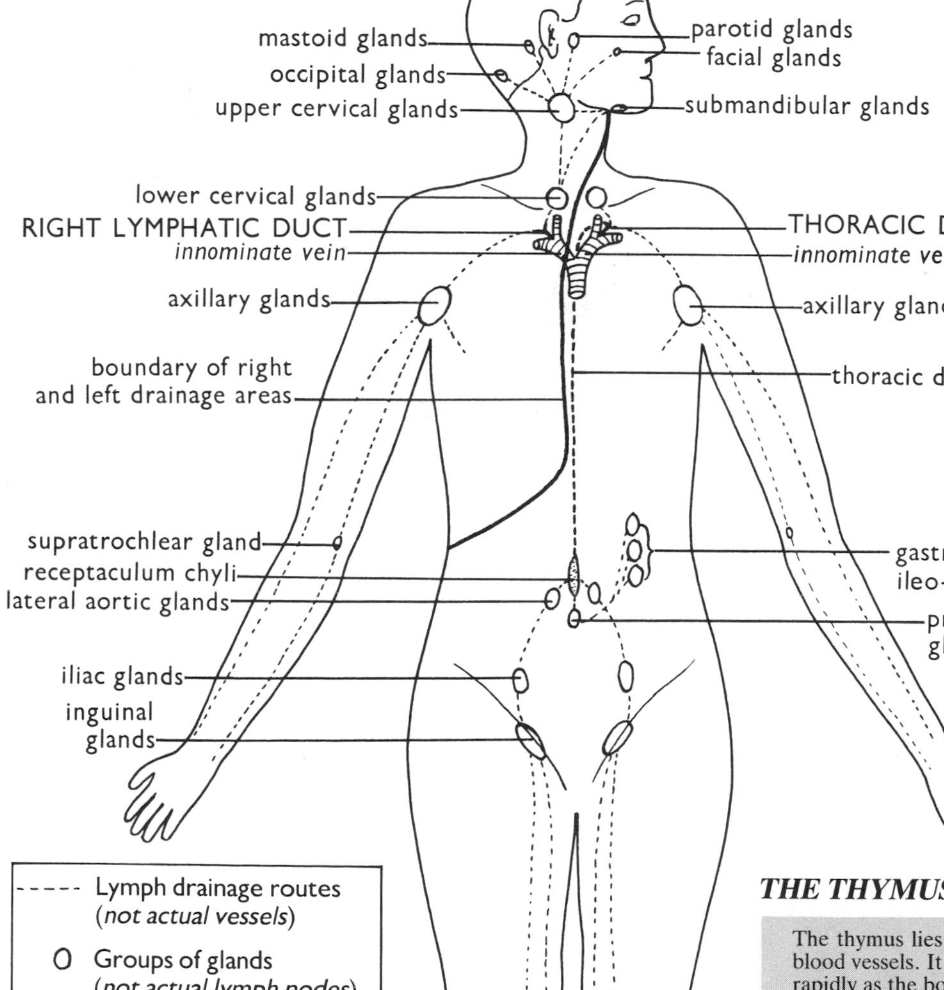

Positions of groups of lymph glands and main lymph vessels

THE THYMUS

The thymus lies behind the sternum, over the roots of the great blood vessels. It is large in infants, but does not increase in size as rapidly as the body. Its maximum weight at puberty is about 40 g, after which it gradually degenerates, most of its characteristic **lymphoidal cells** being replaced by **adipose cells**.

In structure the thymus has a **capsule** and **trabeculae**, dividing it into **lobes**, each consisting almost entirely of **lymphocytes**, supported by a loose net of **endothelio-reticular tissue**. The central **medulla** of each lobe contains **thymic corpuscles** of unknown function.

Functions of the thymus

Before birth small, medium and large lymphocytes migrate to the thymus where they become tightly packed and are processed to acquire **immunological competence**. These T-lymphocytes then go to the lymph nodes where they are stored and clone when required to provide immunity to specific antigens—see page 101. The processing is completed before or very soon after birth, but some lymphocytes remain in the thymus and may be released when there is depletion through stress.

The thymus produces the following hormones:
1. **promine** which stimulates general growth;
2. **thymosine**, **thymic humoral factor**, **thymic factor** and **thymopoietin**, all of which are involved in stimulating the cloning of lymphocytes in lymphoid tissue and the maturation of plasma cells.

All these hormones are important in early life when growth is fastest and immunities are being established.

THE SPLEEN

The spleen lies below the diaphragm, behind and to the left of the stomach, with depressions of its surface indicating the organs with which it is in contact. Its shape and size vary with the movements of these organs and with the degree of distention of the spleen itself, but it averages 12 cm long.

Structure of the spleen

The spleen is covered by peritoneum, within which is a **fibrous capsule**. Fibrous **trabeculae**, similar to those of a lymph node—see page 114, support a pulpy, reticulate mass, forming **splenic cords**, between which blood circulates, coming in direct contact with the phagocytic **splenic cells** (cf. the sinusoid of the liver—see page 93).

Scattered in the **splenic pulp** are **lymph nodes**, each containing an arterial network and surrounded by a marginal zone containing short, straight **penicellate** (tufted) **arterioles**, on each of which is a swelling called an **ellipsoid**. The arterioles open into the venous sinusoids from which blood is eventually removed through **pulp veins**. Because of this blood supply, lymphocytes released from the lymph nodules pass directly into the blood stream.

The splenic vein joins the hepatic portal vein—see pages 92 and 107; thus breakdown products of phagocytic activity of the splenic cells are carried straight to the liver.

Functions of the spleen

The spleen controls the quality and quantity of blood in circulation.

1. It acts as a reservoir for blood cells of all kinds.
2. It destroys worn-out erythrocytes.
3. It produces lymphocytes in its lymph nodules.
4. It produces erythrocytes and granulocytes during foetal life and on certain occasions in adult life, e.g. after severe haemorrhage.

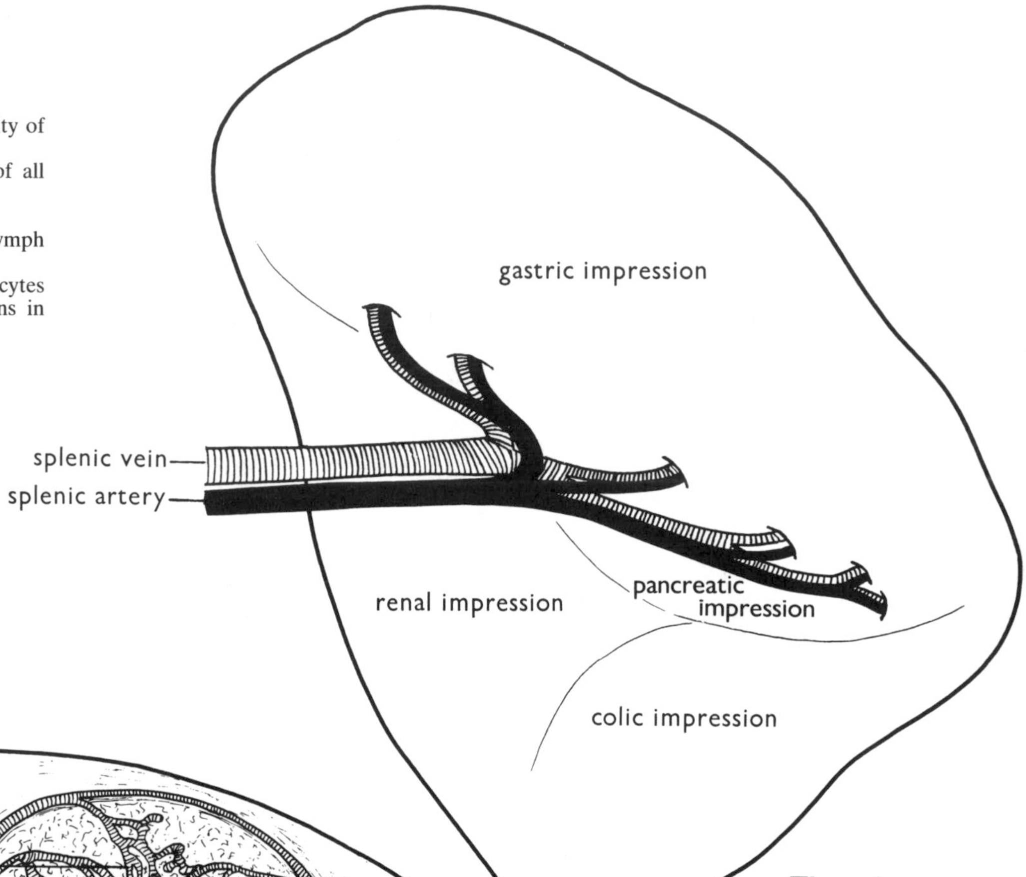

The spleen

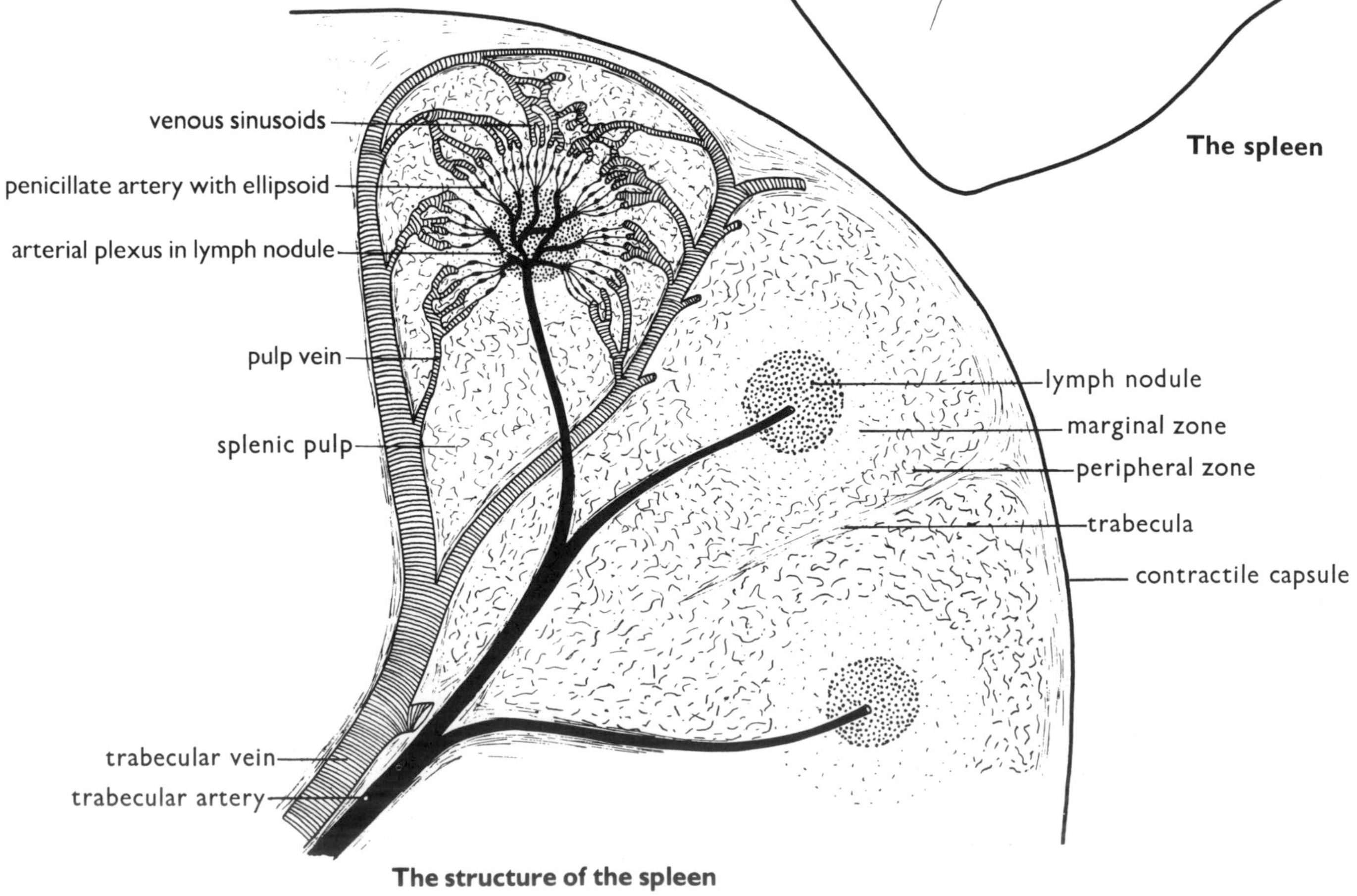

The structure of the spleen

The Excretory System

Excretion to maintain a constant internal environment involves removal from the cells and body fluids of 1. **carbon dioxide**; 2. **nitrogenous waste** from breakdown of proteins and nucleic acids; 3. **toxic materials** including drugs; 4. **excess salts, water** and **heat**. A little of the water, salts and nitrogenous waste from broken-down gut-lining cells and bile leave the body with the **faeces**—see page 90. Most of the unwanted products leave the body through:
(a) the **lungs**, which remove CO_2. Considerable amounts of water and heat may be lost from the respiratory passages—see page 97.
(b) The **skin**, which removes excess heat not lost through breathing. Loss of heat is aided by sweating, which incidentally removes some nitrogenous waste, salts and water—see page 81.

(c) The **kidneys**, which remove nitrogenous and other potentially toxic waste and the excess salts and water not removed in other ways. They cleanse the blood plasma, adjust the salt content (osmotic balance) and pH, and control the overall volume of body fluid. The kidneys with the **ureters**, **bladder** and **urethra** form the **urinary system**, by means of which **urine** is formed, temporarily stored and passed to the exterior on **micturition**.
Note. The faeces and urine are spoken of collectively as the **excreta**.

THE URINARY SYSTEM

The kidneys are bean-shaped, about 11 cm long, 5 cm broad, and 3 cm thick. They lie against the back wall of the abdomen, behind the liver and stomach, the left slightly higher than the right. Each is embedded in fat, covered ventrally by peritoneum and encased in a **fibrous capsule** which is turned in at a notch called the **hilum** to line a **renal sinus**—scc next page. The kidneys are served by the **renal arteries** and **veins** and are connected to the back of the bladder by 25–30 cm long **ureters**, which lie against the abdominal wall and curve round in the pelvic region.

The **urinary bladder** is a distensible sac with three layers of smooth muscle in its walls and folds of epithelial lining which permit stretching. In the average adult 200 ml of urine collected in the bladder stimulates stretch receptors, producing a desire to micturate. The desire fades and recurs as the bladder fills further. Emptying normally takes place well before the limit capacity of about 800 ml is reached. Micturition is a reflex action with contraction of the bladder wall and relaxation of the **smooth muscle sphincter** at its neck, but release of urine can be controlled voluntarily by a **striated muscle sphincter**. Complete emptying is assisted by the abdominal muscles and diaphragm—see pages 55–59.

Passage of urine is through the urethra. That of the female is short (about 4 cm) and opens in front of the vagina—see page 127. That of the male is longer (about 20 cm), is joined by the reproductive ducts and opens on the tip of the penis—see page 125.

Structure of the kidney

The kidney substance is divided into outer **cortex** and inner **medulla**. The latter is striated in appearance and forms about 14 **pyramids**. The cortex extends down the sides of the pyramids as the **renal columns** while the tip of each pyramid lies in a cup-shaped **calyx**, which opens into the **pelvis of the ureter**—see next page.

The urine-forming units of the kidneys are the **nephrons**, of which there are about 1 million. Each nephron starts with a **renal corpuscle** formed of a knot of blood capillaries called a **glomerulus** surrounded by a **glomerular (Bowman's) capsule**, the inner layer of which is closely adherent to the blood vessels. The renal tubule leading from the capsule is differentiated into **proximal** and **distal convoluted tubules** and U-shaped **loop of Henle**. The loops of **cortical nephrons** are short and lie with the convoluted tubules and renal corpuscles in the cortex, while those of the **juxtamedullary nephrons** are long and extend into the pyramids beside common **collecting ducts**. The collecting ducts open at the tips of the pyramids after receiving fluid from 3000–6000 nephrons each.

The epithelial cells lining the glomerular capsules and parts of the loops of Henle are flattened. Those adjacent to the glomerular capillaries have interlocking processes and are called **podocytes**. Most of the cells lining the rest of the tubules are cuboid, with **brush border** of **microvilli** (cf. intestine—see page 87). A few cells in the distal tubules are columnar where they lie next to the walls of the corresponding afferent

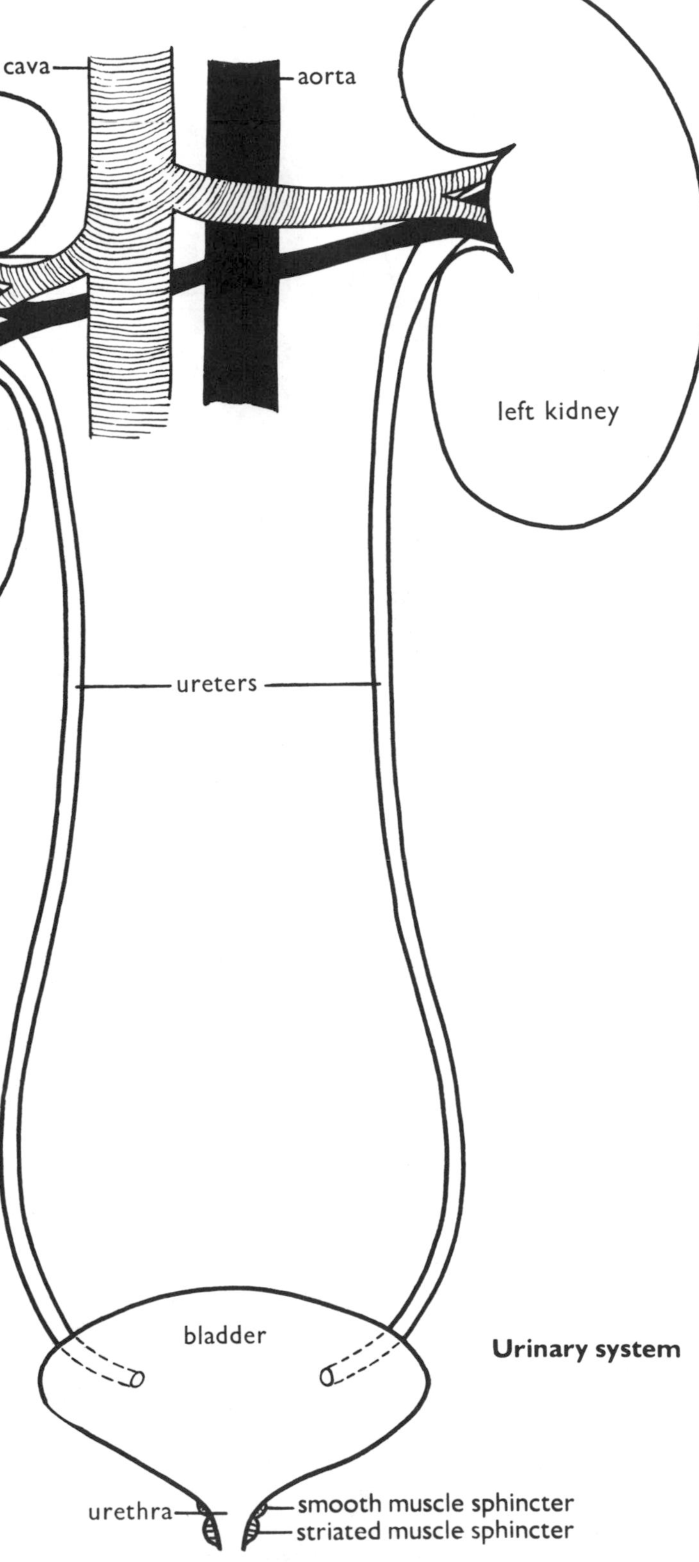

Urinary system

glomerular arterioles, the muscle cells of which are locally differentiated as **juxtaglomerular cells** and form the **juxtaglomerular apparatus**.

The main renal blood vessels traverse the calyx and run up the renal columns to anastomose as the **arcuate arteries** and **veins** between the cortex and the medulla. Branches from the arcuate arteries serve the glomeruli, which have **afferent arterioles** wider than **efferent arterioles**. The latter connect with the **peritubular capillary network** in the cortex and with straighter vessels called **vasa recti** in the medulla—see page 119.

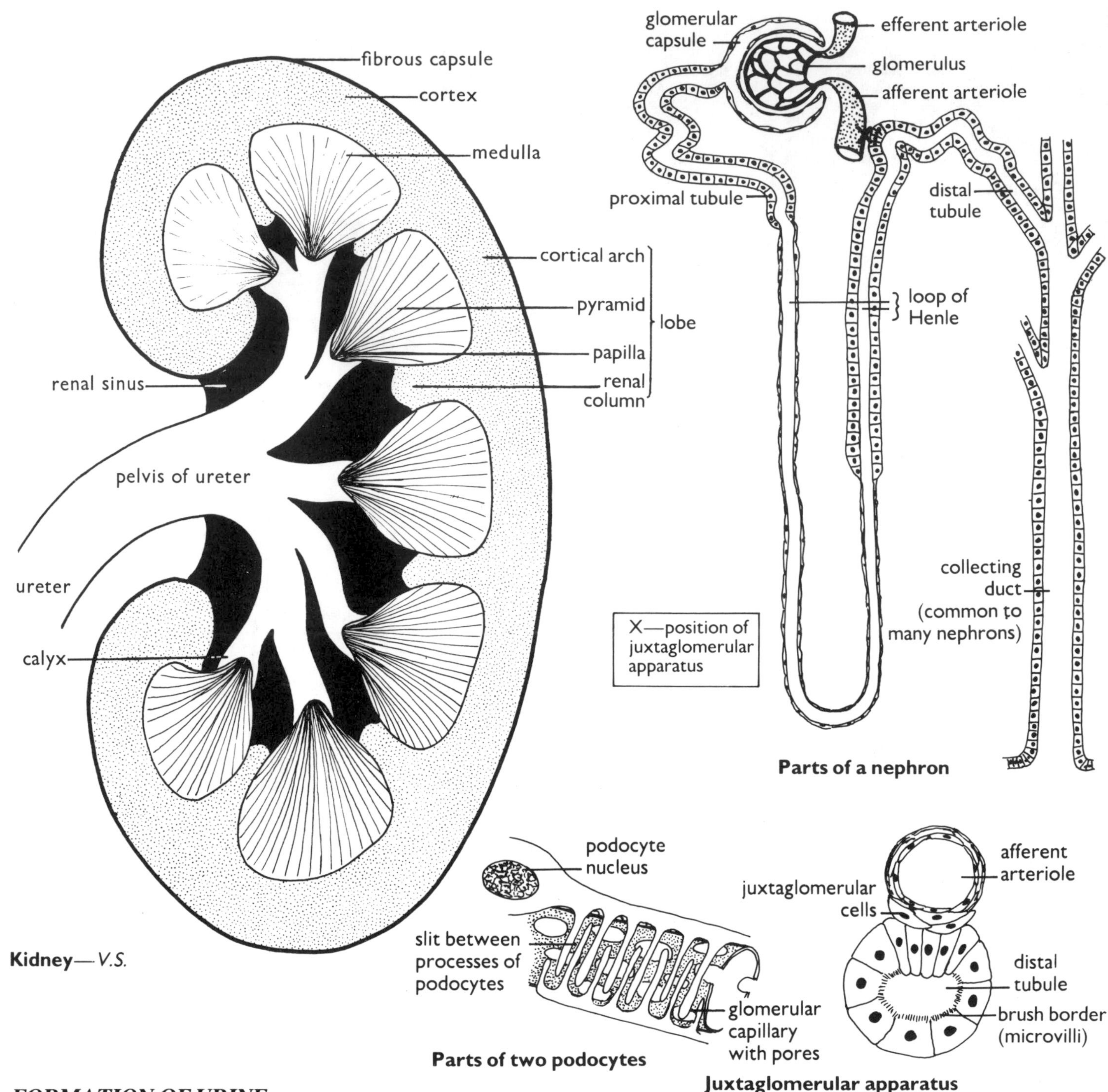

Kidney—*V.S.*

Parts of a nephron

Parts of two podocytes

Juxtaglomerular apparatus

FORMATION OF URINE

Urine has an average composition of 95–96% water, 2% salts, 2% urea and small quantities of other waste materials, especially creatine, uric acid, pigments and potentially toxic drugs. The actual composition fluctuates considerably but urine is usually slightly acid, pH 6.5, and does not normally contain sugar or proteins.

Urine originates as a **filtrate** of the plasma and is **concentrated** during its passage through the renal tubules. Different portions of the tubules have specific functions of selective **reabsorption** and **secretion**, resulting in conservation of useful substances and in **homeostasis** of volume, pH and salinity of blood. Sympathetic nerves and hormones are involved in the regulatory mechanisms.

Filtration

The glomerular capillaries are much more permeable than ordinary capillaries, because their endothelial linings are perforated and the adjacent podocytes have slits between their processes. The intervening basement membrane of glycoprotein with a meshwork of collagen fibres holds back blood cells and most proteins, but permits **dialysis** (filtration) of water carrying the other components of blood plasma including 0.03% of the smallest protein molecules.

For efficient dialysis the hydrostatic pressure of the blood in the glomeruli must exceed the colloidal osmotic pressure of the blood proteins plus the hydrostatic pressure in the renal tubules. Normally the resultant **filtration pressure** is autoregulated:

(a) **Filtration too fast**→increased Cl^- in the distal tubules→constriction of afferent arterioles→reduction in glomerular pressure→reduction in filtration pressure→reduced filtration rate.

(b) **Filtration too slow**→renin secretion by the juxtaglomerular apparatus→formation of angiotensin in the blood→increase in blood pressure and constriction of efferent arterioles→increase in glomerular pressure→increase in filtration pressure→increased filtration rate.

All the nephrons together process about 1200 ml/min of blood, of which 650 ml is plasma. This produces about 125 ml/min of filtrate (= 180 litres/day) but the final quantity of urine produced is only about 1 ml/min (= 1–1.5 litres/day), the rest of the liquid having been reabsorbed. If the blood pressure is high, the amount of urine increases because pressure in the peritubular capillaries resists reabsorption. If the blood pressure is low, a sympathetic reflex response causes vasoconstriction with reduced circulation to the kidneys and reduced filtration, helping to maintain an adequate blood volume, but retaining waste materials, with potentially damaging side effects if the condition is prolonged.

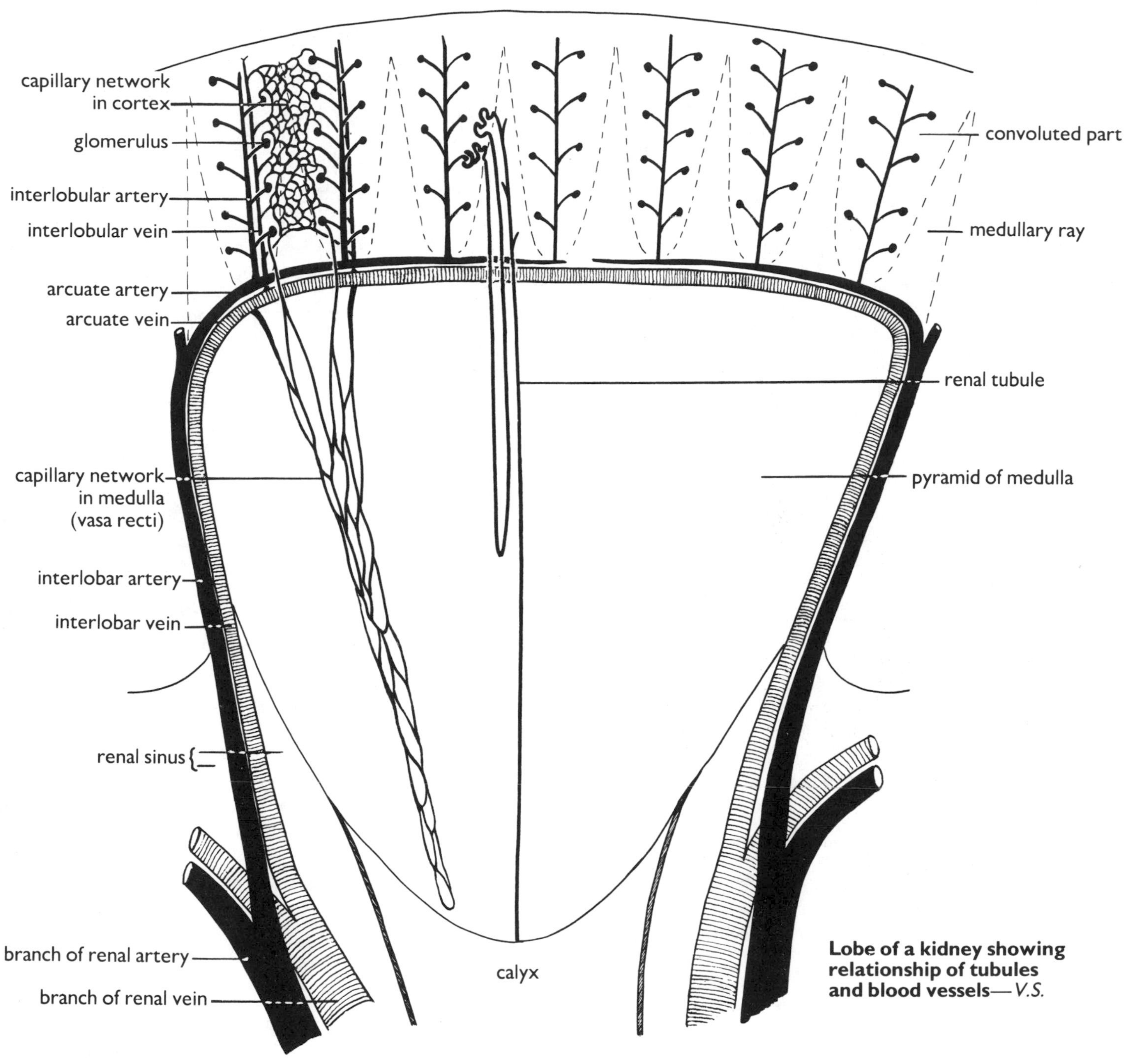

Lobe of a kidney showing relationship of tubules and blood vessels— *V.S.*

Tubular reabsorption

Sugars, **amino acids** and **proteins** are actively reabsorbed in association with carriers in the proximal tubules, normally leaving none in the urine. **Inorganic ions** are selectively reabsorbed, dependent on the amount of each ion in the blood. The process involves diffusion through the exposed borders of the cells with their microvilli and **active onward transport** involving **carriers** and **energy**. Such active transport can take place against the diffusion gradient and is controlled by **aldosterone**—see page 122. Absorption of Na^+ in the proximal tubules is accompanied by Cl^- to match, by up to 80% of the water, which follows the osmotic gradient, and by up to 50% of the **urea**, which, having larger molecules, diffuses more slowly. Urea may also be reabsorbed from the collecting ducts. Some of this is recycled into adjacent loops of Henle, but ultimately only about 13% of the urea filtered leaves the body in the urine. Creatine, phosphates, sulphates, nitrates, uric acid and phenols are not reabsorbed, i.e. they have 100% **renal clearance**.

Tubular secretion

Secretion into the tubules is concerned with regulation of **potassium ions** and **hydrogen ions** and removal of certain drugs which cannot pass as filtrate.

(a) **Potassium**. Accurate levels of blood K^+ are important to the maintenance of polarised membranes—see page 61. Though K^+ is absorbed along with other ions, excess in the blood is prevented by secretion from the distal tubules and collecting ducts under the influence of **aldosterone**, stimulus to produce which is the K^+ itself, so that the mechanism is **autoregulatory**. At the same time the aldosterone stimulates Na^+ absorption to maintain overall salinity of the blood.

(b) **Hydrogen ions**. Certain metabolic processes result in formation of acids, the H^+ of which must be removed to maintain constancy of **pH** in the blood and body fluids. The H^+ is exchanged for Na^+ and the tubular fluid becomes **acid**. Excess acidity is prevented by liberation of **ammonia**, NH_3, chiefly from glutamic acid. This diffuses into the tubule and joins H^+ to form **ammonium ions**, NH_4^+, which are excreted as neutral **ammonium chloride**.

Note. Very occasionally the blood becomes too **alkaline**, in which case **bicarbonates** are excreted, but normally they are retained as useful buffers, particularly in the transport of CO_2 from the tissues to the lungs—see page 97. The pH of urine varies with diet, but is normally about 6.5.

Water regulation

While up to 80% of the water in the original renal filtrate is absorbed by osmotic diffusion in the proximal tubules, further reabsorption is normally necessary to prevent excessive fluid loss (**dehydration**). This reabsorption is controlled by the **antidiuretic hormone**, **ADH**, released from the pituitary body—see page 122, under the control of the hypothalamus. ADH renders the basically impermeable distal tubules and collecting ducts permeable to water and creates concentration gradients so that water can be withdrawn from them. It does the latter by causing Cl^- to be actively pumped out of the thick parts of the loops of Henle. Na^+ is drawn out to match. The ions can pass freely into the permeable thin parts of the juxtamedullary loops of Henle, where they are added to the incoming Na^+ and Cl^-. Though some Na^+ and Cl^- ions pass into the vasa recti, circulation here is slow so that dispersal is also slow. As the process, known as the **counter-current mechanism**, is repeated, a **concentration gradient** builds up between the cortex and the medulla and the fluid leaving the loops is reduced in salinity. This permits up to a further 15% water to be reabsorbed from the distal tubules. On entering the collecting ducts the renal fluid is isosmotic with the intertubular environment, but as it flows towards the tips of the pyramids it passes through the zone of high ion concentration and loses a further 4% water, leaving only about 1% of the original filtrate to carry the soluble waste materials from the body as urine.

The amount of ADH released and thus the concentration of urine is related to the osmotic stimulus by the concentration of ions, chiefly Na^+ and Cl^-, in the blood reaching the hypothalamus. The same stimulus produces the sensation of **thirst**. Thus water intake can be balanced to water output.

Ultimately the **volume of urine** depends on:

1. hydrostatic pressure;
2. concentration of solutes;
3. ADH;
4. water loss by other systems.

In hot weather, when sweating is copious, urine tends to be highly concentrated and small in volume and thirst is greater. A dry atmosphere has the same effects owing to water losses from the respiratory passages.

In cold weather urine is dilute and more copious.

There is also a relationship to the amount of salt eaten. Excess tends to raise blood pressure and result in more copious urine as well as thirst.

Diuretics increase the volume of urine by suppressing Cl^- pumping, thus reducing the build-up of the counter-current and consequently reducing the power to concentrate urine.

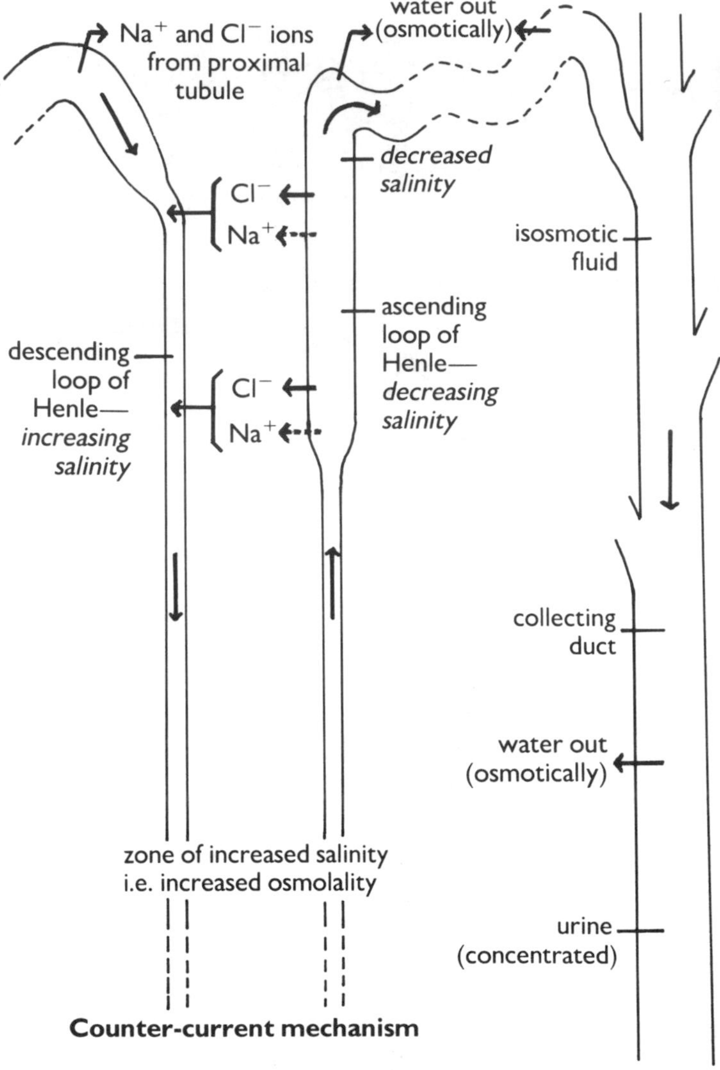

Counter-current mechanism

JUXTAGLOMERULAR APPARATUS

The structure of the apparatus and renin formation have already been described—see pages 117 and 118. The juxtaglomerular cells also secrete the **erythropoietic factor**, which, conjugated to a **globulin**, forms **erythropoietin** and stimulates production of erythrocytes from red bone marrow—see page 100.

The juxtaglomerular cells are thus **endocrine glands** in that they pass their products into the blood stream, as opposed to **exocrine glands** which secrete into ducts or cavities connected to the outside of the body.

Note. The kidneys are not considered to be glands because the urine produced starts as a filtrate rather than a secretion.

EXOCRINE GLANDS

Many exocrine glands have been described earlier in the text in conjunction with the systems in which they occur. To summarise, they may be:

(a) **unicellular**, e.g. goblet cells of the alimentary canal;
(b) **multicellular** of various types shown in the diagrams.

Functions of exocrine glands

Exocrine glands produce secretions which do not affect the metabolism of other cells. They include a variety of substances:

(a) mucus and/or enzymes or enzyme precursors, e.g. digestive glands—see pages 84–94.
(b) special non-enzymatic useful substances, e.g. sebaceous glands (oil)—see pages 80 and 81; ceruminous glands (wax)—see page 74; mammary glands (milk)—see page 131.
(c) waste materials, e.g. sweat glands—see pages 80 and 81; liver—see pages 92 and 93.

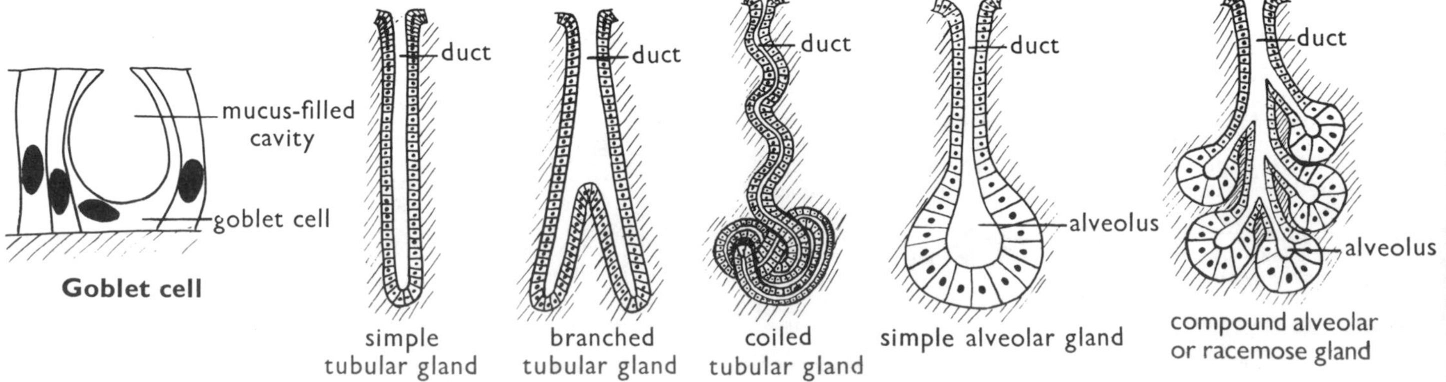

Types of multicellular glands

The Endocrine System

The endocrine system concerns the production and function of **hormones**. These chemical messengers are secreted into the blood stream and circulate to affect the metabolism of cells at a distance from those by which they are produced. They supplement control by the autonomic nervous system—see pages 69 and 70, and help to keep the physiological activities within the limits necessary for maintenance of health while adjusting to special demands. Any disturbance of one part of the endocrine system is liable to lead to malfunctioning of another.

Hormones are chemically varied—see table, and specific (cf. vitamins—see page 99).

Note. Ascorbic acid, required as a vitamin in the diet of man, is formed as a hormone in mice, while calciferol, formed by irradiation of the skin with ultraviolet light, is needed as vitamin D when there is too little sunlight or the skin is kept covered with clothes.

Action of hormones

Hormones change the **rates of metabolic activities** by:
(a) changing the rate of transport of substances in and out of cells;
(b) changing enzyme effectiveness by activating inactive enzyme precursors or causing production of more enzyme.

Sometimes hormones act singly, but more often they form parts of balanced systems, increasing or decreasing a particular function antagonistically. Sometimes the action is indirect, stimulating the release or activity of another hormone. There is no action unless the hormone is specifically recognised by the responding cells. A water-soluble hormone can only react with a surface receptor, thereby releasing a messenger substance into the cell, while a fat-soluble hormone diffuses through the phospholipid plasma membrane—see page 5, to interact with intracellular protein and affect the nucleus.

Small amounts of most hormones are present most of the time, with carefully controlled increases and decreases to maintain the delicate metabolic balances. There is also continual destruction and removal by the liver and kidneys, which prevents large build-up.

PITUITARY BODY

The pituitary body, or hypophysis cerebri, lies in the sella turcica of the sphenoid bone, underneath and attached to the hypophyseal region of the brain—see page 63. It has dual origin.

1. The **neural stalk** or **infundibulum** and the **neural lobe** or **pars nervosa** originate from the brain of the embryo and together make up the **neurohypophysis**. There is close nervous connection between this and the autonomic centres of the hypothalamus.

2. The **pars anterior**, **pars intermedia** and **pars tuberalis** are partially wrapped round the neurohypophysis. They form the **adenohypophysis** which develops from a pouch in the roof of the embryonic nasopharynx, remains of which form the intraglandular cleft. The adenohypophysis is connected to the hypothalamus by portal veins, but not by any nerve connection.

Note. The terms posterior lobe and anterior lobe represent morphological division through the intraglandular cleft and do not correspond to division in origin and function.

Chemical nature of hormones

Polypeptide	Insulin, glucagon, gastrin, secretin, cholecystokinin, ACTH, ADH, oxytocin, calcitonin, parathyroid hormone
Protein	GH, prolactin, renin/angiotensin
Glycoprotein	TSH, LH, FSL, erythropoietin
Iodinated amino acid	Thyroxine
Catecholamine	Adrenaline, noradrenaline
Steroid	Aldosterone, cortisol, corticosterone, androgens, oestrogens, progesterone

Organs which produce hormones

Organs which have endocrine function as well as other functions include:
(a) **stomach**—see page 87;
(b) **small intestine**—see page 88;
(c) **pancreas**—see page 91;
(d) **kidneys**—see page 120;
(e) **thymus**—see page 115;
(f) **testes**—see page 125;
(g) **ovaries**—see page 127;
(h) **placenta**—see page 129;
(i) **hypothalamus**—see page 62.

The hypothalamus is particularly important in that it not only produces hormones, but it has ultimate control over much of the endocrine system.

Organs with purely endocrine function are:
(a) **pituitary body** (hypophysis cerebri);
(b) **adrenal glands** (supra-renal glands);
(c) **thyroid gland**;
(d) **parathyroid glands**;
(e) **pineal body**.

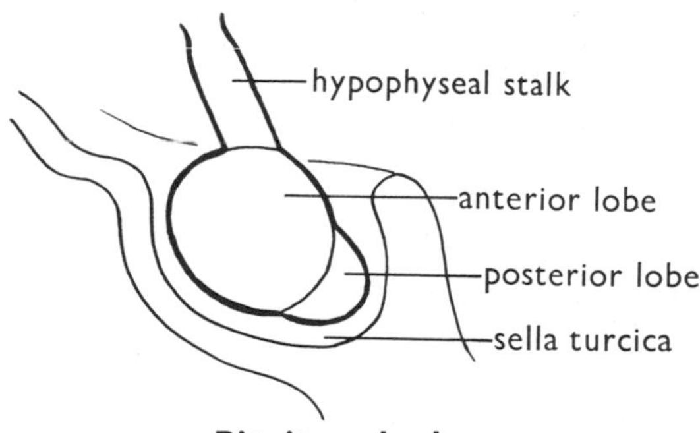

Pituitary body

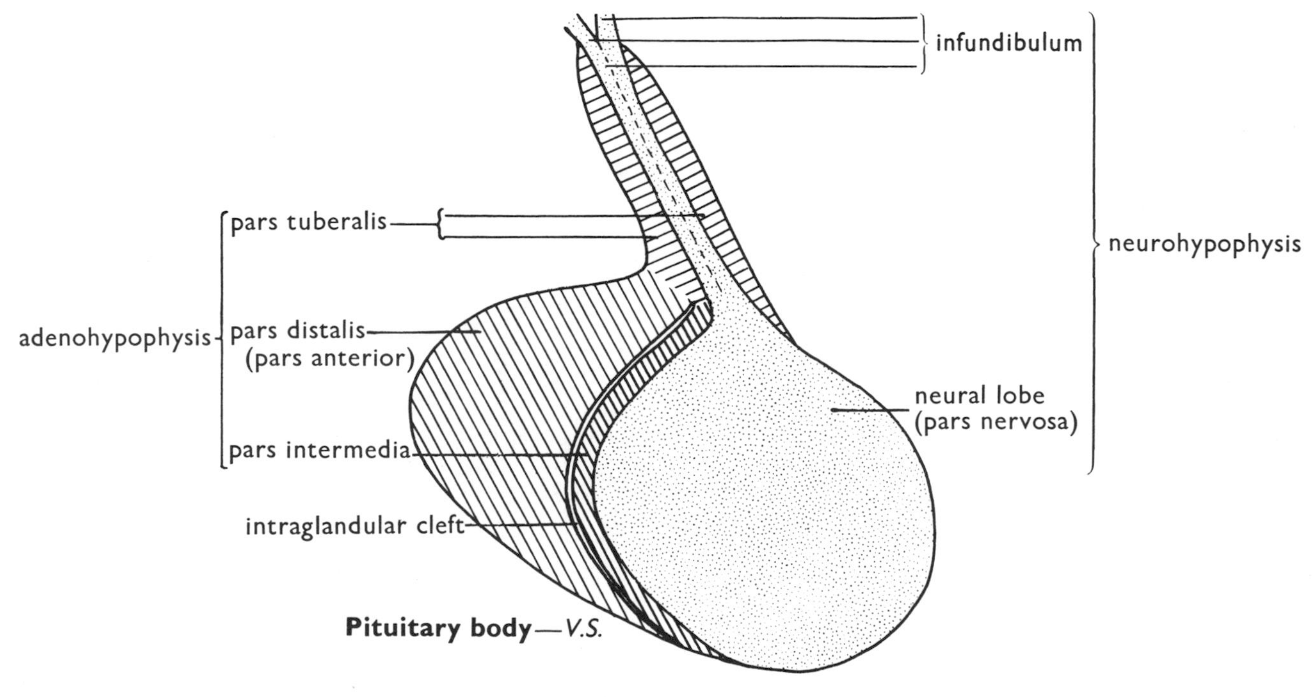

Pituitary body—*V.S.*

Functions of the neurohypophysis

Two hormones, **oxytocin** and the **antidiuretic hormone**, **ADH**, have been extracted from the neurohypophysis, but it is now known that they are manufactured in the **hypothalamus** and passed to the neurohypophysis for storage until stimulation of the hypothalamic nerve centres triggers their release.

Oxytocin stimulates the smooth muscle of the uterus and the breasts. Its production assists childbirth and the ejection of milk during lactation—see page 131.

The **antidiuretic hormone**, **ADH**, increase the permeability of the distal tubules and collecting ducts of the kidneys and controls the counter-current mechanism by which urine is concentrated—see page 120. Deficiency causes diabetes insipidus. ADH is identical with the hormone named **vasopressin** or **pitressin**. The vasopressor effect, raising blood pressure by contraction of arterioles, is negligible in man.

Functions of the adenohypophysis

The adenohypophysis is often called the 'master gland' of the endocrine system, but it is, itself, under the control of hormonal substances called **releasing factors** secreted by the **hypothalamus** and carried to it in the portal blood supply. The adenohypophysis produces at least seven hormones.

1. The **adrenocorticotropic hormone** (**ACTH**) controls the activity of the **adrenal cortex**, particularly the production of glucocorticoids and gonocorticoids.

2. The **thyrotropic** or **thyroid-stimulating hormone** (**TSH**) controls the activity of the **thyroid gland**. Intense emotion or cold increase the production by the hypothalamus of the releasing factor and thus of TSH. The consequent increase in thyroxine—see next page—enhances the effects of adrenaline—see below.

3. The **growth hormone** (**GH**) favours protein synthesis by assisting uptake of amino acids into cells. Its most obvious effect is increase in **skeletal development**. Excess of the hormone during childhood causes gigantism, and in adults a bone thickening called acromegaly. Deficiency in childhood causes dwarfism. Even after growth has ceased, GH is important for maintenance of health and its short-term effects include mobilisation of fat to replace glucose during starvation. GH production is controlled by a balance of many influences: (a) releasing and inhibitory factors from the hypothalamus; (b) thyroxine from the thyroid gland; (c) oestrogens (female sex hormones); (d) somatosatin from the pancreas; (e) glucose in the blood stream. It is normally abundant compared with other hormones. There is rapid turnover, with up to 5% of the gland content secreted each day and half the circulating GH destroyed in 25 minutes.

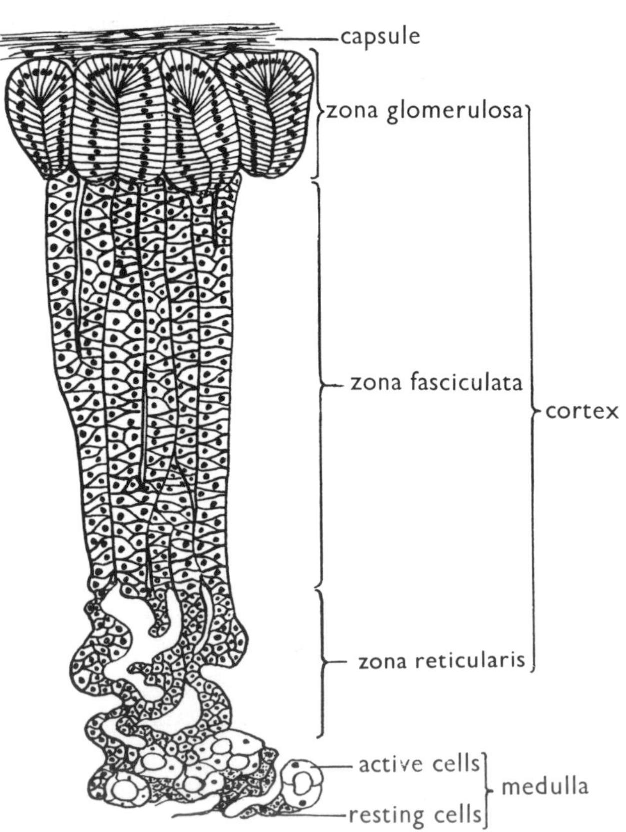

Adrenal gland—*section*

4. The **melanocyte-stimulating hormone** (**MSH**) causes darkening of the skin, but is normally inhibited by the hormones of the adrenal cortex.

5 and 6. Two **gonadotropic hormones** are produced: (a) the **follicle-stimulating hormone** (**FSH**) and (b) the **luteinising hormone** (**LH**). They get their names from their effects on the follicles of the ovaries and the corpora lutea respectively—see page 127. They are, however, also active in the male, where FSH stimulates the seminiferous tissue and LH the interstitial tissue—see page 125.

7. The **luteotropic** or **lactogenic hormone** (**LTH**), also known as **prolactin**, is concerned with production of milk during the lactation period following pregnancy—see page 132. At other times the amounts present are kept small by an inhibiting factor from the hypothalamus. This inhibition is itself inhibited by nipple stimulation, thus suckling helps to maintain the flow of milk. At the same time the LTH helps to maintain the corpus luteum and suppresses FSH production, thus suppressing ovulation—see page 127.

ADRENAL (SUPRA-RENAL) GLANDS

One of these glands lies capping each kidney. Each gland has an outer layer or **cortex** derived from the same tissue as the kidneys and the reproductive organs, and an inner region or **medulla** derived from the same tissue as the sympathetic cords. Related to this dual origin, there is dual function and method of control.

Functions of the adrenal cortex

Over 50 adrenal cortical hormones are known, secretion of which is under the control of other hormones. All the cortical hormones are **steroids** and they are grouped into three categories according to the type of action in which they are involved.

1. **Mineralocorticoids** are formed in the **glomerular zone** of the cortex. They control the salt content of extra-cellular fluids. The best known example is **aldosterone** which increases reabsorption of ions, particularly Na^+ from the renal filtrate, and increases secretion of K^+—see page 119. Secretion of aldosterone is controlled by renin/angiotensin and K^+ levels in the blood. Excess aldosterone leads to retention of body fluid and thus oedema.

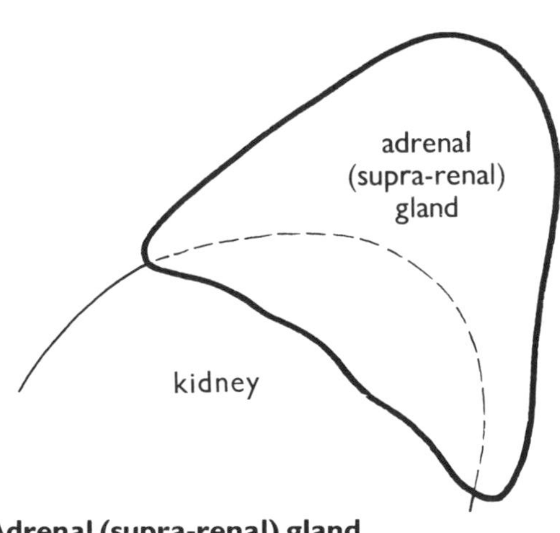

Adrenal (supra-renal) gland

2. **Glucocorticoids** are formed in the **fascicular zone** of the cortex. They affect **metabolism** by increasing utilisation of fats and proteins as a source of energy and decreasing utilisation of carbohydrate, thus raising blood sugar level and increasing deposition of glycogen in the liver. They also increase erythrocyte production but decrease the number of circulating lymphocytes, and thus antibody formation. By stabilising cell membranes they reduce liberation of lysosomes and have an anti-inflammatory effect. They also depress pituitary secretion of ACTH, TSH and MSH—see above. Glucocorticoid activity increases in response to stress. **Cortisol** is an example of this group of hormones.

3. **Gonadocorticoids** are formed in the **reticular zone** of the cortex. They include **androgens** which mimic the effects of the testicular hormones—see page 125—and **oestrogens** and **progesterones**, which mimic the effects of the ovarian hormones—see page 127.

Functions of the adrenal medulla

The secretion of the adrenal medulla contains **adrenaline** and **noradrenaline** (also known as **epinephrine** and **norepinephrine**). Production is stimulated by sympathetic nerves—see page 70, whose action it reinforces. Adrenaline is the more active in increasing breakdown of glycogen in the liver, increasing cardiac output and stimulating ACTH and TSH production. Noradrenaline is also produced in sympathetic ganglia and parts of the brain—see page 69.

THYROID GLAND

The thyroid gland has two main **lobes**, one on either side of the larynx and trachea, joined together by the **isthmus**. In structure it resembles a racemose gland (cf. salivary gland), but has no ducts, the alveoli being replaced by closed **follicles** filled with **colloidal material**. The walls of the follicles have two types of cells: (a) **principal cells** which border onto the follicles, (b) **parafollicular cells** which do not. The gland is very richly supplied with blood vessels.

Functions of the thyroid gland

1. The **principal cells** produce a group of iodine-containing hormones, the best known of which is **thyroxine**. Part of the manufacture and the storage take place in the colloid of the follicles. The product is absorbed back into the cells and secreted into the blood stream, from which it diffuses into the intracellular spaces. Most of the thyroxine in circulation is carried bound to plasma proteins from which it is liberated before it enters the cells. There it stimulates production of **RNA** and therefore **protein synthesis** and subsequently other **metabolic activities** and **growth**. Increase in thyroxine not only increases **metabolic rate**, with consequent release of heat, but increases rate of heart beat, blood pressure, mental activity and fertility. However it decreases secretion of TSH from the pituitary body, and thus is autoregulatory.

2. The **parafollicular cells** produce **calcitonin**. This is released in response to high levels of Ca^{2+} in the blood. It inhibits osteoclast activity and stimulates osteoblast activity so that excess calcium is rapidly deposited in bone—see page 9, and does not form insoluble precipitates in the blood.

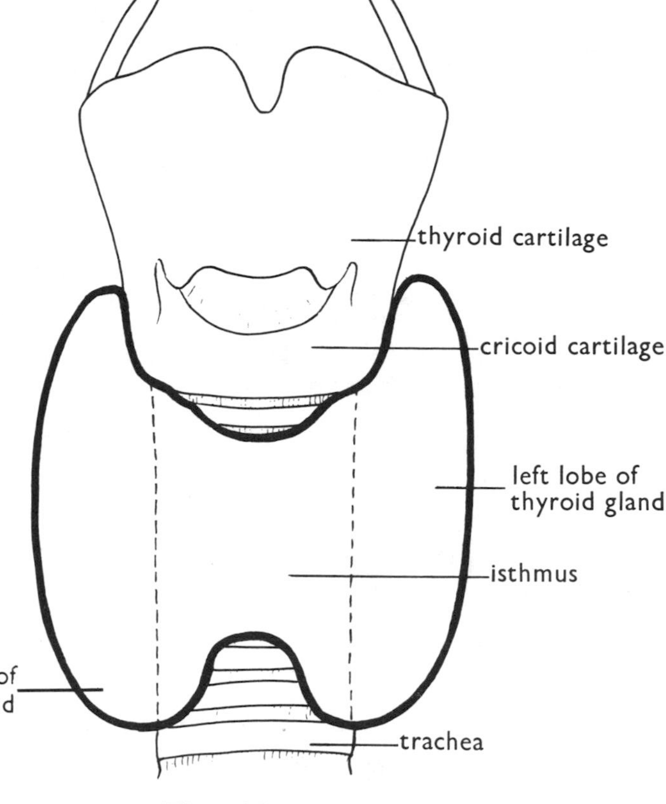

Thyroid gland

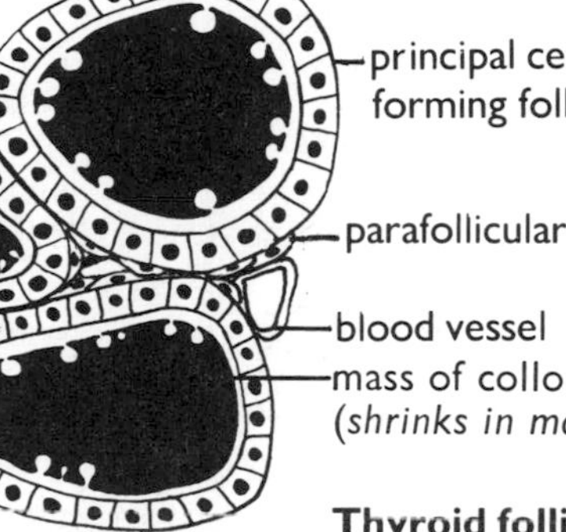

Thyroid follicles—*section*

larynx
thyroid gland
parathyroid glands
oesophagus
trachea

Parathyroid glands

PARATHYROID GLANDS

There are usually four of these glands, behind the thyroid gland. They consist of tightly packed masses of cells, only some of which are active.

Functions of the parathyroid glands

Secretion of the parathyroid glands is **parathormone**. It helps to maintain a constant level of Ca^{2+} in the body fluids. The effects of parathormone are:

(a) release of calcium salts from bone—see page 9;
(b) stimulation of kidneys to remove the excess phosphate also released;
(c) increased reabsorption of Ca^{2+} from renal tubules;
(d) increased absorption of Ca^{2+} from the intestine.

Absorption of Ca^{2+} is also helped by **calciferol** (vitamin D)—see page 99. Ca^{2+} reduces permeability of cell membranes and activates many intracellular enzymes. Lack of circulating Ca^{2+} causes tetany (painful muscular cramps) and collapse. When there is dietary deficiency the bones act as a reservoir from which Ca^{2+} can be drawn to maintain levels in the blood. Parathormone secretion is regulated by Ca^{2+} levels, which are themselves affected by calcitonin—see above.

PINEAL BODY

The pineal body lies above the colliculi—see page 63. In early life it consists of **pinealocytes** supported by **neuroglial cells**.

Functions of the pineal body

The **pinealocytes** produce **melatonin**, a hormone which inhibits reproductive development, by inhibiting release of the gonadotropic hormones. From the start of puberty, the pinealocytes become calcified to form **brain sand**, melatonin production ceases and inhibition of the gonadotropic hormones is removed.

Noradrenaline, serotonin, histamine and an adrenocorticotropic hormone have also been found in the pineal body, but all of these are produced in larger quantities elsewhere.

HYPOTHALAMUS

The hypothalamus is part of the brain—see page 62. It contains centres which regulate numerous functions. These have been described in appropriate places throughout the text and may be summarised as follows.

1. Control, through relay centres in the brain stem, of blood pressure and cardiac output—see page 110.
2. Control, with the aid of the kidneys, of blood and tissue fluid volume and of osmotic pressure, i.e. salt/water balance—see page 120.
3. Excitation to drink (thirst) to replace water losses, to eat (hunger) to maintain supplies of nutrients, and to stop eating when the stomach is full (satiety)—see pages 120, 86 and 70.
4. Control of body temperature when it differs from the core temperature of the hypothalamus itself—see page 81 and above.
5. Control of secretion of many of the pituitary hormones and thus of the activities controlled by these hormones, particularly metabolic rate, growth and sexual development—see page 122 and above.
6. Control, in conjunction with the rest of the limbic system, of the emotions of excitement, fear and rage, and preparation of the body to respond—see pages 62 and 70.
7. Control of diurnal and other biological rhythms, whereby many physiological activities vary regularly within tolerated limits. These rhythms often have related behaviour patterns.

Cell Division

Cell division is essential for **growth**, normal **replacement** and **repair** after damage, and also for production of **ova** and **spermatozoa**—see pages 125 and 127.

Some cells, e.g. those of the blood—see pages 100 and 101, the skin—see page 80, and the lining of the alimentary canal—see pages 84 and 87, are short-lived and replaced continually. Others are renewed less frequently, e.g. smooth muscle and bone. Nerve cells cannot be replaced once development of the nervous system is complete. The pace of renewals may be increased following loss, e.g. bleeding or damage, and some organs, e.g. the liver, have considerable powers of regeneration. Production of new cells is stimulated by certain hormones and vitamins, and kept in check by production of substances called **chalones** from the dividing cells themselves. In general cell division slows down with age, so that replacement and repair becomes less efficient.

Divisions leading to growth and repair are described as **mitotic**. Those leading to maturation of ova and spermatozoa are **meiotic**. The sequence of events in mitosis is shown in the diagrams. Before studying these, reference should be made to cell structure—see pages 5–7, and to the part played in metabolism by DNA, the main ingredient of chromosomes—see page 4.

Note. For simplicity, the behaviour of only two pairs of chromosomes is illustrated. Human cells normally have 23 pairs of chromosomes, identifiable by shape and size, each carrying up to 20 000 genes, which are thus paired. Pairs of genes are called **alleles** and may or may not be identical, thus producing genetic variation.

Growth and differentiation

During the **interphase** or **metabolic phase**, the cell synthesises RNA and proteins and doubles its cellular components. The DNA strands are replicated, uncoiling and attracting complementary nucleotide bases, until each double strand becomes two double strands, i.e. the chromatids are formed ready for the next cell division. Histones and other proteins around the DNA help twisting and shortening in the next prophase.

Note. Mitochondria contain their own DNA and are replicated independently—see page 7. They are shared at cell division.

In mitosis each daughter cell receives DNA with the same arrangement of base pairs, i.e. the same genes, and thus has the same developmental potential as the parent cell. What actually happens to the cell then depends on its position in the body and relationship to other cells.

Primary differentiation occurs very early in embryonic life—see page 128. The mechanism for this is not yet fully understood, but hormone-like organiser substances direct otherwise undifferentiated cells to become the precursors of specific tissues and organs. In tissue culture (outside the body) many types of highly specialised cells can revert to a simpler state and multiply, but they never give rise to cells of a different tissue group, thus indicating subtle changes in genetic constitution. Neurones cannot divide because they lose their centrosomes and therefore cannot form spindles, while mature red blood cells cannot divide because they lack nuclei entirely.

MITOSIS

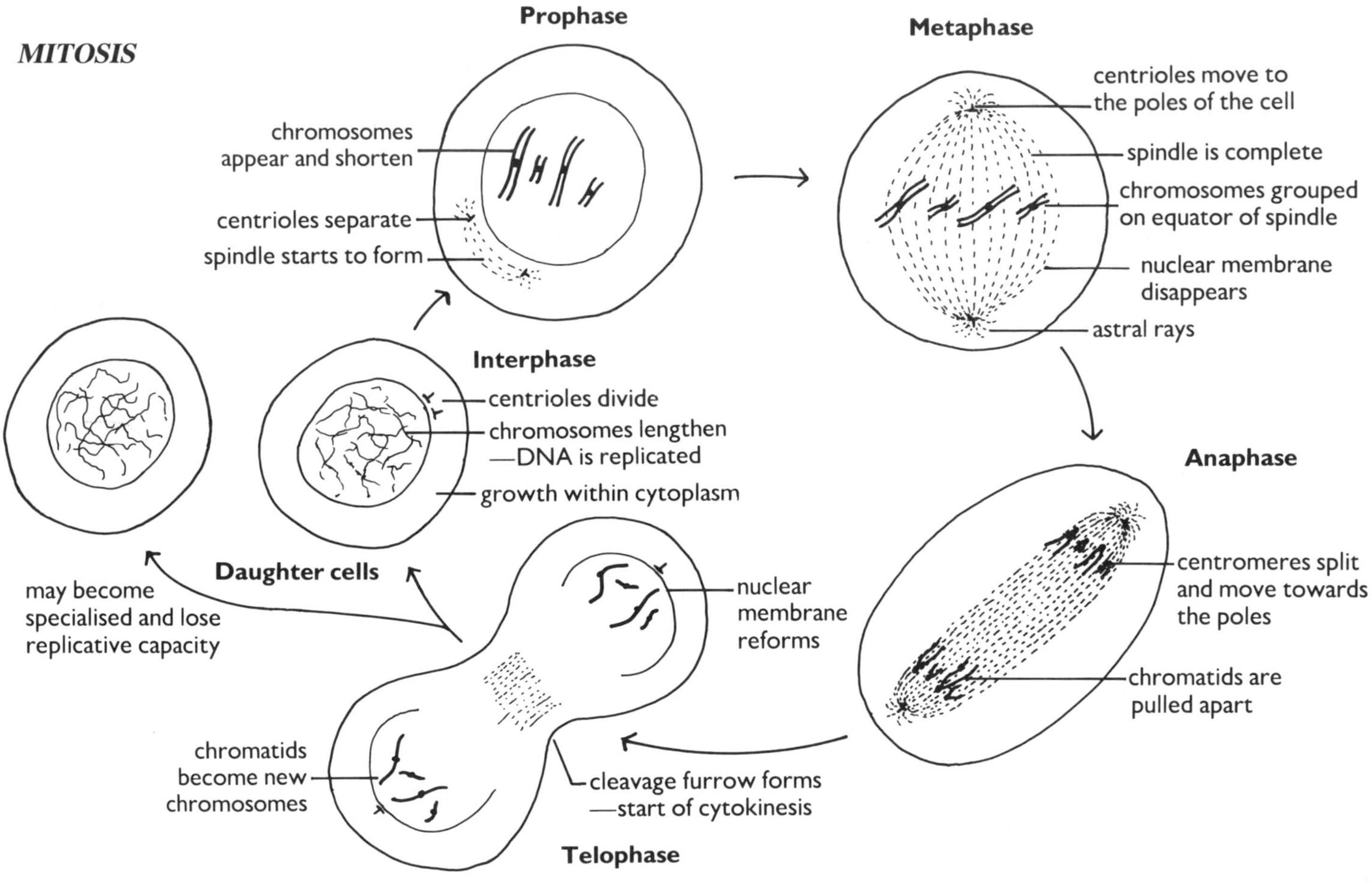

Prophase
chromosomes appear and shorten
centrioles separate
spindle starts to form

Metaphase
centrioles move to the poles of the cell
spindle is complete
chromosomes grouped on equator of spindle
nuclear membrane disappears
astral rays

Anaphase
centromeres split and move towards the poles
chromatids are pulled apart

Telophase
nuclear membrane reforms
chromatids become new chromosomes
cleavage furrow forms —start of cytokinesis

Interphase
centrioles divide
chromosomes lengthen —DNA is replicated
growth within cytoplasm

Daughter cells
may become specialised and lose replicative capacity

MEIOSIS

Meiosis occurs during the formation of gametes (eggs and sperm, i.e. mature ova and spermatozoa). It involves the same stages as mitosis, with the following important differences:

1. During **prophase** the homologous (like) chromosomes pair, forming **tetrads**, in which **crossing over** can occur between sections of the constituent chromatids, thus producing genetic recombination in the DNA.

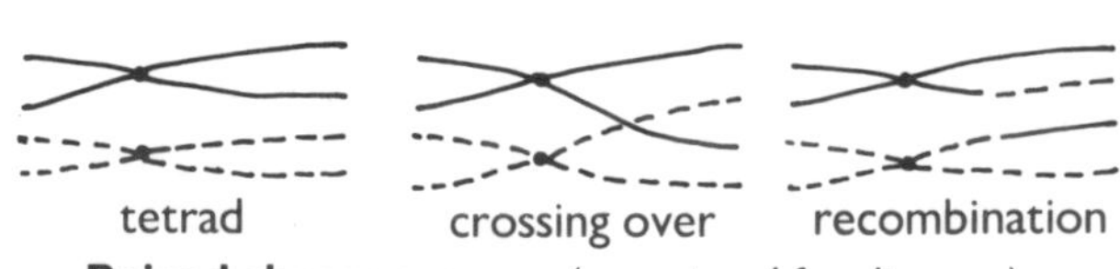

Paired chromosomes (*untwisted for diagram*)

2. During **metaphase** the chromosomes remain paired as they line up on the equator of the spindle.
3. At **anaphase** the pairs separate so that half the original number of chromosomes goes to each pole.
4. Each chromosome of the telophase nuclei consists of two chromatids so that further replication of DNA does not occur during the subsequent interphase.
5. The meiotic division is followed rapidly by a second maturation division, sometimes called the **meiotic mitosis**, when the chromatids separate and form the chromosomes of the gamete nuclei.

The gametes, with their single set of chromosomes, are said to be **haploid**. When an ovum is fertilised by a spermatozoon—see page 127, the **diploid** (paired) number of chromosomes is restored. The new individual thus receives half his/her genes from each parent. Different combinations of genes are possible so that the offspring is similar to but still different from its parents.

The Reproductive System

The differences between men and women are due to genes carried on the so-called sex chromosomes X and Y, which are not a perfect pair. A man has one X and one Y chromosome, making him XY, while a woman has two X chromosomes, making her XX. The primary effect of the appropriate sex chromosomes is to cause the reproductive organs to develop as **testes** in the male and **ovaries** in the female. The testes and ovaries produce the appropriate germ cells or **gametes** and also **hormones** responsible for secondary sexual characteristics.

The male reproductive system is designed to produce numerous very minute **spermatozoa (sperm)**, to store them and to transfer them to the reproductive passages of the female.

The female reproductive system is designed to produce a smaller number of **ova (eggs)**, to provide the young with nourishment and a suitable place to grow during the early part of life.

THE MALE REPRODUCTIVE SYSTEM

The male reproductive system consists of two testes, associated glands and a system of ducts which eventually open on the tip of the penis.

The testes

The testes develop in the abdomen, but just before birth they descend to the groin and come to lie outside the abdominal cavity in bags of skin called the **scrotal sacs**. Each testis is an ovoid body containing many **seminiferous tubules** bound into lobules by connective tissue in which lie **interstitial cells**—see diagram on next page. Under the influence of gonadotropic hormones—see page 122, the walls of the tubules produce spermatozoa while the interstitial cells produce **androgens** (male hormones), of which testosterone is the most important. The androgens are responsible for development of **secondary sexual characteristics** including growth of beard and deepening of voice.

Ducts and glands

Leading from each testis are numerous fine ducts, the **vasa efferentia**, which join a mass of coiled tubes called the **epididymis** in which the immature and **non-motile** spermatozoa become mature and are stored till required. From the epididymis there is a single duct, the **vas deferens**. The vasa deferentia pass through the **inguinal canals**—see page 55, where they are accompanied by the spermatic arteries and veins and form the **spermatic cords**. The vasa deferentia are joined by ducts from secretory pouches called the **seminal vesicles** and open into the urethra through short ejaculatory ducts. **Prostate** and **bulbo-urethral glands**—see diagram, open into the urethra directly. These glands and the seminal vesicles produce secretions which contribute to the volume of the seminal fluid and increase the motility and viability of the spermatozoa.

The penis

The urethra, acting as common urino-genital duct, runs through the penis, which is short and soft when not in use but contains special strands of **cavernous tissue** which, on sexual stimulation (**arousal**), can be distended with blood and thus stiffened. This erection enables the semen to be injected into the vagina of the female—see page 127—during **copulation**.

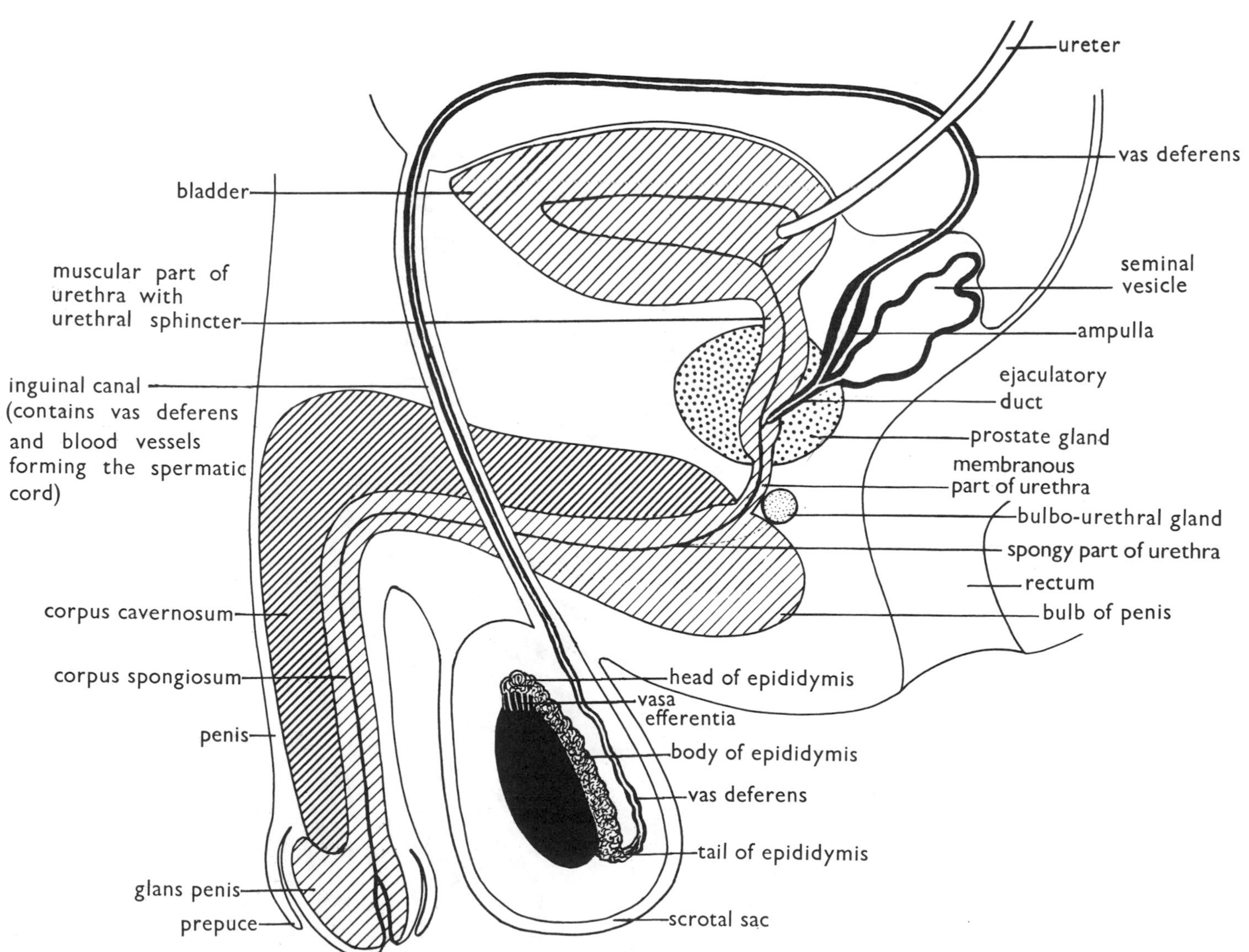

Male reproductive system

Number of chromosomes

Primary spermatocyte	2 × 23
Secondary spermatocyte	23
Spermatid	23

Therefore the spermatozoon has 23, of which one will be a sex chromosome, X or Y but not both. All the ova carry X chromosomes only, therefore on fertilisation (union of gametes) X-bearing sperm will give rise to daughters XX and Y-bearing sperm to sons XY.

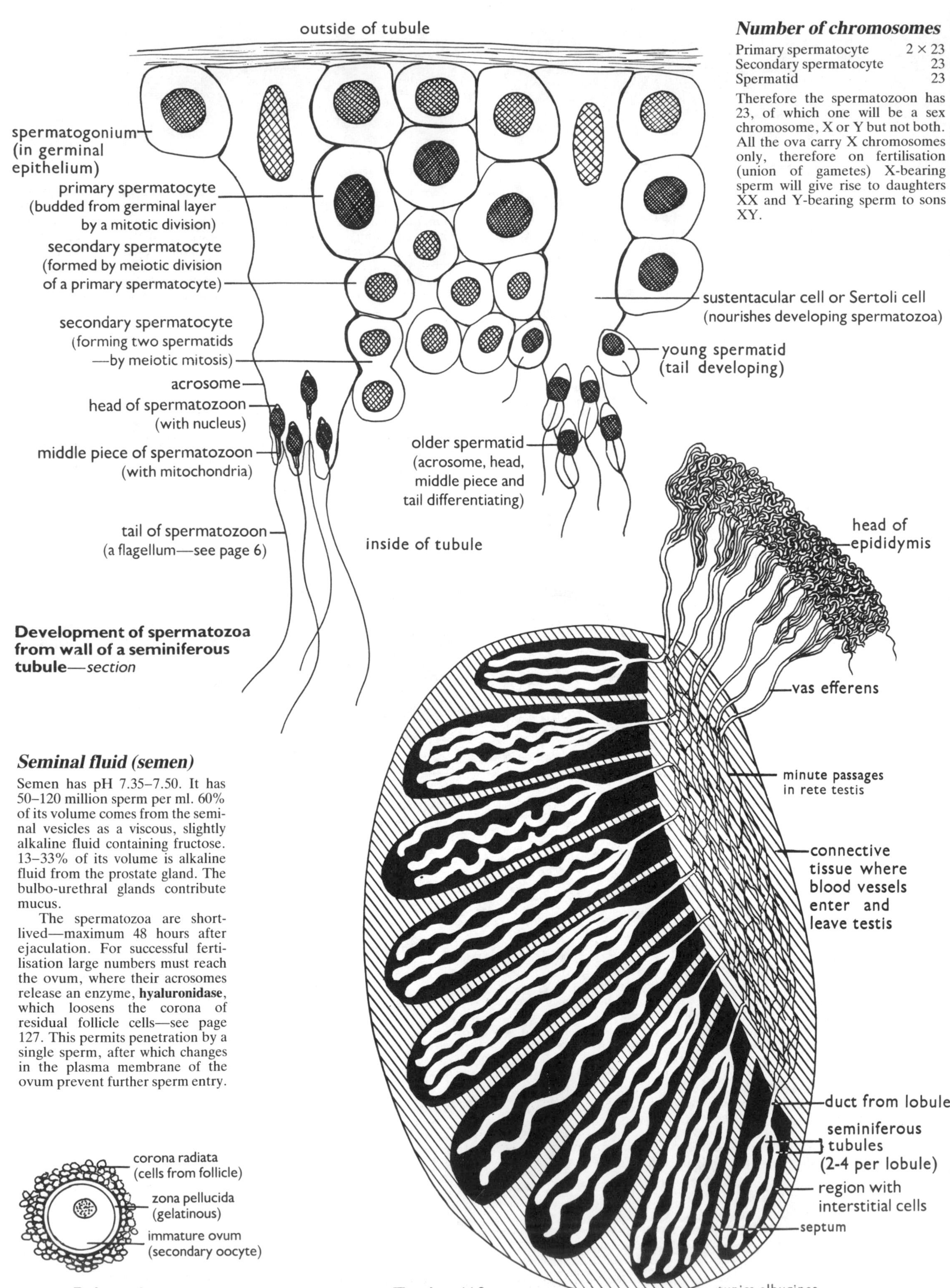

Development of spermatozoa from wall of a seminiferous tubule—*section*

Seminal fluid (semen)

Semen has pH 7.35–7.50. It has 50–120 million sperm per ml. 60% of its volume comes from the seminal vesicles as a viscous, slightly alkaline fluid containing fructose. 13–33% of its volume is alkaline fluid from the prostate gland. The bulbo-urethral glands contribute mucus.

The spermatozoa are short-lived—maximum 48 hours after ejaculation. For successful fertilisation large numbers must reach the ovum, where their acrosomes release an enzyme, **hyaluronidase**, which loosens the corona of residual follicle cells—see page 127. This permits penetration by a single sperm, after which changes in the plasma membrane of the ovum prevent further sperm entry.

Released ovum

Testis—*V.S.*

THE FEMALE REPRODUCTIVE SYSTEM

The female has two **ovaries** with their **ducts**. In place of a penis she has a small knob, the **clitoris**, which can be similarly erected during sexual stimulation and in front of which is a fatty pad, the **mons Veneris**. Her secondary sexual characteristics include a wider pelvis—see page 29, and larger breasts.

The ovaries

The ovaries lie in the pelvic basin, held in position by **ligaments** formed by folds of peritoneum. Each almond-shaped ovary (about 3 cm long) has **germinal epithelium** surrounding a mass of connective tissue called stroma. Before birth, cells are budded from this, which have potential to become **ova** or their surrounding **follicle cells**. Up to 300 000 **primary oocytes** are formed, but most of these never develop.

From **puberty** (about age 14) to **menopause** (about age 45), successive batches of follicles enlarge as shown in the diagrams. At **ovulation**, the ripe follicles rupture, shedding the **secondary oocytes** (immature ova). After ovulation the follicles fill with yellowish cells and clotted blood and become the **corpora lutea**, which normally grow for 10–15 days and then atrophy, but at least one is retained through pregnancy.

Maturation of each ovum involves nuclear division with minimal loss of cytoplasm. The **first maturation division** is **meiotic** and reduces the number of chromosomes to 23. It occurs before ovulation, giving rise to the **secondary oocyte** and the **first polar body**. The **second maturation division** occurs after sperm entry, but before fusion of sperm and **ootid** (mature ovum) nuclei. A **second polar body** is shed.

The uterine tubes, uterus and vagina

The narrow, delicate **uterine** (**Fallopian**) **tubes** are connected to the **uterus**, which opens into the **vagina**, which in turn opens into the **vestibule** behind the urethra and between the folds of the **labia**—see diagram. Each uterine tube has a funnel-shaped opening surrounded by **ciliated** processes called fimbriae. These partially enfold the ovary so that shed ova are normally wafted into the tubes instead of straying into the abdominal cavity. Cilia within and muscular movements of the walls of the tubes aid passage of ova.

The **uterus** is pear-shaped, about 7.5 cm long. It enlarges considerably during pregnancy—see page 129. Its thick muscular wall is lined with **uterine mucosa**, which thickens and breaks down alternately during the **menstrual cycle**. The muscular **cervix** (neck) of the uterus protrudes into the **vagina**. A pair of **vaginal glands** and the vaginal wall produce lubricant with pH about 5. This eases the insertion of the penis during copulation.

The spermatozoa of the ejaculate swim through the cervix and uterus at a rate of 1–4 mm/min. **Fertilisation** normally takes place in the upper parts of the tubes and the embryo starts its development during the 3–4 days it takes to pass down to the uterus.

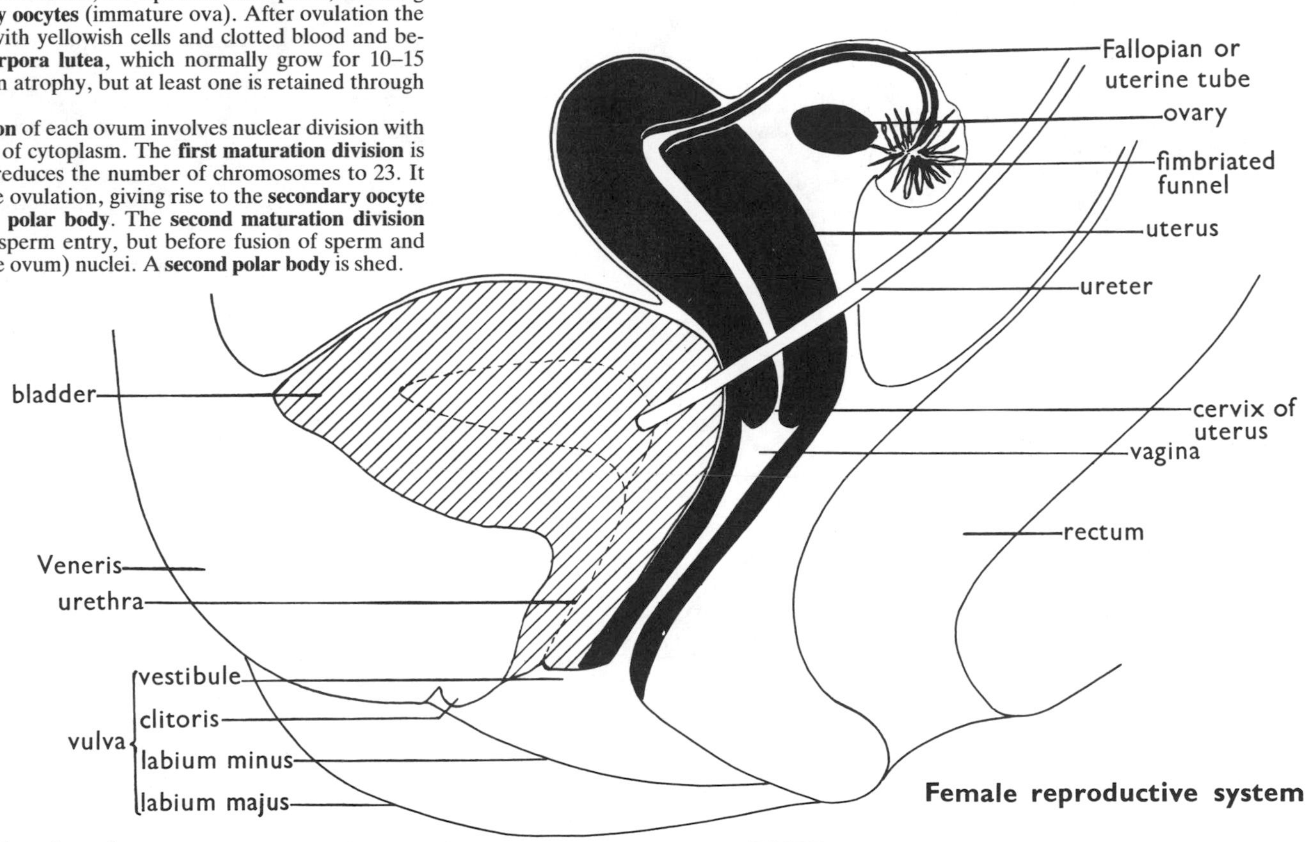

Fallopian or uterine tube
ovary
fimbriated funnel
uterus
ureter
bladder
cervix of uterus
vagina
rectum
Veneris
urethra
vulva
vestibule
clitoris
labium minus
labium majus

Female reproductive system

The menstrual cycle

The ovaries produce at least six hormones called collectively **oestrogens**, which are responsible for female sexual characteristics. The oestrogens, plus **progesterone** from the corpora lutea, stimulate thickening of the uterine mucosa ready for possible implantation of an embryo and prepare the mammary glands for milk production. From puberty to menopause, except during pregnancy, ovulation may occur at monthly intervals, governed by alternating increase of the two **gonadotropic hormones**—see page 122. **FSH** stimulates ripening of the follicles and secretion of oestrogens. The oestrogens then suppress secretion of FSH and increase secretion of **LH**, which then promotes development of the corpora lutea. If no pregnancy follows, the corpora lutea soon atrophy, secretion of both progesterone and oestrogens is reduced, secretion of FSH increases again, the uterine mucosa breaks down and **menstrual bleeding** occurs. Ovulation can be expected about 16 days after **menstruation** starts, but does not always take place.

During pregnancy large quantities of oestrogens and progesterone are produced by the placenta—see page 129, ovulation is inhibited, corpora lutea and thickening of the uterine mucosa are maintained, and there is no menstruation.

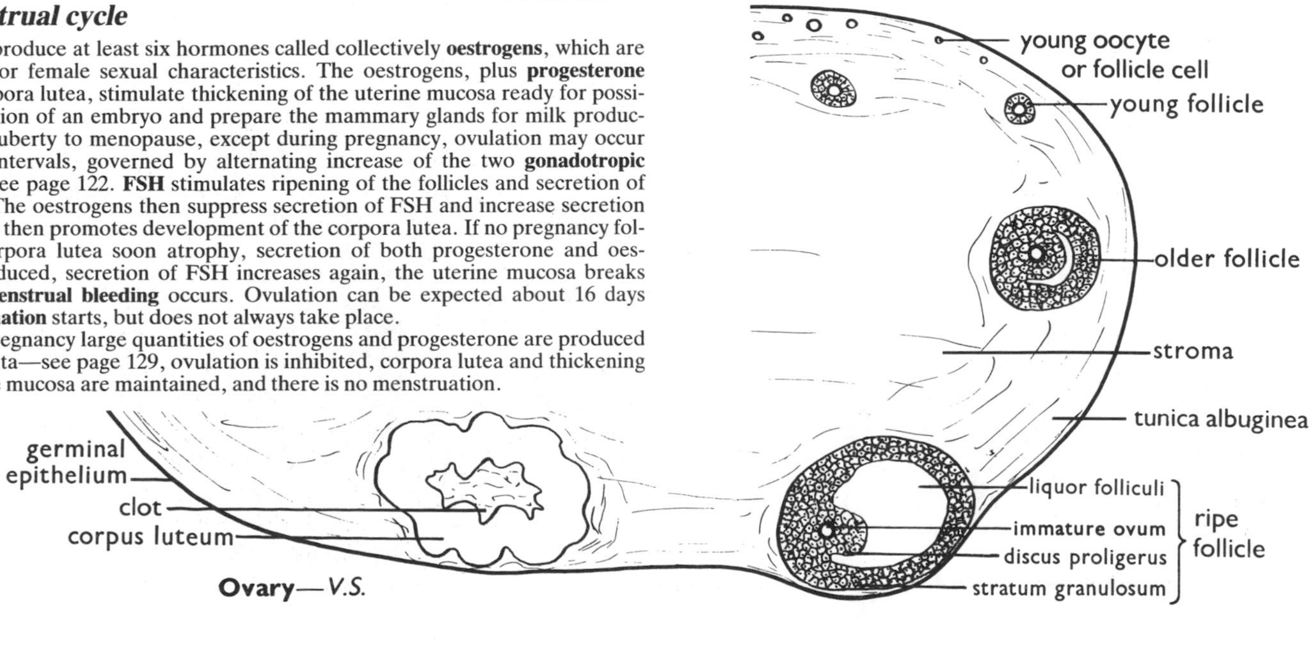

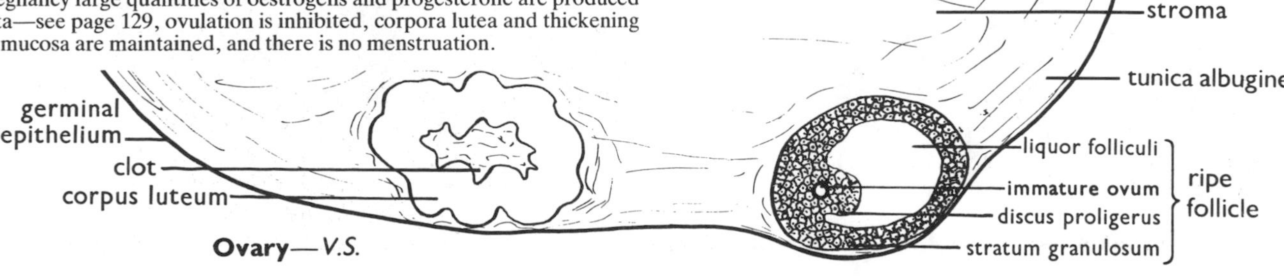

Ovary—*V.S.*

EMBRYOLOGY

The prenatal development of the baby is termed **embryology**.

The fertilised egg is a single cell, but this very soon divides to form a sphere of cells. The sphere becomes hollow and is called the **blastocyst**. Its outer wall is the **trophoblast** and on one side is a mass of cells, the **germinal disc**, from which the embryo develops under the influence of hormone-like organisers. The germinal disc is originally on the surface but becomes enclosed by growth of the **amnion** over it. The cells become differentiated into three tissue layers called **ectoderm**, **endoderm** and **mesoderm**.

The ectoderm of the germinal disc forms the skin and its derivatives and the nervous tissue of the embryo. Ectoderm also forms the trophoblast and the lining of the **amniotic cavity**.

Endoderm forms the lining of the **yolk sac** (a vestige of the sac which encloses the yolk in the eggs of birds and reptiles) and later forms the lining of the alimentary canal and its associated glands.

Mesoderm lies between the ectoderm and the endoderm and gives rise to all connective tissues including blood and bones, all muscle tissue and reproductive tissue.

The blastocyst becomes implanted in the uterine mucosa and forms villi from the trophoblast to increase the surface of contact.

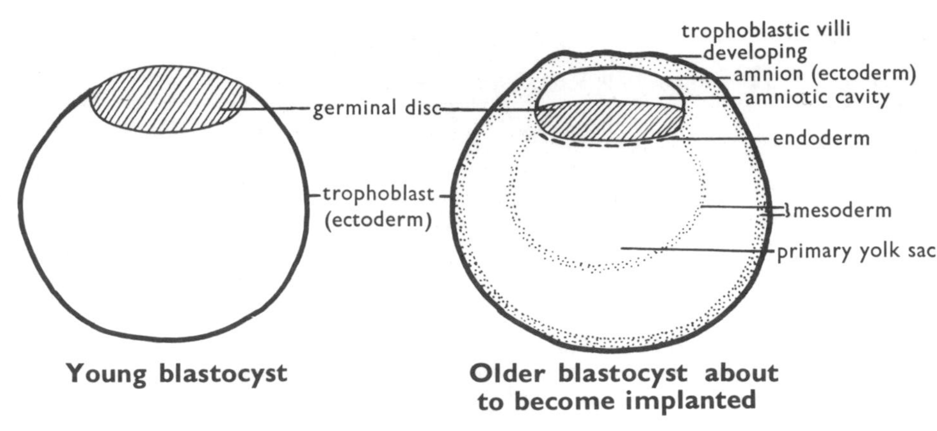

Young blastocyst

Older blastocyst about to become implanted

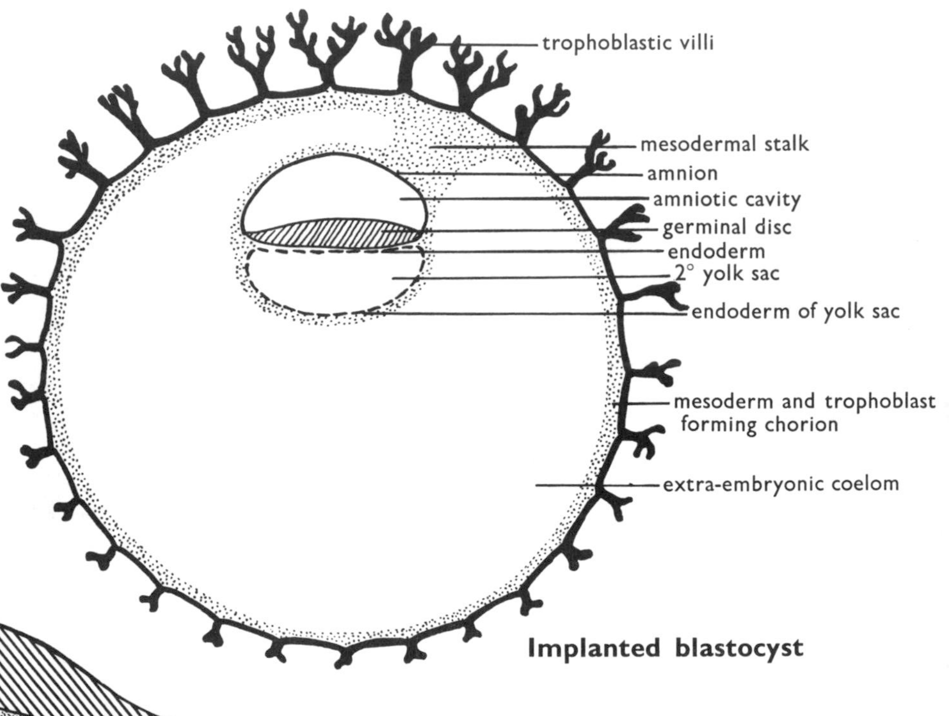

Implanted blastocyst

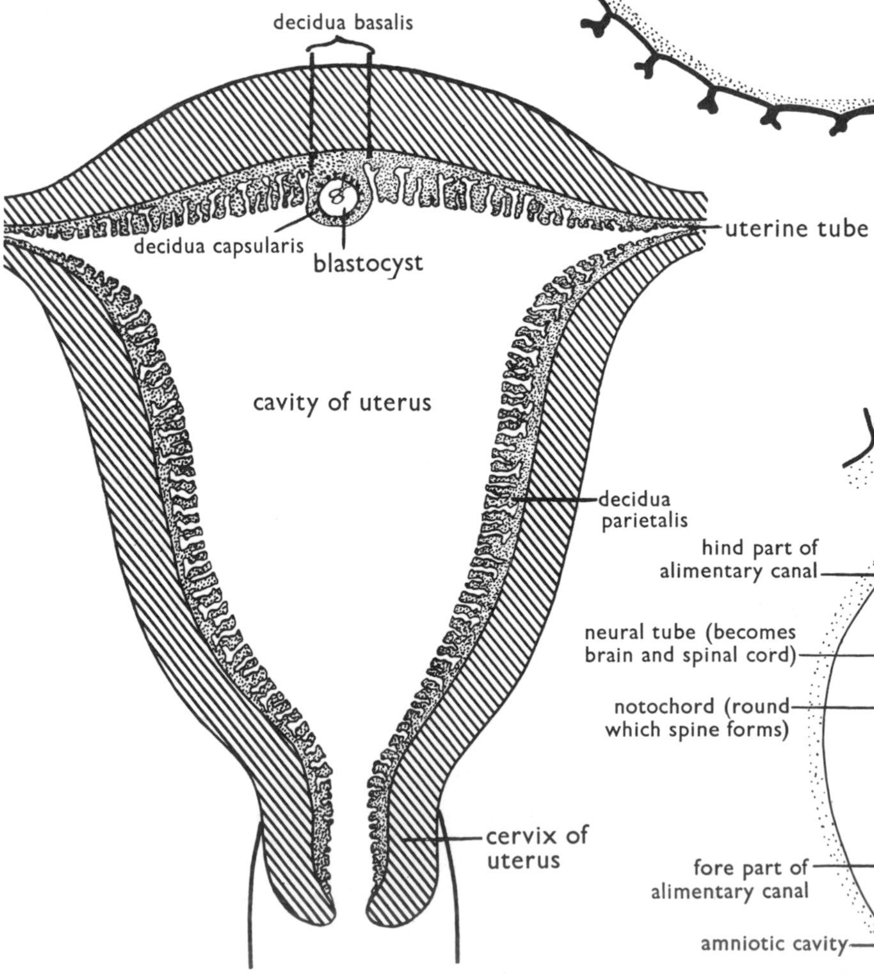

decidua basalis
uterine tube
decidua capsularis
blastocyst
cavity of uterus
decidua parietalis
cervix of uterus

Uterus with implanted blastocyst

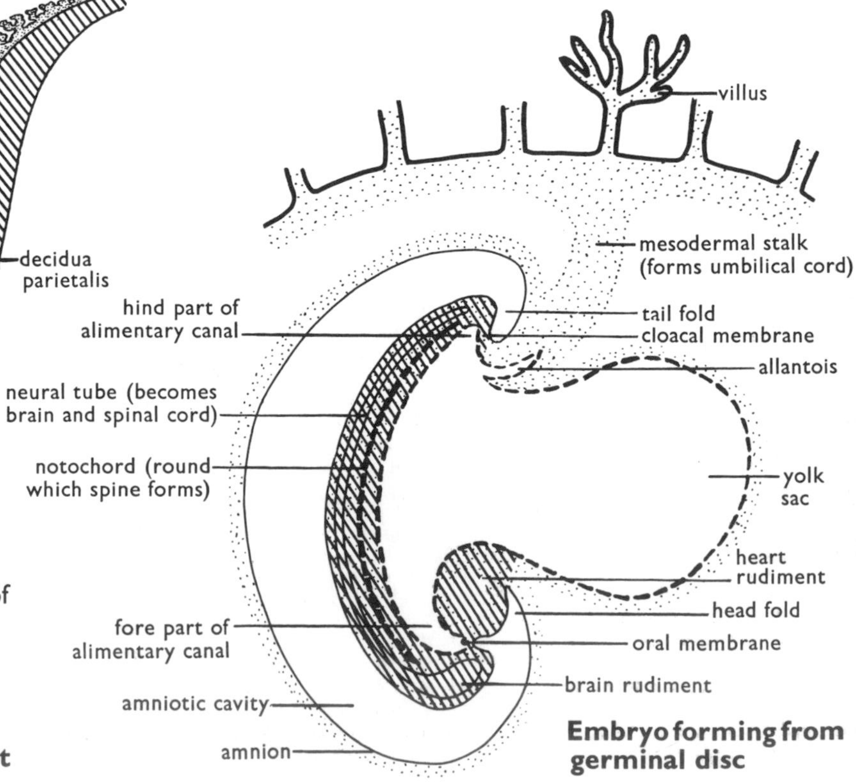

villus
mesodermal stalk (forms umbilical cord)
tail fold
cloacal membrane
allantois
hind part of alimentary canal
neural tube (becomes brain and spinal cord)
notochord (round which spine forms)
yolk sac
heart rudiment
head fold
oral membrane
fore part of alimentary canal
brain rudiment
amniotic cavity
amnion

Embryo forming from germinal disc

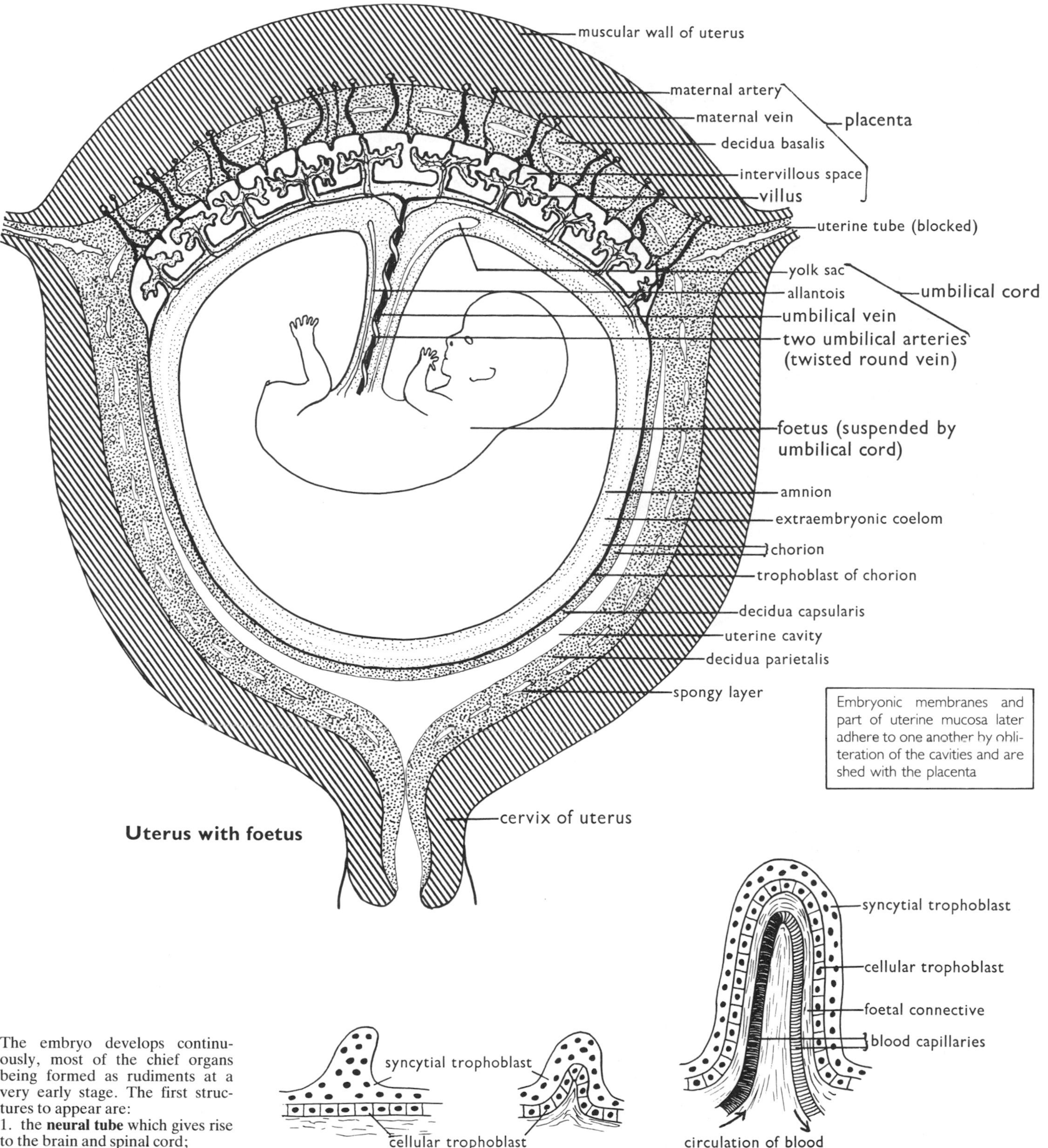

Uterus with foetus

1° villus **2° villi forming by invasion of mesoderm and blood vessels**

The embryo develops continuously, most of the chief organs being formed as rudiments at a very early stage. The first structures to appear are:

1. the **neural tube** which gives rise to the brain and spinal cord;
2. the **notochord** which forms a support and is later replaced by the vertebrae;
3. the **mesodermal blocks** which give rise to the chief muscles;
4. the **head** and **tail folds** which shut in the alimentary canal and form the under side of the embryo;
5. the **heart** and **blood vessels** which form a functioning circulation at a very early stage.

Note. The outgrowth of the hind part of the alimentary canal called the allantois is a vestige of an important structure found in birds and reptiles.

The delicate developing embryo is suspended in, and protected by, fluid in the cavity enclosed by the amnion. The fluid in the extra-embryonic coelom between the amnion and the outer embryonic membrane or chorion has the same effect until the cavity is obliterated, later in development, by the union of amnion and chorion.

As the embryo grows bigger, it and its membranes project into the uterine cavity. The villi are invaded by connective tissue and later by blood vessels. The compound structure of the villi and the uterine mucosa forms the **placenta** through which the embryo, now called the **foetus**, is fed, receives oxygen and gets rid of waste materials. Blood sinuses form in the maternal portion of the placenta and surround the villi, but the maternal blood is always completely separated from the foetal blood by the cellular layer of the trophoblast. The trophoblast has a selective filtering action which prevents the entry of maternal hormones into the foetus (especially important when the foetus is male) and of many toxins, drugs and germs. In addition to **oestrogens** and **progesterone**—see page 127, the placenta secretes **chorionic gonadotropin**, which reinforces the lactogenic action of the lactogenic (luteotropic) hormone—see page 122.

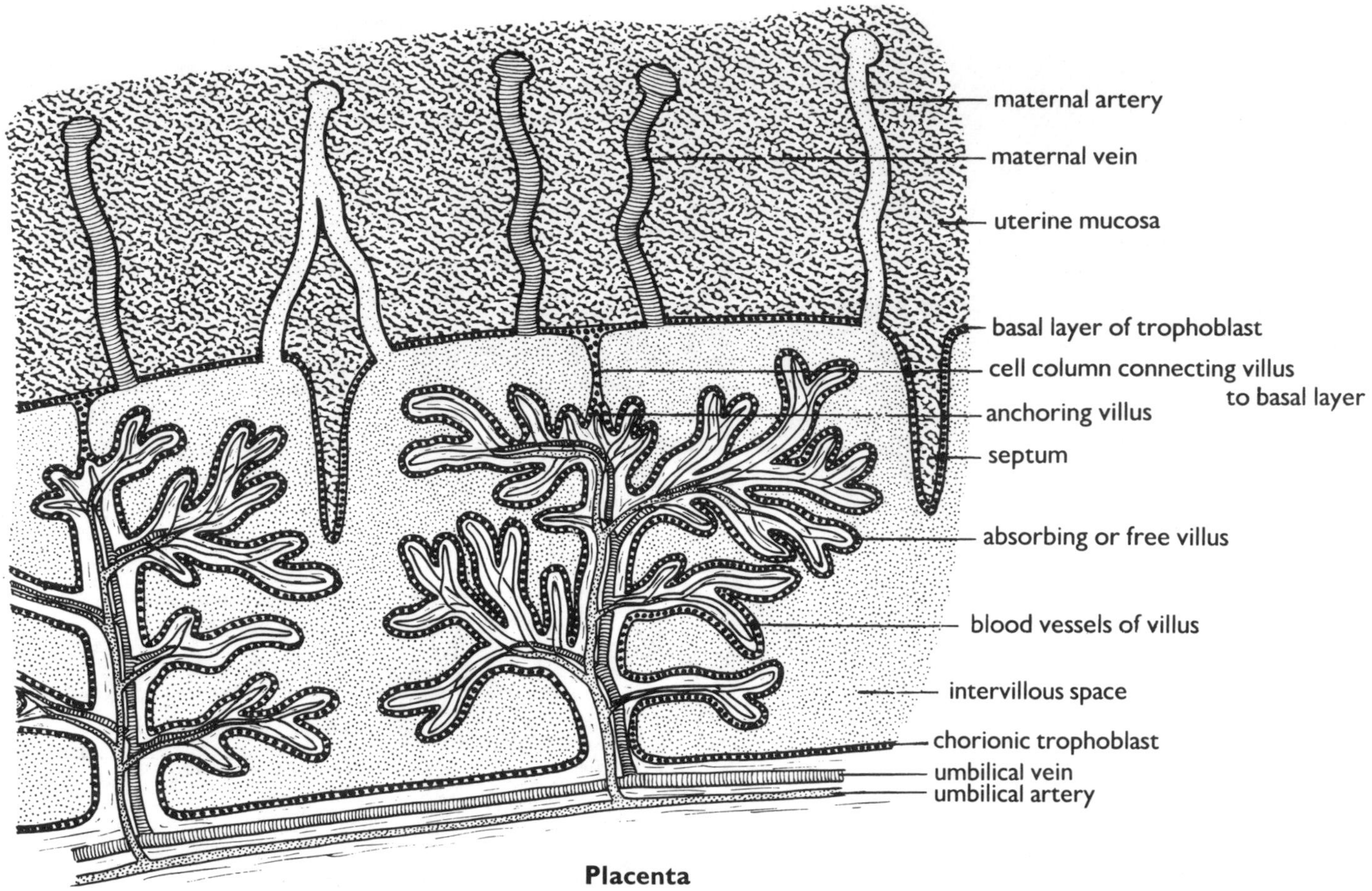

Placenta

The connection between the foetus and the placenta is by the **umbilical cord**, in which there are two arteries, one vein and the allantoic canal. The **umbilical arteries** are branches of the iliac arteries. The blood from the **umbilical vein** reaches the inferior vena cava chiefly through the **ductus venosus**, but some also travels indirectly through the liver.

Before birth the **foramen ovale** (in the septum between the atria of the heart) and the **ductus arteriosus** (between the left pulmonary artery and the aorta) allow short-circuiting of the blood so that very little goes to the, as yet, non-functioning lungs. The valve of the inferior vena cava directs the blood from this vessel through the foramen ovale without mixing with the blood from the superior vena cava.

At birth, or **parturition**, the foramen ovale and the ductus arteriosus close so that the normal double circulation is established (except in some 'blue' babies). The umbilical cord is cut so that the baby is dependent on its own lungs for a supply of oxygen and on food taken through the mouth and must get rid of its own waste materials.

The placenta and the embryonic membranes are shed as the **afterbirth**.

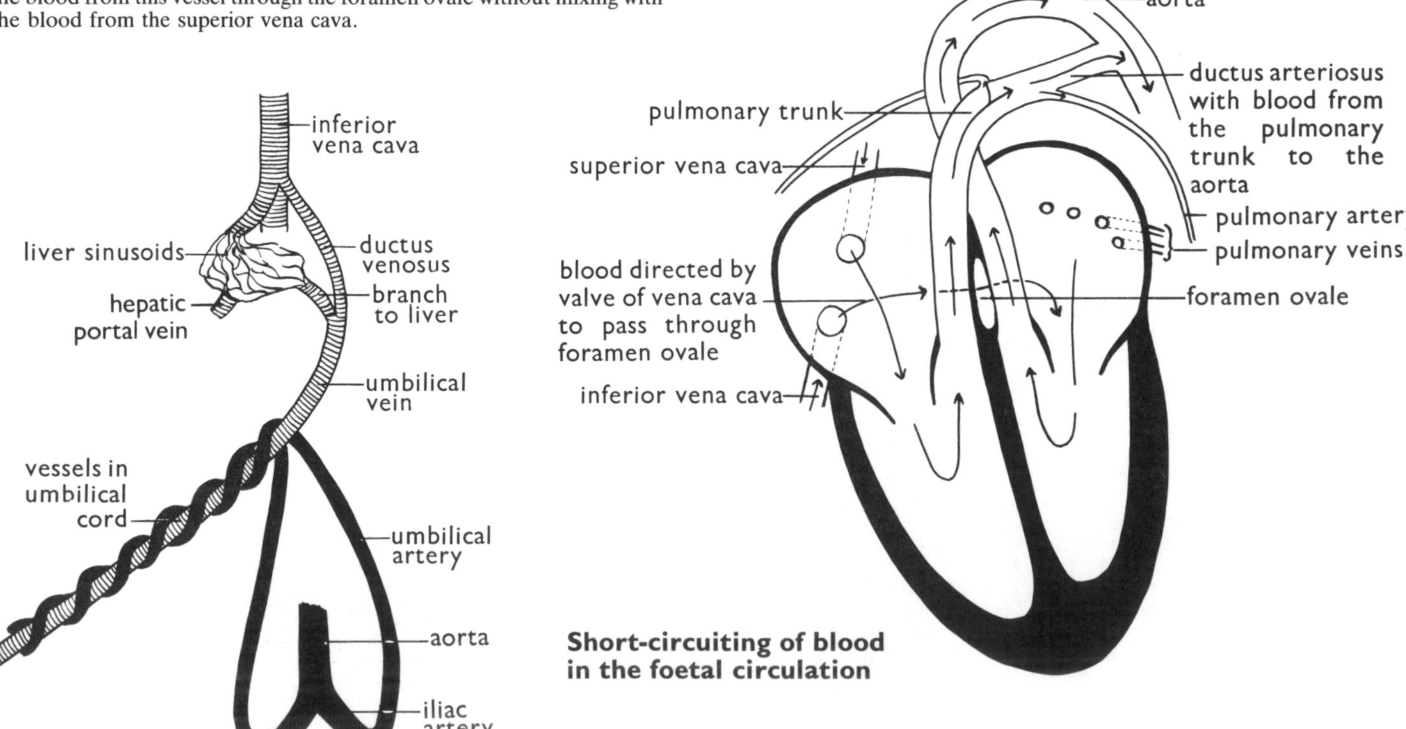

Umbilical circulation

Short-circuiting of blood in the foetal circulation

MAMMARY GLANDS

Mammary glands are present in both sexes, and lie outside the wall of the thorax, over the 2nd to 6th ribs. Those of the female swell at puberty under the influence of the sex hormones.

Each fully developed breast has 12–20 **lobes**, subdivided into **lobules**, the walls of which bear very numerous glandular **alveoli**. The lobules and lobes are bound together by fatty connective tissue. The **lactiferous ducts** from the lobules open into wide **lactiferous sinuses** from each of which a short, straight duct leads to the nipple. Around the nipple, the corrugated area of skin called the **areola** becomes darkly pigmented after pregnancy.

Function of the mammary glands

The mammary glands do not become functional till after childbirth. During pregnancy, the large quantities of oestrogens formed by the placenta inhibit formation of lactogenic hormones, but they and progesterone give rise to additional growth of the breasts and sensitise their secretory tissue. At parturition, oestrogen secretion ceases and secretion of the **luteotropic (lactogenic) hormone** (LTH, prolactin) and the **glucocorticoids**—see page 122—is increased. The lactogenic action of these hormones causes the alveoli to become active.

The initial secretion from the mammary glands is a yellowish fluid containing protein and sugar but no fat. This is called **colostrum** and is replaced by **milk** within 2–3 days. Milk is the perfect food for the young infant and its composition varies with the age of the baby, becoming gradually more concentrated. At the same time the child's digestion improves. The overall average composition of human milk is:

protein 1.5–2.0% (⅓ lactalbumen, ⅔ caseinogen)
fat 3.5%
milk sugar (lactose) 6.5%
mineral salts, including calcium, 0.3%

(*Note*. Milk is deficient in iron but normally enough iron is stored in the liver of the foetus to last the baby till it begins to take a mixed diet.)

vitamins A, B, C and D
water 87.7–88.2%

Provided suckling continues, lactation will be fully maintained for 7–9 months, or during the time that the corpus luteum is still yielding progesterone, but if suckling is discontinued, capacity for milk production is soon lost.

The reaction to suckling is a complex chain of events, starting with stimulation of the nerve endings in the nipple. Nerve impulses reaching the brain affect the hypothalamus and produce two responses:

(a) decrease in formation of the prolactin-inhibiting hormone so that the pituitary body releases additional LTH which stimulates milk production;

(b) release of oxytocin which affects the muscle in the walls of the alveoli, forcing milk through the ducts into the lactiferous sinuses from which it is easily available to the nursing baby.

At the height of lactation, up to 1.5 litres of milk may be formed each day. To supply the nutrients for this it is important for the mother to have adequate supplies in her own diet. It is particularly important for her to take supplementary calcium, phosphate, and vitamin D to guard against decalcification of her bones and teeth.

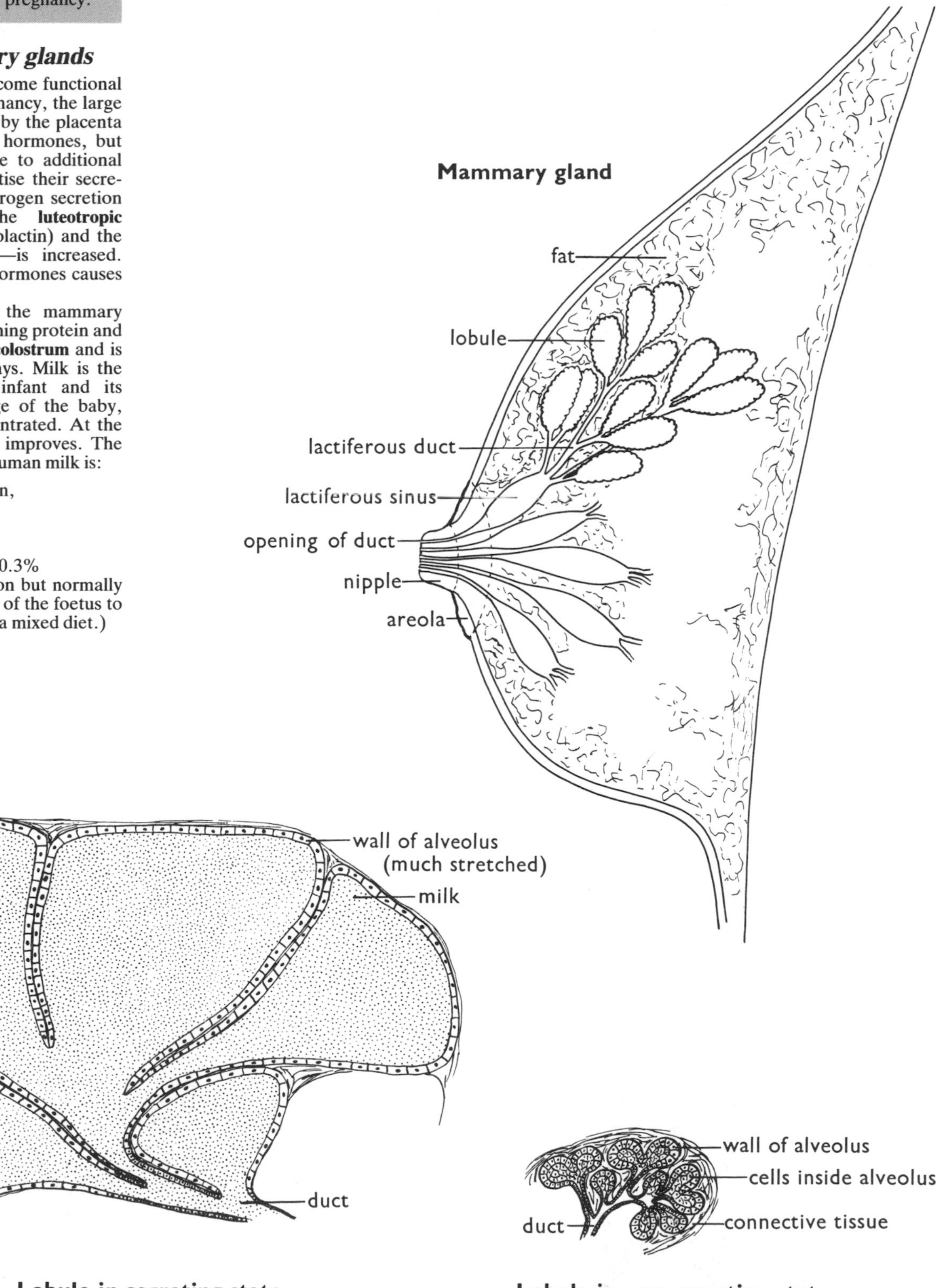

Lobule in secreting state

Lobule in non-secreting state

References

This book is intended as an overview, giving basic facts with the maximum conciseness for clarity. Further information can be obtained from more detailed texts. It is important always to use those which have been recently revised, because modern techniques make possible the identification of additional detail of fine structure and better understanding of the complex chemistry of life. Theories become obsolete in the light of new knowledge and terminology changes.

Note. Sometimes terms used in American textbooks and American spellings of technical words differ from those more commonly used in Britain, but this is generally obvious from context.

Though not an exhaustive bibliography, the following list of books may prove helpful:

Principles of Anatomy and Physiology, Tortora and Anagnostakos (Harper and Row, New York)
Textbook of Anatomy and Physiology, Anthony and Thibodeau (Mosby, St Louis)
Essentials of Physiology, Lamb, Ingram, Johnston and Pitman (Blackwell, Oxford)
Physiology of the Human Body, Guyton (Holt-Saunders, Japan)
Physiology of the Human Body, McClintic (John Wiley, New York)
Basic Human Physiology, Weller and Wiley (Pridle, Weber and Schmidt, USA)
An Introduction to Human Physiology, Green (Oxford Medical Publications)

Classic textbooks which have been updated include:

Kimber, Gray and Stackpole's *Anatomy and Physiology*, revised by Miller and Leavell (Macmillan, New York)
Best and Taylor, *Physiological Basis of Medical Practice*, revised by West (Williams and Wilkins, Baltimore)
Samson Wright's *Applied Physiology*, revised by Keele, Neil and Joels (Oxford University Press)
Cunningham's *Textbook of Anatomy*, revised by Romanes (Oxford University Press)
Gray's Anatomy, revised by Williams and Warwick (Churchill Livingstone, Edinburgh)

Also useful:

McGraw-Hill Dictionary of Chemistry (McGraw-Hill, New York)

Index

Notes

1. Formation of the plural of Latin (or Greek) words obeys complicated rules. Broadly:

-um	becomes -a	-is	becomes	-es	-ex	becomes	-ices or -iges
-a	-ae or -ata	-en		-ina	-ix		-ices
-us	-i or -uses or -ora	-ens		-entia	-x		-xes or -iges
-ur	-ora	-ans		-antes			

Adjectives decline along with nouns, but when the possessive (of) form is used, this does not change.

The appropriate endings are indicated in the index.

With all other nouns the formation of the plural obeys ordinary English usage. Occasionally the English-type plural is an acceptable alternative to the Latin one, e.g. tibias instead of tibiae; femurs instead of femora.

2. In some cases the noun used in the index does not appear in the text because the corresponding adjective, verb or adverb fits the sentence construction, e.g. anabolism/anabolic (p. 3); hydrolysis/to hydrolyse (p. 90); reflex/reflexly (p. 90).